Recent Advances in
Micro/Nano-Fabrication

Recent Advances in Micro/Nano-Fabrication

Guest Editors

Yao Liu
Jinjie Zhou

Basel • Beijing • Wuhan • Barcelona • Belgrade • Novi Sad • Cluj • Manchester

Guest Editors

Yao Liu
Mechanical Engineering
North University of China
Taiyuan
China

Jinjie Zhou
Mechanical Engineering
North University of China
Taiyuan
China

Editorial Office
MDPI AG
Grosspeteranlage 5
4052 Basel, Switzerland

This is a reprint of the Special Issue, published open access by the journal *Micromachines* (ISSN 2072-666X), freely accessible at: www.mdpi.com/journal/micromachines/special_issues/2K545IF546.

For citation purposes, cite each article independently as indicated on the article page online and using the guide below:

Lastname, A.A.; Lastname, B.B. Article Title. *Journal Name* **Year**, *Volume Number*, Page Range.

ISBN 978-3-7258-3002-2 (Hbk)
ISBN 978-3-7258-3001-5 (PDF)
https://doi.org/10.3390/books978-3-7258-3001-5

© 2025 by the authors. Articles in this book are Open Access and distributed under the Creative Commons Attribution (CC BY) license. The book as a whole is distributed by MDPI under the terms and conditions of the Creative Commons Attribution-NonCommercial-NoDerivs (CC BY-NC-ND) license (https://creativecommons.org/licenses/by-nc-nd/4.0/).

Contents

Yao Liu and Jinjie Zhou
Editorial for the Special Issue on Recent Advances in Micro/Nano-Fabrication
Reprinted from: *Micromachines* 2024, 15, 1485, https://doi.org/10.3390/mi15121485 1

Yao Liu, Shaobo Zhai, Jinzhu Guo, Shiling Fu, Bin Shen and Zhigang Zhao et al.
Study on the Abrasive Blasting Mechanism of Solder Welded 304V Wire in Vascular Intervention
Reprinted from: *Micromachines* 2024, 15, 1405, https://doi.org/10.3390/mi15121405 7

Junyu Duan, Gui Long, Xu Xu, Weiming Liu, Chuankun Li and Liang Chen et al.
Hierarchical Micro/Nanostructures with Anti-Reflection and Superhydrophobicity on the Silicon Surface Fabricated by Femtosecond Laser
Reprinted from: *Micromachines* 2024, 15, 1304, https://doi.org/10.3390/mi15111304 21

Yongjie Shi, Min Su, Qianqian Cao and Di Zheng
A Normal Displacement Model and Compensation Method of Polishing Tool for Precision CNC Polishing of Aspheric Surface
Reprinted from: *Micromachines* 2024, 15, 1300, https://doi.org/10.3390/mi15111300 34

Ravi Pratap Singh and Yaolong Chen
Integration of Metrology in Grinding and Polishing Processes for Rotationally Symmetrical Aspherical Surfaces with Optimized Material Removal Functions
Reprinted from: *Micromachines* 2024, 15, 1276, https://doi.org/10.3390/mi15101276 49

Haoqi Luo, Xue Wang, Lin Qin, Hongxin Zhao, Deqing Zhu and Shanyi Ma et al.
Investigation on the Machinability of Polycrystalline ZnS by Micro-Laser-Assisted Diamond Cutting
Reprinted from: *Micromachines* 2024, 15, 1275, https://doi.org/10.3390/mi15101275 66

Qingxu Zhang, Yanyan Yang, Shijie Huo, Shucheng Duan, Tianao Han and Guang Liu et al.
Laser Direct Writing of Setaria Virids-Inspired Hierarchical Surface with TiO_2 Coating for Anti-Sticking of Soft Tissue
Reprinted from: *Micromachines* 2024, 15, 1155, https://doi.org/10.3390/mi15091155 81

Sang-Kon Kim
Contact Hole Shrinkage: Simulation Study of Resist Flow Process and Its Application to Block Copolymers
Reprinted from: *Micromachines* 2024, 15, 1151, https://doi.org/10.3390/mi15091151 97

Haonan Li, Muyang Zhang, Yeqian Liu, Shangneng Yu, Xionghui Li and Zejingqiu Chen et al.
Off-Stoichiometry Thiol-Ene (OSTE) Micro Mushroom Forest: A Superhydrophobic Substrate
Reprinted from: *Micromachines* 2024, 15, 1088, https://doi.org/10.3390/mi15091088 112

Yeeu-Chang Lee, Hsu-Kang Wu, Yu-Zhong Peng and Wei-Chun Chen
The Synthesis and Assembly Mechanism of Micro/Nano-Sized Polystyrene Spheres and Their Application in Subwavelength Structures
Reprinted from: *Micromachines* 2024, 15, 841, https://doi.org/10.3390/mi15070841 121

Quanjing Wang, Ru Zhang, Qingkui Chen and Ran Duan
A Review of Femtosecond Laser Processing of Silicon Carbide
Reprinted from: *Micromachines* 2024, 15, 639, https://doi.org/10.3390/mi15050639 133

Man Yang, Santosh Kumar Parupelli, Zhigang Xu and Salil Desai
Understanding the Effect of Dispersant Rheology and Binder Decomposition on 3D Printing of a Solid Oxide Fuel Cell
Reprinted from: *Micromachines* **2024**, *15*, 636, https://doi.org/10.3390/mi15050636 159

Hao Cao, Zhishuang Xue, Hongfeng Deng, Shuo Chen, Deming Wang and Chengqun Gui
Micromirror Array with Adjustable Reflection Characteristics Based on Different Microstructures and Its Application
Reprinted from: *Micromachines* **2024**, *15*, 506, https://doi.org/10.3390/mi15040506 176

Jonghyeok Kim, Byungjoo Kim, Jiyeon Choi and Sanghoon Ahn
The Effects of Etchant on via Hole Taper Angle and Selectivity in Selective Laser Etching
Reprinted from: *Micromachines* **2024**, *15*, 320, https://doi.org/10.3390/mi15030320 189

Huiyun Zhang, Zhigang Zhao, Jiaojiao Li, Linzheng Ye and Yao Liu
Review on Abrasive Machining Technology of SiC Ceramic Composites
Reprinted from: *Micromachines* **2024**, *15*, 106, https://doi.org/10.3390/mi15010106 199

Fei Peng, Chao Sun, Hui Wan and Chengqun Gui
An Improved 3D OPC Method for the Fabrication of High-Fidelity Micro Fresnel Lenses
Reprinted from: *Micromachines* **2023**, *14*, 2220, https://doi.org/10.3390/mi14122220 228

Denys Moskal, Jiří Martan and Milan Honner
Scanning Strategies in Laser Surface Texturing: A Review
Reprinted from: *Micromachines* **2023**, *14*, 1241, https://doi.org/10.3390/mi14061241 245

Editorial

Editorial for the Special Issue on Recent Advances in Micro/Nano-Fabrication

Yao Liu * and Jinjie Zhou *

School of Mechanical Engineering, North University of China, Taiyuan 030051, China
* Correspondence: liuyao@nuc.edu.cn (Y.L.); zhoujinjiechina@126.com (J.Z.)

This Special Issue of *Micromachines*, named "Recent Advances in Micro/Nano-Fabrication", comprehensively dedicates the latest research on micro/nano-fabrication in various fields. This Special Issue constitutes a synthesis of sixteen papers that investigate the most recent advancements and challenges in micro- and nanoscale fabrication processes. Micro/nanostructures have gained significant attention and wide applications in photocatalysis, coated fabrics, microchips, and sensors [1]. With the continuous increase in modern production demands, the capability to prepare microscopic patterns and structures is crucial for the advancement of many technologies and devices [2]. Removing micro- and nanoscale materials to achieve ultrahigh precision is also challenging. Surface micro/nanostructures generally possess unique textural structures [3] and exhibit certain periodicity [4], which makes them the preferred choice for achieving important performance characteristics in various devices. In the past decade, with technological advancements and the emergence of cost-effective manufacturing methods, the successful implementation of engineered surfaces with self-cleaning properties in industrial applications has been achieved [5]. The pursuit of smaller, multifunctional microdevices and integrated microsystems with superior electrical and mechanical performances stimulated micro/nano-fabrication technology breakthroughs [6]. By advancing the manufacturing technology of micro/nanostructures and their interfaces with industrial applications, this Special Issue presents the latest research progress in the micro/nano-fabrication process. The commonly used micro/nano-fabrication methods include electrochemical machining (ECM) [7], electrical discharge machining (EDM) [8], conventional mechanical machining [9], and laser beam machining (LBM) [10]. For example, ECM and EDM have been applied to fabricate the brain electrode in a microscale to reduce the damage to the tissue and nervous system during the invasion process [11]. Ultrasonic machining is also used to generate structure surfaces by using high-frequency vibration [12]. The microstructure was fabricated by the laser or chemical machining for the coating to find the super hydrophilic or superhydrophobic surface [13]. Laser transfer printing technology was used to make the wire in the photovoltaic panel with a thickness of less than 40 μm for high photoelectric conversion efficiency [14]. The micro/nano-fabrication process tackles challenges like high power consumption, low integration, and poor performance. Despite complexities and high costs, micro/nano-fabrication facilitates high-performance products and structures, which expand into ultra-large-scale integrated circuit manufacturing, nanoscale electronics, optoelectronics, high-density magnetic storage devices, microelectromechanical systems, biochips, and nanotechnology. Those applications offer even greater development opportunities for micro/nanofabrication.

The following contributions in this Special Issue cover a wide range of applications and processing techniques, providing valuable insights for the fabrication of various complex material surfaces:

A review article by Denys et al. [15] compared classical and recently developed scanning strategies using laser surface texturing (LST). Plasma and particle shielding effects and heat accumulation were described as fundamental physical limitations of short and

Citation: Liu, Y.; Zhou, J. Editorial for the Special Issue on Recent Advances in Micro/Nano-Fabrication. *Micromachines* 2024, 15, 1485. https://doi.org/10.3390/mi15121485

Received: 3 December 2024
Accepted: 6 December 2024
Published: 11 December 2024

Copyright: © 2024 by the authors. Licensee MDPI, Basel, Switzerland. This article is an open access article distributed under the terms and conditions of the Creative Commons Attribution (CC BY) license (https://creativecommons.org/licenses/by/4.0/).

ultrashort LSTs. Several methods of laser beam shifting, including the scanning strategy, straight shadow lines, path archiving, cross-coupled texture, and Lissajous or Hilbert curve filling, were discussed. The final discussion shows that a combination of several techniques, such as multi-beam processing, asynchronously shifted LST strategies, and Laser-Induced Periodic Surface Structure formation, can provide a way to achieve higher processing rates in LST. The subsequent phase in attaining elevated standards of Industry 4.0 may be the implementation of dynamic non-contact infrared temperature control in conjunction with high-speed scanning technology in laser surface treatment.

The latest abrasive processing techniques for SiC ceramic composites were reviewed by Zhang et al. [16]. Both conventional grinding (CG) and non-conventional grinding (non-CG) techniques were introduced and compared the advantages and disadvantages. Subsequently, four simulation methods, namely the finite element method (FEM), smooth particle hydrodynamics (SPH), molecular dynamics (MD) analysis, and discrete element method (DEM) analysis, are presented briefly, with an overview of the fundamental theory, the current status of the research, and the scope of application. It is proposed that the application of multiscale simulation methods for the study of ceramic matrix composite (CMC) grinding is a promising method. Furthermore, the particular structural and physical characteristics of CMCs are examined, and the findings indicate that reinforcing fibers not only address the inherent deficiencies in CMC materials but also alter the removal mechanism. Ultimately, it is recommended that the utilization of novel technologies to enhance the quality of grinding, comprehend the influence of fiber orientation and distribution on the removal mechanism and removal mode of the material, and develop a suitable machining model are crucial avenues of future research.

In a review of femtosecond laser processing on silicon carbide, Wang et al. [17] provided a comprehensive analysis of the machining techniques and applications of femtosecond laser on SiC. Various femtosecond laser hybrid fabrication methods and the utilization of femtosecond laser machining SiC structures in microelectromechanical systems (MEMS) were discussed. The femtosecond laser can machine pits, corrugated structures, nanoscale structures, grooves, and large surface areas directly on SiC. Additionally, femtosecond lasers can be used in conjunction with other processing techniques to machine SiC, such as water jet processing, underwater processing, and chemical selective etching. The advantages of femtosecond laser machining SiC for sensors, microcavities, and micromechanical systems were also presented. The femtosecond laser hybrid fabrication techniques address the limitations of high hardness, excellent thermal conductivity, and the chemical inertness of SiC semiconductors, which promote femtosecond laser machining technique development and application. This provides an important direction for exploring advances in femtosecond laser micro- and nanofabrication, with the goal of diversifying the processes used to fabricate SiC devices.

Peng et al. [18] developed a photolithography model based on optical imaging and photochemical reaction profiles. The study proposed a sub-domain division method with statistical principles to enhance the efficiency and precision of 3D Optical Proximity Correction (3D OPC). This method was then applied to fabricate high-fidelity micro-Fresnel lenses. The proposed 3D OPC method for fabricating Fresnel lenses demonstrated a significant reduction of 79.98% in profile error. Additionally, the average peak signal-to-noise ratio (PSNR) of the images utilized in the imaging system exhibited an 18.92% enhancement, while the average contrast of the images showed a 36% improvement. Those outcomes validated the reliability of the algorithm. Furthermore, the potential of this technique in future work to use 3D OPC algorithms to optimize achromatic lenses to improve the PSNR and the contrast of imaging was discussed.

Kim et al. [19] employed a selective laser etching (SLE) technique to fabricate a connection hole on the glass-mediated layer (TGV). The objective was to identify the most effective etchant to improve the etching rate. Four different etchants (HF, KOH, NaOH, and NH_4F) were compared. The results demonstrated that the most effective etchant was NH_4F. The TGV could be generated within 3 h through etching with the 8 M NH_4F solution at

85 °C. The HF, NH$_4$F, NaOH, and KOH solutions generated cone angles of 1–53°, 47–60°, 53–62°, 53–62°, and 58–66°, respectively. The optimal process parameters were found to enhance the productivity of TGVs in glass-mediated layers.

A study on micromirror arrays with tunable reflective properties based on different microstructures and their applications was conducted by Cao et al. [20]. The results demonstrated that microstructures prepared using 3D lithography can be used to tune the reflective properties without the need for an additional power supply. The microstructures were also applied to the projection display segment, which showed that the maximum surface gain of the concave–convex aspheric micromirror arrays (MMA) was up to 2.66. The maximum surface gain of less than 2.72 had a small color difference, minimal geometric distortion, and high-gain reflective properties, which provided enhanced performance and reduced processing costs compared to existing mainstream optical screens. This study addressed the limitation of the modulation of reflective property adjustment due to the complex structure and preparation process in power systems, which provided new opportunities for the development of advanced reflective optical surfaces.

Man et al. [21] fabricated parallel interconnects using a digital design and investigated the effect of Triton X-100 dispersant on the rheology of La$_{0.6}$Sr$_{0.2}$Fe$_{0.8}$Co$_{0.2}$O$_3$-δ (LSCF) slurries. This was achieved by using a direct-write micromanipulation technique to fabricate micro-single-compartment SOFCs (u-SC-SOFCs). The results indicated that the optimal concentration of Triton X-100 for various slurries was approximately 0.2–0.4% of the LSCF solid in weight. In detail, the slurry composition of 50% solids, 12% binder, and 0.2% dispersant was identified as the optimal formulation for SOF fabrication via the direct-write method. This study established a foundation for the development of optimal process parameters and sintering cycles for the deposition of high-fidelity cathode electrodes for miniature single-compartment solid oxide fuel cells for green energy.

Li et al. [22] employed dispersion polymerization to synthesize micro/nanoscale polystyrene (PS) spheres, which were subsequently deposited onto a silicon substrate via a floating assembly method to form a long-range monolayer. Sub-wavelength structures were subsequently manufactured using dry etching techniques. The polystyrene spheres with diameters ranging from 500 nm to 5.6 μm were produced by making adjustments to the polyvinylpyrrolidone (PVP) stabilizer and variations in component concentration. The optical limitations inherent in colloidal lithography in the conventional lithography process were resolved.

In a study by Li et al. [23], a superhydrophobic substrate, developed via a novel method using a bottom-up scheme, was fabricated using stoichiometric-grade thiol-ene (OSTE) micro-mushroom forests. In a relatively short time period, a large area of OSTE micro-mushroom forests was fabricated, which exhibited excellent superhydrophobicity and low adhesion properties. Those properties were evidenced by a contact angle of 152.9 ± 0.2°, a sliding angle of 4.1°, and a contact angle hysteresis of less than 0.5°. The deficiencies of the traditional fabrication methods for superhydrophobic surfaces were solved in this study. Those properties indicated that this substrate has great potential for superhydrophobic applications.

Kim et al. [24] employed the evolutionary, finite element, machine learning, and deep learning methods to model the photoresist flow process (RFP) under various reflow conditions. Additionally, the self-consistent field theory (SCFT) was utilized to describe the self-assembly of a cylindrical block copolymer (BCP), which was confined in the pre-patterning of the photoresist reflow process (RFP) to produce small contact hole (C/H) sizes. The study established a foundation for shrinkage modeling and the generation of smaller contact holes, which enhanced the photoresist reflow process (RFP) with random contact holes (C/Hs).

The nanosecond laser direct-write ablation technique was employed to create a micro- and nano-textured surface on a high-frequency electrosurgical knife [25]. The surface was designed to mimic the structure found on the leaves of Setaria viridis. Additionally, the TiO$_2$ coating was deposited on the knife surface through magnetron sputtering. Then, the plasma-induced hydrophobic modification and octadecyl trichlorosilane (OTS) treatments were conducted to enhance the hydrophilic nature of the silicone oil. These can construct a self-lubricating and anti-stick surface. The study successfully reduced the tendency of human tissues to adhere to electrosurgical knives, which can reduce the risk of surgical complexity and increased medical complications.

Luo et al. [26] investigated the mechanism of the in situ laser-enhanced ultra-precision cutting machinability of ZnS. The physical properties of ZnS were characterized by high-temperature nanoindentation experiments. The material removal mechanism of ZnS during in situ laser-assisted cutting was also investigated. The results revealed that the initiation depth of the groove damage produced by in situ laser-assisted cutting increased by 57.99%. The formation of micrometer-scale pits was effectively mitigated through the use of in situ laser-assisted cutting. Following the planning of cutting experiments, it was successfully demonstrated that a smooth and uniform surface with a Sa of 3.607 nm was achieved at a laser power of 20 W, which was 73.58% higher than that achieved through normal cutting. The study provided a promising method for ultra-precision cutting of ZnS.

Singh et al. [27] developed CAM software (version 1) for aspherical surface machining by integrating the measurement technology with an optimized material removal mechanism in grinding and polishing. Computer numerical control (CNC) technology, computer technology, and data analysis were also combined. The software employs compensating correction algorithms to process error data and generate numerical control programs for machining. Thie study addressed the existing manufacturing challenges and improved the performance of the optical system.

Shi et al. [28] developed a normal displacement model based on the Hertz contact theory. A normal displacement compensation method was subsequently proposed based on the decoupled polishing system. The experimental results demonstrated that compared to no displacement compensation, the utilization of the normal displacement compensation during the polishing process resulted in a notable reduction in the surface roughness from 0.4 μm to 0.21 μm. Additionally, the unevenness coefficient of surface roughness exhibited a considerable decline from 112.5% to 19%. This study solved the polishing tool location error caused by the contact deformation of the tool, abrasive, and workpiece.

Duan et al. [29] fabricated layered micro/nanostructures, which comprise periodic microstructures, laser-induced periodic surface junctions (LIPSSs), and nanoparticles, through femtosecond laser processing (LP). Furthermore, a hydrophobic layer was formed on the micro/nanostructures through perfluorosilane modification (PM). The experiments were conducted under the optimized parameters, and the results demonstrated that the average reflectivity on the silicon surface in the visible light band was reduced to 3.0%. Additionally, the surface exhibited a large contact angle of $172.3 \pm 0.8°$ and a low sliding angle of $4.2 \pm 1.4°$. This approach has potential applications in several fields, including optics, detectors, and photovoltaics.

In the last contribution, Liu et al. [30] conducted sandblasting experiments to investigate the effects of process parameters (air pressure, lift-off height, abrasive volume, and abrasive type) on the processing time, surface roughness, and mechanical properties of wire component. The goal was to reveal the different cutting mechanisms between the SAC solder and 304 V wire. The results demonstrated that resin abrasives can effectively remove SAC burrs while maintaining the integrity of 304 V due to the proper hardness and young modulus. The processing time decreased with increasing air pressure, while the surface roughness increased with increasing abrasive volume. Additionally, sandblasting led to a decrease in the yield strength of the wire. The tensile strength was influenced by Young's modulus and the hardness of the abrasive. The study provided a comprehensive understanding of sandblasting and the different cutting mechanisms.

In conclusion, this Special Issue of *Micromachines* offers a comprehensive and insightful overview of the latest developments in micro/nano-fabrication. The processing methods and combinations of techniques presented in these papers provide innovative ideas for the future micro/nano-fabrication of materials and promote the application of micro/nano-fabrication technologies in a wide range of fields. The findings of the above studies provide researchers with valuable insights that can be employed to further develop and enhance the utilization of micro/nano-fabrication techniques in electronic devices, semiconductors, and bio-devices. This will lead to improved product performance and value.

Conflicts of Interest: The authors declare no conflict of interest.

References

1. Weiling, W.; Jinlin, C.; Lei, C.; Weng, D.; Yu, Y.; Hou, Y.; Yu, G.; Wang, J.; Wang, X. A laser-processed micro/nanostructures surface and its photothermal de-icing and self-cleaning performance. *J. Colloid Interface Sci.* **2024**, *655*, 307–318.
2. Felicis, D.D.; Mughal, Z.M.; Bemporad, E. A method to improve the quality of 2.5-dimensional micro-and nano-structures produced by focused ion beam machining. *Micron* **2017**, *101*, 8–15. [CrossRef]
3. Huili, H.; Minglin, H.; Hao, L.; Bing, Z.; Chong, Z. Damage evolution and crystalline orientation effects in ultrafast laser micro/nano processing of single-crystal diamond. *Opt. Laser Technol.* **2024**, *169*, 110120.
4. Yang, Y.; Min, F.; Wang, Y.; Guo, L.; Long, H.; Qu, Z.; Zhang, K.; Wang, Y.; Yang, J.; Chen, Y.; et al. Solution-Processed Micro-Nanostructured Electron Transport Layer via Bubble-Assisted Assembly for Efficient Perovskite Photovoltaics. *Adv. Mater.* **2024**, *36*, e2408448. [CrossRef] [PubMed]
5. Shivaprakash, N.K.; Banerjee, P.S.; Banerjee, S.S.; Barry, C.; Mead, J. Advanced polymer processing technologies for micro and nanostructured surfaces: A review. *Polym. Eng. Sci.* **2023**, *63*, 1057–1081. [CrossRef]
6. Ma, Z.; Wang, W.; Xiong, Y.; Long, Y.; Shao, Q.; Wu, L.; Wang, J.; Tian, P.; Khan, A.U.; Yang, W.; et al. Carbon Micro/Nano Machining toward Miniaturized Device: Structural Engineering, Large-Scale Fabrication, and Performance Optimization. *Small* **2024**, *19*, e2400179. [CrossRef]
7. Zhao, D.; Yu, X.; Yang, B.; Liu, Y.; Xiu, W. Electrochemical behavior and microstructural characterization of nano-SiC particles reinforced aluminum matrix composites prepared via friction stir processing. *Ceram. Int.* **2024**, *50*, 35525–35536. [CrossRef]
8. Jiajun, L.; Sanjun, L.; Yonghua, Z. Enabling Jet-Electrochemical Discharge Machining on Niobium-Like Passivating Metal and the Single Step Fabrication of Coated Microstructures. *J. Electrochem. Soc.* **2023**, *170*, 093508.
9. Yang, D.; Zhao, L.; Cheng, J.; Chen, M.; Liu, H.; Wang, J.; Han, C.; Sun, Y. Unveiling sub-bandgap energy-level structures on machined optical surfaces based on weak photo-luminescence. *Nanoscale* **2023**, *15*, 18250–18264. [CrossRef]
10. Mita, H.; Mizuno, Y.; Tanaka, H.; Fujie, T. UV laser-processed microstructure for building biohybrid actuators with anisotropic movement. *Biofabrication* **2024**, *16*, 025010. [CrossRef]
11. Gore, P.; Chen, J.Y.; Sundaram, M. Unsupervised detection and mapping of sparks in the Electrochemical Discharge Machining (ECDM) process. *Manuf. Lett.* **2024**, *41*, 135–441. [CrossRef]
12. Wang, Y.; Wang, Y.; Cao, X.; Cheng, Z.; Zhang, M.; Yin, J.; Dong, Y.; Fu, Z. Study on the formation mechanism of functional surface microstructure by ultrasonic vibration assisted laser processing. *Opt. Commun.* **2023**, *546*, 29805. [CrossRef]
13. Zuo, P.; Liu, T.; Li, F.; Wang, G.; Zhang, K.; Li, X.; Han, W.; Tian, H.; Hu, L.; Huang, H.; et al. Controllable Fabrication of Hydrophilic Surface Micro/Nanostructures of CFRP by Femtosecond Laser. *ACS Omega* **2024**, *9*, 20988–20996. [CrossRef] [PubMed]
14. Drabczyk, K.; Sobik, P.; Kulesza-Matlak, G.; Jeremiasz, O. Laser-Induced Backward Transfer of Light Reflecting Zinc Patterns on Glass for High Performance Photovoltaic Modules. *Materials* **2023**, *16*, 7538. [CrossRef] [PubMed]
15. Moskal, D.; Martan, J.; Honner, M. Scanning Strategies in Laser Surface Texturing: A Review. *Micromachines* **2023**, *14*, 1241. [CrossRef]
16. Zhang, H.; Zhao, Z.; Li, J.; Ye, L.; Liu, Y. Review on Abrasive Machining Technology of SiC Ceramic Composites. *Micromachines* **2024**, *15*, 106. [CrossRef]
17. Wang, Q.; Zhang, R.; Chen, Q.; Duan, R. A Review of Femtosecond Laser Processing of Silicon Carbide. *Micromachines* **2024**, *15*, 639. [CrossRef]
18. Peng, F.; Sun, C.; Wan, H.; Gui, C. An Improved 3D OPC Method for the Fabrication of High-Fidelity Micro Fresnel Lenses. *Micromachines* **2023**, *14*, 2220. [CrossRef]
19. Kim, J.; Kim, B.; Choi, J.; Ahn, S. The Effects of Etchant on via Hole Taper Angle and Selectivity in Selective Laser Etching. *Micromachines* **2024**, *15*, 320. [CrossRef]
20. Cao, H.; Xue, Z.; Deng, H.; Chen, S.; Wang, D.; Gui, C. Micromirror Array with Adjustable Reflection Characteristics Based on Different Microstructures and Its Application. *Micromachines* **2024**, *15*, 506. [CrossRef]
21. Yang, M.; Parupelli, S.K.; Xu, Z.; Desai, S. Understanding the Effect of Dispersant Rheology and Binder Decomposition on 3D Printing of a Solid Oxide Fuel Cell. *Micromachines* **2024**, *15*, 636. [CrossRef] [PubMed]
22. Lee, Y.-C.; Wu, H.-K.; Peng, Y.-Z.; Chen, W.-C. The Synthesis and Assembly Mechanism of Micro/Nano-Sized Polystyrene Spheres and Their Application in Subwavelength Structures. *Micromachines* **2024**, *15*, 841. [CrossRef] [PubMed]
23. Li, H.; Zhang, M.; Liu, Y.; Yu, S.; Li, X.; Chen, Z.; Feng, Z.; Zhou, J.; He, Q.; Chen, X.; et al. Off-Stoichiometry Thiol-Ene (OSTE) Micro Mushroom Forest: A Superhydrophobic Substrate. *Micromachines* **2024**, *15*, 1088. [CrossRef] [PubMed]
24. Kim, S.-K. Contact Hole Shrinkage: Simulation Study of Resist Flow Process and Its Application to Block Copolymers. *Micromachines* **2024**, *15*, 1151. [CrossRef] [PubMed]

25. Zhang, Q.; Yang, Y.; Huo, S.; Duan, S.; Han, T.; Liu, G.; Zhang, K.; Chen, D.; Yang, G.; Chen, H. Laser Direct Writing of Setaria Virids-Inspired Hierarchical Surface with TiO_2 Coating for Anti-Sticking of Soft Tissue. *Micromachines* **2024**, *15*, 1155. [CrossRef]
26. Luo, H.; Wang, X.; Qin, L.; Zhao, H.; Zhu, D.; Ma, S.; Zhang, J.; Xiao, J. Investigation on the Machinability of Polycrystalline ZnS by Micro-Laser-Assisted Diamond Cutting. *Micromachines* **2024**, *15*, 1275. [CrossRef]
27. Singh, R.P.; Chen, Y. Integration of Metrology in Grinding and Polishing Processes for Rotationally Symmetrical Aspherical Surfaces with Optimized Material Removal Functions. *Micromachines* **2024**, *15*, 1276. [CrossRef]
28. Shi, Y.; Su, M.; Cao, Q.; Zheng, D. A Normal Displacement Model and Compensation Method of Polishing Tool for Precision CNC Polishing of Aspheric Surface. *Micromachines* **2024**, *15*, 1300. [CrossRef]
29. Duan, J.; Long, G.; Xu, X.; Liu, W.; Li, C.; Chen, L.; Zhang, J.; Xiao, J. Hierarchical Micro/Nanostructures with Anti-Reflection and Superhydrophobicity on the Silicon Surface Fabricated by Femtosecond Laser. *Micromachines* **2024**, *15*, 1304. [CrossRef]
30. Liu, Y.; Zhai, S.; Guo, J.; Fu, S.; Shen, B.; Zhao, Z.; Ding, Q. Study on the Abrasive Blasting Mechanism of Solder Welded 304V Wire in Vascular Intervention. *Micromachines* **2024**, *15*, 1405. [CrossRef]

Disclaimer/Publisher's Note: The statements, opinions and data contained in all publications are solely those of the individual author(s) and contributor(s) and not of MDPI and/or the editor(s). MDPI and/or the editor(s) disclaim responsibility for any injury to people or property resulting from any ideas, methods, instructions or products referred to in the content.

Article

Study on the Abrasive Blasting Mechanism of Solder Welded 304V Wire in Vascular Intervention

Yao Liu [1,2,3], Shaobo Zhai [1,4], Jinzhu Guo [1], Shiling Fu [3,4], Bin Shen [3,4], Zhigang Zhao [1] and Qingwei Ding [5,*]

1. Shanxi Key Lab of Advanced Manufacturing Technology, North University of China, Taiyuan 030051, China; liuyao@nuc.edu.cn (Y.L.); z13834038164@163.com (S.Z.); 18234146352@163.com (J.G.); zhao105820@163.com (Z.Z.)
2. Guangdong Provincial Key Laboratory of Minimally Invasive Surgical Instruments and Manufacturing Technology, Guangdong University of Technology, Guangzhou 510006, China
3. Jiaxing Jiangxin Medical Technology Co., Ltd., Jiaxing 314200, China; fu18823128633@hotmail.com (S.F.); binshen@zgjczg.com (B.S.)
4. Yangtze Delta Region Institute of Tsinghua University, Jiaxing 314000, China
5. Department of Vascular Surgery, General Surgery Clinical Center, Shanghai General Hospital, Shanghai Jiao Tong University School of Medicine, 100 Haining Road, Shanghai 200080, China
* Correspondence: qingwei.ding@shgh.cn

Abstract: The solder burrs on the 304V wire surface can easily scratch the vascular tissue during interventional treatment, resulting in complications such as medial tears, bleeding, dissection, and rupture. Abrasive blasting is often used to remove solder burr and obtain a smooth surface for the interventional device. This study conducted an abrasive blasting experiment to explore the effects of process parameters (air pressure, lift-off height, abrasive volume, and abrasive type) on processing time, surface roughness, and mechanical properties to reveal the material removal mechanism. The results indicated that the resin abrasive can remove the SAC burr and keep the 304V integrity due to the proper hardness and Young's module. Impaction pits are the main material removal mode in abrasive blasting. The processing time decreases with the increase in air pressure. The surface roughness increases with the increase in abrasive volume. The primary and secondary factors affecting the surface roughness of the 304V wire after abrasive blasting are the abrasive type and air pressure, followed by the abrasive volume and lift-off height. Blasting leads to a decrease in yield strength, and Young's modulus and the hardness of the abrasive will affect the tensile strength. This study lays a foundation for understanding abrasive blasting and different cutting mechanisms.

Keywords: abrasive blasting; orthogonal experiment; surface roughness; interventional devices

1. Introduction

Guidewire [1,2], stent [3], filter [4], interventional coil [5], and braided catheter, used in vascular intervention, are made of precision wire. The connection between the wire and other components, such as steel tube, endoscopic lens house, radiopaque marker bands, and coil, is usually soldered. Due to the natural solidification of the solder after melting and the effect of gravity, burrs is easy to generate. In the vascular intervention, the burrs may scratch the intima and media of the blood vessels, resulting in hemorrhagic thrombosis, intima endometrial, media tearing, and dissection. In severe cases, it may cause vascular perforation and fracture, which may cause myocardial infarction and death. It is necessary to remove the residual solder burr to generate a smooth surface. The grinding and abrasive blasting are the two main ways to deburr the solder. Grinding is usually used for a regular shape with uniform dimensions. However, for heterogeneous structures, large dimensional variation, and sharp corner surfaces such as vascular filters, stents, and ventricular occlude, burrs on the corners cannot be reached by the wheel and can only be removed by abrasive blasting.

The abrasive blasting process uses compressed air to drive an abrasive to impact and scratch the workpiece surface at high speed [6]. Many factors affect the performance of the workpiece after abrasive blasting, including the size and shape of the abrasive particles, the jet pressure, the lift-off height, the impact velocity, and so on. The parameters for evaluating the performance of the workpiece include surface roughness, surface properties, residual stress, mechanical properties, etc. The existing literature focuses on the relationship between them.

Some scholars have conducted in-depth research on the specific influence mechanisms of the individual factors. Li et al. [7] found that abrasive blasting destroys the surface integrity of single-crystal superalloys, resulting in irregular pits on the surface caused by the cutting by abrasive particles while also changing the surface morphology; the surface roughness and microhardness increase as sand particle size increases. Masanoa et al. [8] evaluated the bending strength of four zirconia materials after abrasive blasting using the Al2O3 abrasive and found that microcracks appeared on the subsurface with high compressive stress. Cruz et al. [9] pointed out that pressure is the key factor in workpiece modification; it accounts for 57.0%, 72.8%, and 61.0% of the change in roughness, hardness, and corrosion resistance, respectively. Ding et al. [10] found that the surface layer of titanium after 200–300 μm SiO_2 abrasive blasting was transformed into a nanocrystalline structure, which improved corrosion resistance and fatigue strength. Based on Abaqus, Cao et al. [11] pointed out that the higher the shot peening intensity, the greater the depth corresponding to the residual stress field of the target, and the more serious the surface roughness. Some scholars have analyzed the effect of two different factors on the performance of the workpiece. Anna [12] pointed out that the abrasive type has a larger influence on the surface roughness and adhesion of the workpiece than the air pressure in abrasive blasting. Lundgren et al. [13] evaluated the tooth enamel damage degree caused by abrasive blasting and pointed out that time a has stronger influence than that of lift-off height. Some scholars have carried out multi-factor analysis by carrying out experiments and constructing related models. Laureniu et al. [14] constructed a mathematical model to predict the abrasive blasting surface roughness, which is affected by lift-off height, abrasive size, and impact angle. Kanesan et al. [15] studied the erosion characteristics of abrasive blasting on 316 stainless steel wire screen meshes. The wear of the workpiece becomes more serious with the increase in the abrasive particle size and impact speed, and there was a shift in the erosion mechanism from micro-plowing to deeper plowing and in the pitting mechanism when the impact angle increased from 30° to 45°. However, the magnitude of plowing and pitting reduced as the impact angle increased to 60° and 90°. Budi et al. [16] studied the influence mechanism of three different abrasives (metallic shot, slag balls, and silica particles) on the surface morphology of 316L stainless steel in abrasive blasting, and the surface roughness value caused by the silica particles that were angular and had an irregular surface morphology was the largest. Guo et al. [17] used six different abrasives with different sizes to blast titanium plates, and they exhibited different characteristics. Carlos et al. [18] pointed out that the residual stress generated during ceramic abrasive blasting is the main factor that causes surface defects.

In the removal of solder burr by abrasive blasting, the abrasive also removes the substrate material, which causes a decrease in the strength and fatigue life. Therefore, the process of different cutting, which is to remove the solder and keep the integrity of the substrate as much as possible, should be investigated via a systematic analysis of the interaction between multiple factors of sandblasting, which has not been reported in the studies reviewed above. Orthogonal tests are usually used for multi-factor analysis, which is another design method to study multiple factors and multiple levels. It selects some representative points from the comprehensive test according to the orthogonality and can reduce the test time and cost, while ensuring the reliability and repeatability of the experimental results.

In this study, abrasive blasting experiments were conducted to reveal the differential cutting mechanism of 304V wire and SAC305. Firstly, the abrasive blasting experimental setup was introduced, and Taguchi experiments were conducted to investigate the effect of air pressure, lift-off height, abrasive volume, and the abrasive (including type and size). Then, the surface topography was observed to reveal the removal mechanism in abrasive blasting. The processing time for the SAC removal, the surface roughness, and the tensile and yield strength of the wire component after abrasive blasting were measured to reveal the removal mechanism indirectly. Finally, the conclusions were drawn regarding the significance of this study.

2. Experimental Setup

The wire component is a spring shaft (outer diameter: 0.65 mm) coiled by a 304V stainless steel wire (diameter: 0.15 mm), which is the most used metal for medical devices. The solder is SAC305, which is 96.5% tin, 3% silver, and 0.5% copper in weight, and has been proven to have no toxicity for the human body. The melted SAC305 adheres to the wire component surface and further realizes the connection to other components. Table 1 presents the size and mechanical properties of the wire component and solder.

Table 1. Material parameters used in the experiment.

Name	Material	Size	Density (g·cm^{-3})	Young's Modulus (MPa)	Vickers Hardness (HV)	Figure
Wire component	304V	Ø0.65 mm	7.93	1.7×10^5	210	
Solder	SAC305	0.6 mm^3	7.37	46	14.1	
Resin	Urea formaldehyde resin	#120, #150, #200	1.2	3358	145–175	
Plastic abrasive	TPU	#300	1.1	30	90–95	
	PVC	#200	1.38	2900	120–125	
	PTFE	#300	2.1–2.3	500	58	
Biomass	Corn	#200	0.37	-	-	
	Walnut	#200	0.24	-	-	
Industry	Alumina	#200	3.85	3.5×10^5	900	

The device used in this experiment is a single-pen abrasive blasting machine (BP-1, Tianjin Lizhong Huier Technology Development Co., Ltd., Tianjin, China), which uses compressed air to carry the abrasive in dry conditions. The abrasive tank is a cylindrical container with a diameter of 70 mm and a height of 170 mm. There is an abrasive delivery pipeline with a height of 110 mm in the tank. The height of the abrasive should not exceed the exit tube of the container. The compressed air from the pipeline inlet impacts on and mixes with the abrasives. The mixture was taken from the outlet to the nozzle. The abrasive in the mixture flow with the air. Then, the abrasive impacts the workpiece, punching and scratching it, as shown in Figure 1.

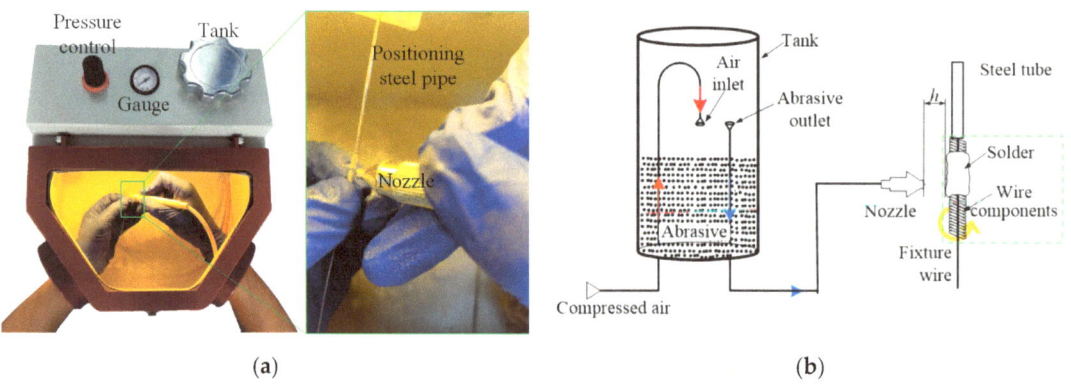

Figure 1. Abrasive blasting experimental setup. (**a**) Overview and (**b**) configuration.

During the experiment, to figure out the effect of compressed air, the lift-off height of the nozzle, and the abrasive volume and type on the removal process, the Taguchi experiments were designed and conducted. The resin abrasive used was thermosetting urea-formaldehyde (UF) resin [19]. The plastic abrasives used were thermoplastic polyurethane rubber (TPU), polyvinyl chloride (PVC), and polytetrafluoroethylene (PTFE). The ground corn kernels and walnut shells were used as representations of the biomass material. An industry alumina abrasive with high hardness was chosen as the last one. Based on the number and level of the factors, the standard L27 ($3^9 \times 9^1$) experiment table was designed, as shown in Table 2.

The range analyses are often used to assess the effect of the factors. The larger the range, the greater the effect of the factors. The calculation of the range R is as follows:

$$R = \overline{K}_{max} - \overline{K}_{min}, \quad (1)$$

where $\overline{K}_{max}$ and $\overline{K}_{min}$ are the maximum and the minimum meaning of results at specified factors with different levels. The $\overline{K}$ can be calculated as follows:

$$\overline{K} = \frac{K}{i}, \quad (2)$$

where K is the sum of the experimental results of factor level, and i is the number of factor levels in the experiment; i = 3 or 9 for this study.

In this study, the air pressures were selected as 0.3, 0.4, and 0.5 MPa. The lift-off heights were 2, 4, and 6 mm, which is controlled by a rule before the blasting. The abrasive volumes in height were 60, 80, and 100 mm. A total of 9 types of abrasive were selected in this experiment, as shown in Table 1, which were easily obtained from the market. To achieve the uniform removal of the solder, the 304V wire component was rotated at a constant speed of 80°/s during abrasive blasting. A magnifying glass was set on the 304V wire to check the remaining solder, and the minimum time to remove all solder was recorded with a stopwatch. To reduce the potential damage to the wire component, the maximum abrasive blasting time is set at 300 s. When the abrasive size or type was changed, the tank and pipeline were washed with water and dried with compressed air.

Table 2. Orthogonal experiment.

No.	Factor Level	Pressure (MPa)	Lift-Off Height (mm)	Abrasive Volume (mm)	Abrasive	Times (s)	Surface Roughness (μm)	Tensile Strength (MPa)	Yield Strength (MPa)
1	$A_1B_1C_1D_1$	0.3	2	60	#120 resin	160	0.19	926.1	77.5
2	$A_1B_1C_1D_2$	0.3	2	60	#150 resin	120	0.21	946.3	82.3
3	$A_1B_1C_1D_3$	0.3	2	60	#200 resin	150	0.25	896.7	83.9
4	$A_1B_2C_3D_4$	0.3	4	100	TPU	300	0.16	947.9	82.5
5	$A_1B_2C_3D_5$	0.3	4	100	PVC	300	1.02	917.4	79.4
6	$A_1B_2C_3D_6$	0.3	4	100	PTFE	300	0.16	931.5	84.8
7	$A_1B_3C_2D_7$	0.3	6	80	Corn	300	0.54	984.0	96.1
8	$A_1B_3C_2D_8$	0.3	6	80	Walnut	300	0.22	944.4	89.1
9	$A_1B_3C_2D_9$	0.3	6	80	Alumina	10	0.83	721.3	68.8
10	$A_2B_1C_3D_9$	0.4	2	100	Alumina	5	0.44	881.2	86.8
11	$A_2B_1C_3D_7$	0.4	2	100	Corn	180	0.77	940.7	81.5
12	$A_2B_1C_3D_8$	0.4	2	100	Walnut	180	0.86	929.8	89.1
13	$A_2B_2C_2D_3$	0.4	4	80	#200 resin	50	0.29	922.7	82.9
14	$A_2B_2C_2D_1$	0.4	4	80	#120 resin	170	0.62	933.3	90.4
15	$A_2B_2C_2D_2$	0.4	4	80	#150 resin	90	0.34	909.7	80.3
16	$A_2B_3C_1D_6$	0.4	6	60	PTFE	300	0.67	938.0	83.1
17	$A_2B_3C_1D_4$	0.4	6	60	TPU	300	0.37	941.4	90.7
18	$A_2B_3C_1D_5$	0.4	6	60	PVC	300	0.16	944.7	75.5
19	$A_3B_1C_2D_5$	0.5	2	80	PVC	300	0.74	909.2	90.9
20	$A_3B_1C_2D_6$	0.5	2	80	PTFE	300	0.33	927.7	78.8
21	$A_3B_1C_2D_4$	0.5	2	80	TPU	300	0.16	920.2	73.8
22	$A_3B_2C_1D_8$	0.5	4	60	Walnut	245	0.74	939.9	78.8
23	$A_3B_2C_1D_9$	0.5	4	60	Alumina	5	0.60	641.4	60.6
24	$A_3B_2C_1D_7$	0.5	4	60	Corn	190	0.64	929.5	86.1
25	$A_3B_3C_3D_2$	0.5	6	100	#150 resin	40	0.44	935.4	87.1
26	$A_3B_3C_3D_3$	0.5	6	100	#200 resin	35	0.42	986.9	85.7
27	$A_3B_3C_3D_1$	0.5	6	100	#120 resin	100	0.32	939.4	87.8

A laser confocal microscope LEXT OLS4000 (Olympus, Tokyo, Japan) was used to observe the surface after abrasive blasting. The solder's removal condition was double-checked. The experiment samples that could not remove the solder completely are excluded in the following discussion. The surface roughness of the wire was measured by a laser confocal microscope LEXT OLS4000 (Olympus, Tokyo, Japan) before and after the abrasive blasting. Polynomial filtering was used to remove the wire curvature for surface roughness measurement. Five points on each sample were measured, and the average value was calculated. The original surface roughness of the 304 V wire before the abrasive blasting was 0.16 μm, as shown in Figure 2a. Six photos uniformly distributed on the wire surface were taken for damage checking, as shown in Figure 2b. The pits with a diameter larger than 10 μm and the scratches with a length greater than 50 μm on the surface were counted to evaluate the wire damage after abrasive blasting, as shown in Figure 2b. The mechanical properties of the wire components after abrasive blasting were measured with a tensile test. As shown in Figure 2c, the initial length of the wire was 100 mm, and the feed rate was 200 mm/min. Ten tensile tests were conducted on the wire component after abrasive blasting.

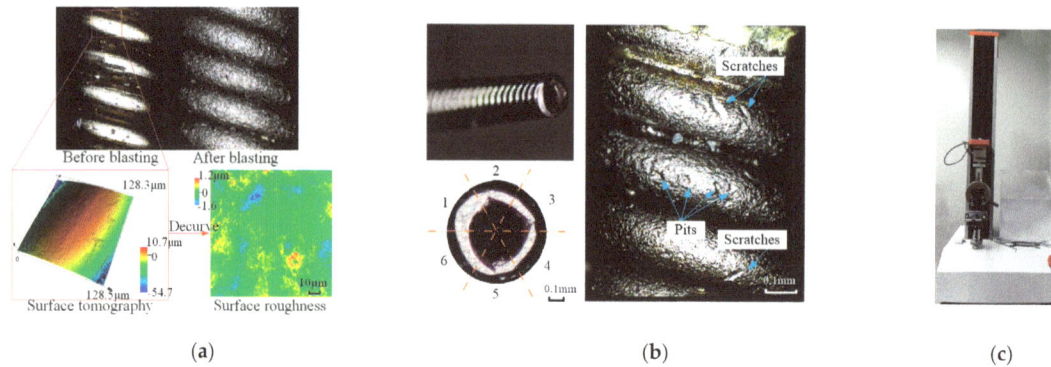

Figure 2. Measurement of test results: (**a**) surface roughness, (**b**) surface damage, and (**c**) mechanical properties.

3. Results

3.1. Surface Topography

Figures 3 and 4 show the 304V wire components after abrasive blasting. The resin abrasive in nine tests can effectively remove all the SAC 305 solder on the surface, as shown in Figure 3a. The size of the wire is the same before and after abrasive blasting, which shows the poor removal ability of the resin abrasive on the 304V wire. The smooth wire surface after abrasive blasting shows many pits, which cause the loss of metallic luster. The plastic abrasives in nine tests fail to remove all the solder on the wire surface, as shown in Figure 3b. The solder after abrasive blasting by a plastic abrasive was not subjected to any obvious impact, which indicates a poor removal ability. The PVC (D5) abrasive showed a stronger ability to remove solder than the TPU (D4) and PTFE (D6) abrasives after 300 s, which also partially changed the wire surface topography obviously.

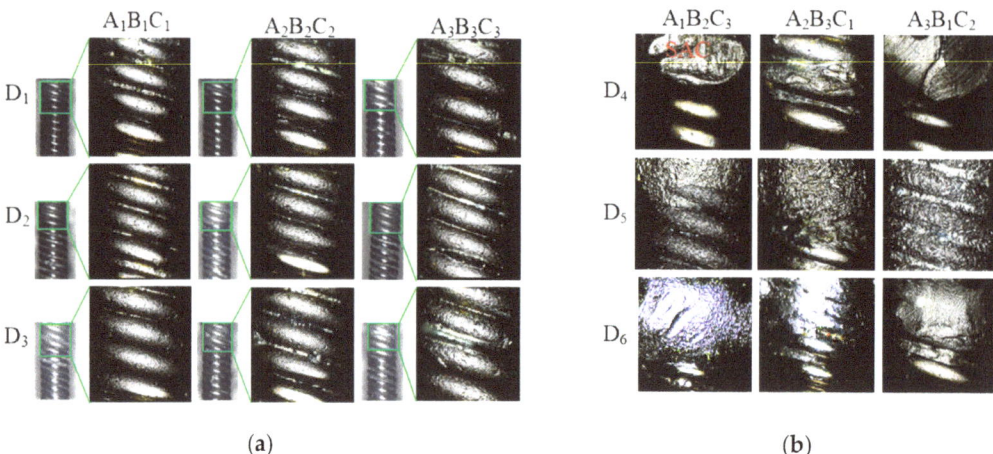

Figure 3. Blasting surface topography by (**a**) resin and (**b**) plastic abrasive.

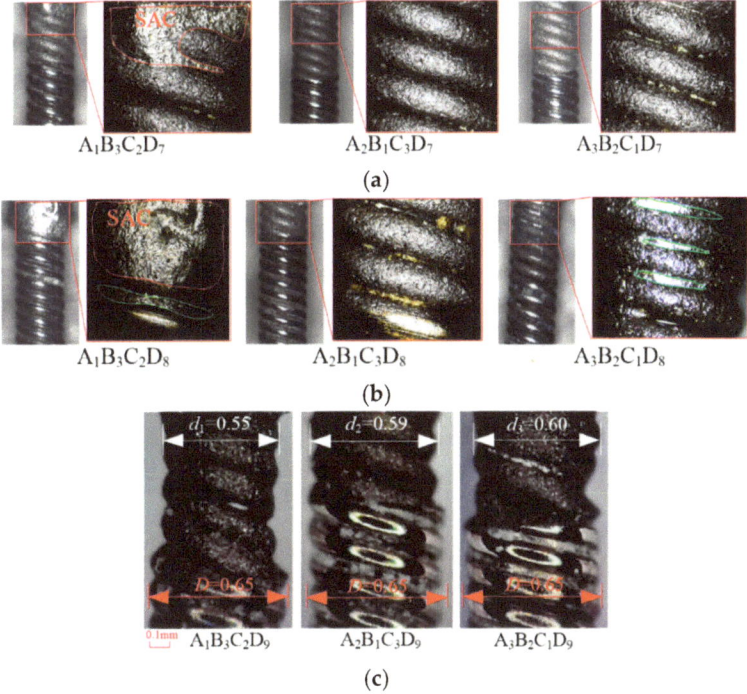

Figure 4. Blasting surface topography by (**a**) corn, (**b**) walnut, and (**c**) alumina abrasive.

The biomass abrasive (corn (D7) and walnut (D8)) cannot remove the SAC 305 on the wire surface completely, as shown in Figure 4a,b. Due to the poor material removal ability, all the abrasive blasting times are 300 s. Under the same processing time, the remaining SAC305 area on the wire surface after processing by corn is smaller than that of walnut. The wire diameter after processing by the resin, plastic, and biomass abrasives is not changed, except for the many impact pits. The impact from the abrasive can generate compressive stress on the wire surface, which can improve the fatigue life of the wire theoretically. The SAC305 was partially or completely removed for those kinds of abrasive, even if the efficiency may not have been satisfactory. However, the wire diameter did not change, which showed the different cuttings. A reduction in the diameter of the wire components was observed after abrasive blasting by the alumina (D9) abrasive, after the SAC305 was removed. As shown in Figure 4c, the diameters of the 304V components after abrasive blasting were 0.55 mm, 0.59 mm, and 0.60 mm, which is smaller than the original 0.65 mm. This diameter reduction will weaken the wire strength and result in a potential fracture in future applications.

3.2. Processing Time

Table 2 shows the processing time in each test. The TPU (D4), PVC (D5), and PTFE (D6) plastic abrasives did not remove all the SAC305 on the wire surface within 300 s, indicating a weak removal ability. The average processing time of the alumina abrasive (D9) was 6.67 s, which is much less than that of the other eight abrasives, indicating a strong material removal ability. However, the short processing time of the alumina abrasive was detrimental to the control of the removal process, inducing great damage to the wire components. The processing times of the resin abrasives (D1, D2, D3) under different mesh sizes and process parameters were in the range of 35 to 170 s. The time distribution range is 135 s. The large time range makes adjustment easy. The processing time decreases with the decrease in abrasive size. In the six tests (No. 7, 8, 11, 12, 22, and 24) using corn (D7) and walnut (D8) abrasives, the SAC305 can be removed, except in No. 7 and 8, which failed to remove all SAC305 in 300 s. The processing time was between 180 and 300 s. This shows that the removal material ability of the corn and walnut abrasives was greatly affected by the process parameters. The overall processing time was long, resulting in low production efficiency.

The factor analysis by using the processing time is calculated and shown in Figure 5. The processing time decreases with the increase in air pressure. The high air pressure results in high abrasive velocity and impact energy. The lift-off height shows no effect on the processing time. The abrasive volume under 80 mm shows no difference in the processing time. However, when the abrasive volume increases to 100 mm, the processing time decreases by 20% to 160 s. A higher abrasive volume can obtain a shorter processing time, due to an increase in the impact frequency, which can remove solder more efficiently. The abrasive types have a great effect on the processing time. The resin can remove the solder in an acceptable range. However, the plastic cannot remove the solder within 300 s. The biomass abrasive can remove the solder after 200 s, which shows lower efficiency than the resin abrasive. The alumina abrasive can eliminate the solder very quickly, which makes it hard to control the time and result in the wire removal.

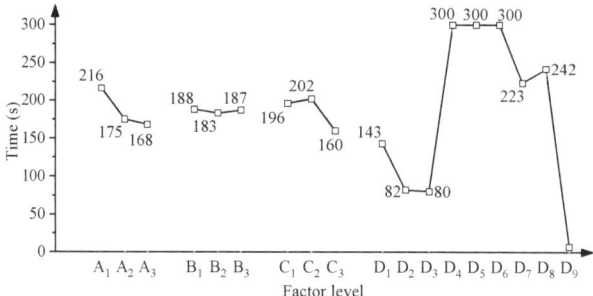

Figure 5. Effect of factor level on processing time.

3.3. Surface Roughness

After the SAC305 is removed, the 304V wire surface is exposed and directly impacted by the abrasive, which leads to the deformation and change in surface roughness. The surface roughness of the wire after sandblasting is shown in Table 2. The roughness of testing No. 5, 7, 8, 16, 17, 19, 20, and 21, which can only partially remove SAC305 solder within 300 s, is measured on the exposure area. Tests No. 4, 6, and 18 have an insufficient ability to remove soldering, and the wire is still buried in the SAC305, due to the poor removal ability in the limited time. Therefore, the surface roughness adopts the initial roughness. The resin abrasive in the nine tests shows the smallest surface roughness. The maximum 304V surface roughness after sandblasting is 0.62 µm, indicating little damage to the wire substrate. The use of the resin abrasive can effectively remove solder without damaging the 304V wire, which is more suitable for differential cutting. The surface roughness caused by the alumina abrasive ranges from 0.44 to 0.83 µm, which is higher than that of the resin abrasive. The corn and walnut abrasives are ground from biomass, resulting in an irregular shape, which in turn leads to higher surface roughness. The surface roughness of tests No. 5 and 19, which use the PVC abrasive, is 1.02 µm and 0.74 µm, respectively.

Figure 6 shows the relationship between factor levels and surface roughness. The surface roughness of the 304V wire increases with the increase in abrasive volume. The high abrasive volume causes more abrasive to impact the wire, which results in an increase in surface roughness. The surface roughness increases first and then decreases with the increase in air pressure and lift-off height. As the mesh size of the abrasive decreases, the surface roughness decreases first and then remains stable. The small abrasive size has a low impact energy, which generates a small pit size. However, the small size causes more abrasive to impact the wire and causes the pits number increase in the surface, which keeps the surface roughness stable.

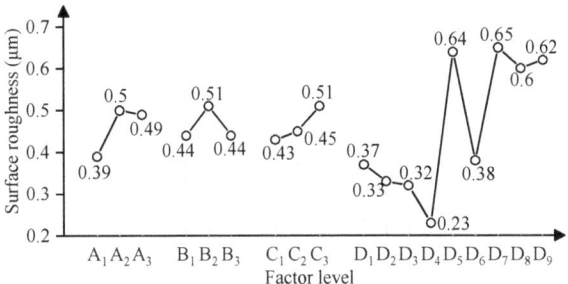

Figure 6. Effect of factor level on surface roughness.

3.4. Wire Mechanical Properties After Abrasive Blasting

Figure 7 presents the relationship between the wire component stress (σ_T) and strain (ε_T) in the tensile process, which goes through an elastic deformation stage and a plastic deformation stage with stress hardening and fracturing. The yield strength and tensile strength of the wire component are 87.4 MPa–96.2 MPa and 879.9 MPa–933.7 MPa before the abrasive blasting.

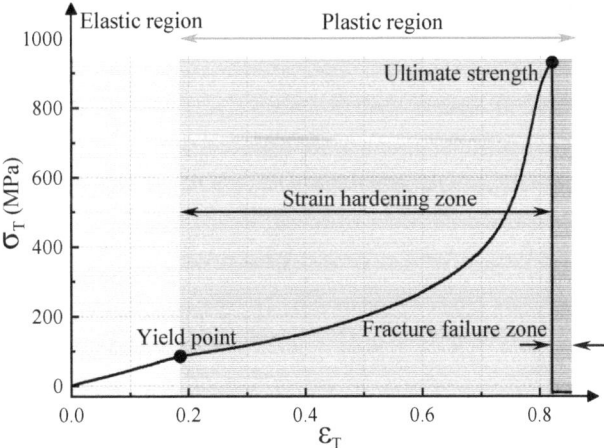

Figure 7. The stress-strain curve of the wire component.

The tensile strength and yield strength of the wire after sandblasting are given in Table 2. Figure 8a shows the relationship between the factor level and the wire tensile strength. The air pressure, lift-off height, and abrasive volume do not affect the tensile strength of the wire. The use of the alumina abrasive reduces the tensile strength by about 28% through the removal of the wire material. Other abrasives do not affect the wire tensile strength. The relationship between the wire yield strength and factor level is shown in Figure 8b. Obviously, the process parameters of abrasive blasting reduce the yield strength and weaken the ability of the wire to resist plastic deformation.

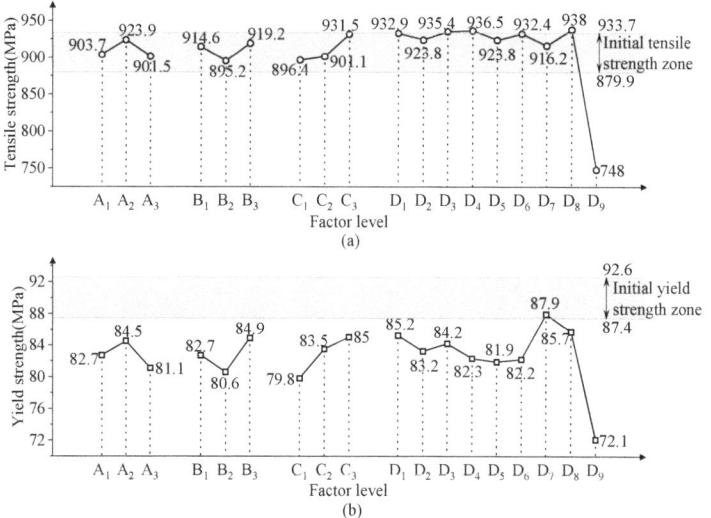

Figure 8. Effect of factor level on mechanical properties: (**a**) tensile and (**b**) yield strength.

3.5. Range Analysis

According to Equations (1) and (2), the K, $\overline{K}$, and R of the processing time, surface roughness, tensile strength, and yield strength are calculated, as presented in Table 3. According to the order of the R, the primary and secondary factors affecting the processing time and surface roughness of the 304V wire are the abrasive type and air pressure, followed by the abrasive volume and lift-off height. The primary and secondary factors that affect the mechanical properties are the abrasive type and abrasive volume, followed by the lift-off height and air pressure.

Table 3. The range of experimental results by factor level.

Factor Level	Times (s)			Surface Roughness (μm)			Tensile Strength (MPa)			Yield Strength (MPa)		
	K	$\overline{K}$	R	K	$\overline{K}$	R	K	$\overline{K}$	R	K	$\overline{K}$	R
A1	1940	215.56		3.55	0.39		8132.9	903.7		744.4	82.7	
A2	1575	175.00	47.22	4.53	0.50	0.11	8314.9	923.9	22.4	760.3	84.5	3.4
A3	1515	168.33		4.39	0.49		8113.3	901.5		729.6	81.1	
B1	1695	188.33		3.95	0.44		8231.3	914.6		744.6	82.7	
B2	1650	183.33	5.00	4.56	0.51	0.07	8057.0	895.2	24.0	725.8	80.6	4.3
B3	1685	187.22		3.97	0.44		8272.8	919.2		763.9	84.9	
C1	1770	196.67		3.83	0.43		8067.7	896.4		718.5	79.8	
C2	1820	202.22	42.22	4.06	0.45	0.08	8109.8	901.1	35.1	751.1	83.5	5.2
C3	1440	160.00		4.59	0.51		8383.6	931.5		764.7	85.0	
D1	430	143.33		1.12	0.37		2798.8	932.9		255.7	85.2	
D2	245	81.67		0.99	0.33		2771.4	923.8		249.7	83.2	
D3	240	80.00		0.96	0.32		2806.3	935.4		252.5	84.2	
D4	900	300.00		0.68	0.23		2809.5	936.5		247	82.3	
D5	900	300.00	293.33	1.92	0.64	0.42	2771.3	923.8	190.0	245.8	81.9	15.8
D6	900	300.00		1.15	0.38		2797.2	932.4		246.7	82.2	
D7	670	223.33		1.96	0.65		2748.6	916.2		263.7	87.9	
D8	725	241.67		1.81	0.60		2814.1	938.0		257.0	85.7	
D9	20	6.67		1.87	0.62		2243.9	748.0		216.2	72.1	

3.6. Damage Quantity

From the surface observation, the main damage after blasting comprises pits and scratches. After sandblasting, ultrasonic cleaning is used to clean the wire assembly to ensure that there is no abrasive inside the pits and scratche. The number of pits and scratches on the 304V wire surface after abrasive blasting were counted, as shown in Figure 9. The quantity of pits is significantly higher than that of scratches under all conditions, which indicates that the impact pit is the primary removal mode in abrasive blasting.

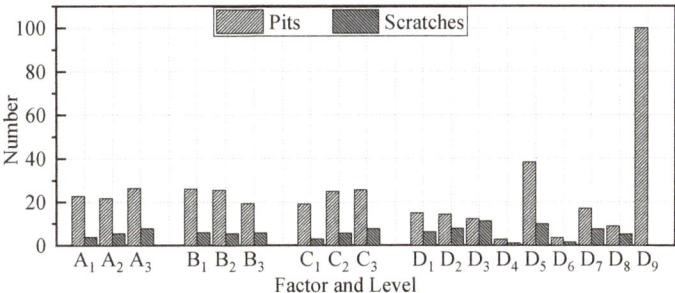

Figure 9. Surface damage quantity on stainless steel wire components.

With the increase in air pressure, the number of scratches increases due to the increased energy of the abrasive. The pits also show an increasing trend, even if the A2 has the lowest quantity. The lift-off height shows that the number of pits on the 304V surface is negative in proportion to the lift-off height and positive in proportion to the abrasive volume. The number of scratches is positively correlated to the air pressure, lift-off height, and abrasive volume and negatively correlated to the abrasive size. The larger mesh size can reduce pit quantity and increase scratch quantity due to the high velocity in the smaller mass, based on the resin abrasive results (D_1, D_2, and D_3). For the plastic abrasive, the removal ability of the TPU (D_4) and PTFE (D_6) abrasive is poor, which left much fewer pits and scratches on the surface. The pits and scratches on the surface processed by the PVC (D_5) abrasive are much larger than D_4 and D_6, which shows the high removal efficiency. The scratches for corn (D_7) are larger than those for walnut (D_8), which results in a high material removal rate. However, the roughness of the workpiece is larger. The alumina abrasive (D_9) shows the most pits and least scratches on the wire surface among all the abrasive materials, which indicates the high material removal ability.

3.7. Material Removal Mechanism

During the abrasive blasting process, the abrasive impacts the solder and the 304V components. Under different impact parameters and angles, different material removal mechanisms can be observed, as shown in Figure 10.

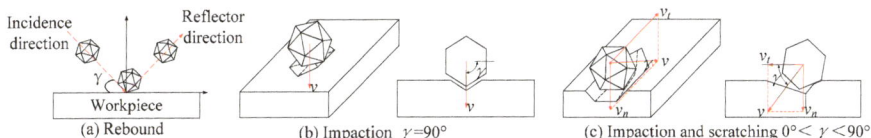

Figure 10. Material removal mechanism in abrasive blasting.

(1) When the hardness and Young's modulus of the abrasive are much smaller than the workpiece, a rebound occurs after the abrasive impacts the surface.

(2) With the increase in the abrasive hardness and Young's modulus, the impact causes plastic deformation of the workpiece. Under $\gamma = 90°$, impact pits are left on the surface. Under $0° < \gamma < 90°$, the v_t mat induces scratch marks on the surface.

Figure 11 presents Young's modulus and hardness comparison of the five abrasive types, SAC305, and the 304V wire. Due to the mechanical properties of the biomass material changing with the moisture and biostructure, which is hard to determine, the corn and walnut properties are excluded in Figure 11. Young's modulus is a physical parameter that describes the ability to resist elastic deformation. Hardness is the ability to resist plastic deformation. Young's modulus and hardness of SAC305 are 14.1 HV and 46×10^5 MPa. Young's modulus and hardness of 304V are 210 HV and 1.7×10^5 MPa. To achieve the different cuttings of the SAC305 and 304V, the plastic deformation should occur in SAC305, and elastic deformation should occur in 304V. Based on the above analysis, the ranges for the different cuttings are marked by the light gray rectangle. Young's modulus of TPU is lower than that in SAC305, which indicated that more elastic deformation occurred in the TPU after impact, which consumes the most energy and results in poor removal ability. The hardness of TPU is between the SAC305 and 304V, which shows the plastic deformation of the SAC305, which is validated in Figure 3b with many pits on the SAC305 surface. The alumina abrasive has a higher Young's modulus and hardness than 304V, which directly results in the removal of the wire, as shown in Figure 4c.

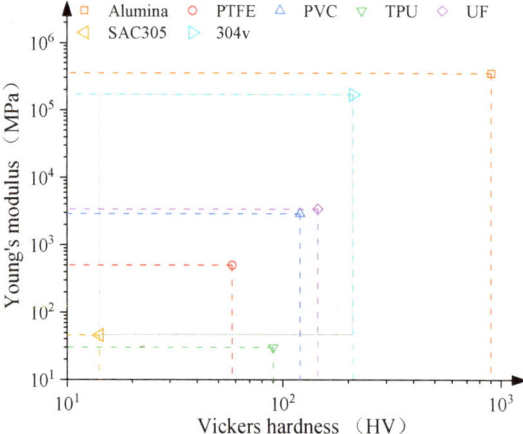

Figure 11. Young's modulus and hardness of the experimental materials.

The properties of UF resin, PVC, and PTFE are located in the gray rectangle, which means all of them can obtain a different cutting. The UF resin has the highest Young's modulus and hardness, which shows the fastest removal rate. The properties of PVC are close to the UF resin, which shows that the solder can also be removed. Young's modulus and hardness of PVC is lower than that of the UF resin, resulting in a lower efficiency, as shown in Figure 3b. The PTFE has a much lower Young's modulus and hardness, which results in a much lower removal rate and longer processing time. Based on the results above, the high abrasive hardness and Young's modulus decrease the processing time, resulting in a higher removal efficiency.

When Young's modulus and the hardness of the abrasive are lower than that in SAC305, a high degree of elasticity occurs in the abrasive, which consumes the majority of the impact energy, causing the abrasive rebound. The effect on SAC is weak and cannot easily cause the removal. When Young's modulus and hardness are in the gray rectangle, the impact of the abrasive damages the SAC directly without causing failure of the stainless steel wire. Under this range, the higher Young's modulus and hardness can achieve a higher material removal rate and lower processing time. If Young's modulus and the hardness of the abrasive are higher than 304V, the removal of 304V is inevitable, which causes the tensile strength to be reduced. Abrasive blasting causes defects such as pits and scratches on the wire, which change the wire's integrity and reduce the yield strength. During blasting, the abrasive directly acts on the surface of the workpiece to achieve material removal. The differences in abrasive material properties, size, and shape lead to different material removal mechanisms, which results in different processing times and surface roughness. The air pressure determines the abrasive energy. The abrasive volume and the lift-off height affect the abrasive quantity per unit time impacted on the 304V wire. The effect of abrasive volume is much larger than the lift-off height, which results in a stronger processing time and surface roughness change.

4. Conclusions

In this study, experiments were conducted to reveal the abrasive blasting mechanism. The effects of air pressure, lift-off height, abrasive volume, and abrasive type on the processing time and surface roughness are systematically discussed. The conclusions are as follows:

(1) The different cutting can be achieved as the abrasive hardness and Young's modulus are in the middle of the SAC305 solder and 304V stainless steel. The high hardness and Young's modulus can improve the blasting efficiency. Under the six types of abrasive, the resin is the best for removing the SAC305 solder and maintaining the integrity of the 304V with high efficiency.

(2) The primary and secondary factors affecting the surface roughness and processing time are the abrasive type and air pressure, followed by the abrasive volume and lift-off height. The primary and secondary factors that affect the mechanical properties are the abrasive type and abrasive volume, followed by the lift-off height and air pressure.

(3) The processing time decreases with the increase in air pressure, abrasive hardness, and Young's modulus. Abrasive blasting surface roughness increases with the increase in the abrasive volume. The tensile strength of the wire is affected by the abrasive Young's modulus and hardness, and the yield strength decreases due to the blasting.

(4) When using the resin abrasive, the solder can be removed without damaging the stainless steel wire substrate by adjusting the process parameters.

(5) The impact pits are the primary material removal mode in abrasive blasting. The quantity of pits on the surface has a negative relationship with the lift-off height and abrasive mesh size. The quantity of scratches has a positive correlation to air pressure, lift-off height, and abrasive volume, and a negative correlation to the abrasive size.

Author Contributions: Conceptualization, Y.L., B.S., and Q.D.; validation, J.G.; formal analysis, Y.L., S.Z. and J.G.; investigation, S.Z., J.G., S.F. and Z.Z.; resources, B.S.; data curation, S.F. and Z.Z.; writing—original draft preparation, S.Z.; writing—review and editing, Y.L.; supervision, Q.D.; project administration, Q.D.; All authors have read and agreed to the published version of the manuscript.

Funding: This research was funded by the Fundamental Research Program of Shanxi Province (CN) (Grant No. 202303021211146), the Zhejiang Provincial Natural Science Foundation of China (Grant No. LY23E050002), the Opening Project of Guangdong Provincial Key Laboratory of Minimally Invasive Surgical Instruments and Manufacturing Technology (Grant No. MISIMT-2022-2), and the Medical Engineering Cross Research Foundation of Shanghai Jiaotong University (Grant No. YG2024QNB18).

Data Availability Statement: The original contributions presented in the study are included in the article, further inquiries can be directed to the corresponding authors.

Conflicts of Interest: The authors declare no conflicts of interest. Shiling Fu and Bin Shen are employees of Jiaxing Jiangxin Medical Technology Co., Ltd. The paper reflects the views of the scientists, and not the company.

References

1. Tummala, S.; Patel, M. Basics of Guidewire Technology and Peripheral Artery Disease. *Semin. Interv.Radiol.* **2023**, *40*, 129–135. [CrossRef] [PubMed]
2. Schafer, S. *Guidewire Path Reproducibility and Guidewire Path Simulation*; State University of New York at Buffalo: Getzville, NY, USA, 2007.
3. Jiang, W.; Zhao, W.; Zhou, T.; Wang, L.; Qiu, T. A review on manufacturing and post-processing technology of vascular stents. *Micromachines* **2022**, *13*, 140. [CrossRef] [PubMed]
4. Bisbal, F.; Sarrias, A.; Villuendas, R. Ventricular Tachycardia: Inferior Vena Cava Filter in a Case of Peripheral Vascular Disease. *Rev. Esp. Cardiol. (Engl. Ed.)* **2016**, *69*, 517. [CrossRef] [PubMed]
5. Li, W.; Liu, Y.; Chu, F.; Wang, Y. Interventional Embolization of Unilateral Cavernous Sinus with ONXY Glue Combined with Coils for the Treatment of Bilateral Dural Arteriovenous Fistula. *J. Craniofacial Surg.* **2024**, *35*, e451–e454. [CrossRef] [PubMed]
6. Applications of sandblasting and shot peening. *Plat. Finish.* **2016**, *38*, 24.
7. Li, J.; Dong, J.; Han, M.; Liu, S. Effects of Sand Blasting on Surface Integrity and High Cycle Fatigue Properties of DD6 Single Crystal Superalloy. *Acta Metall. Sin.* **2023**, *59*, 1201–1208.
8. Inokoshi, M.; Shimizubata, M.; Nozaki, K.; Takagaki, T.; Yoshihara, K.; Minakuchi, S.; Vleugels, J.; Van Meerbeek, B.; Zhang, F. Impact of sandblasting on the flexural strength of highly translucent zirconia. *J. Mech. Behav. Biomed. Mater.* **2021**, *115*, 104268. [CrossRef] [PubMed]
9. Peñuela-Cruz1, C.E.; Márquez-Herrera, A.; Aguilera-Gómez, E.; Saldaña-Robles, A.; Mis-Fernández, R.; Peña, J.L.; Caballero-Briones, F.; Loeza-Poot, M. The effects of sandblasting on the surface properties of magnesium sheets: A statistical study. *J. Mater. Res. Technol.* **2023**, *23*, 1321–1331. [CrossRef]
10. Ding, L.; Amir, P. The impact of sandblasting as a surface modification method on the corrosion behavior of steels in simulated concrete pore solution. *Constr. Build. Mater.* **2017**, *157*, 591–599. [CrossRef]
11. Cao, Y.; Niu, T.; Gai, P. Numerical simulation of shot peening based on surface coverage and shot peening intensity. *J. Cent. South Univ. Sci. Technol.* **2024**, *55*, 69–79.
12. Rudawska, A.; Danczak, I.; Müller, M.; Valasek, P. The effect of sandblasting on surface properties for adhesion. *Int. J. Adhes. Adhes.* **2016**, *70*, 176–190. [CrossRef]
13. Lundgren, T.; Samuelson, A.; Clase, C.; Naoumova, J. How sandblasting on lingual surfaces can be carried out with minimum enamel damage: An in vitro study on human teeth. *Int. Orthod.* **2020**, *18*, 820–826. [CrossRef] [PubMed]
14. Slătineanu, L.; Potârniche, Ș.; Coteață, M.; Grigoraș, I.; Gherman, L.; Negoescu, F. Surface roughness at aluminium parts sand blasting. *Proc. Manuf. Syst.* **2011**, *6*, 69–74.
15. Kanesan, D.; Mohyaldinn, M.E.; Ismail, N.I.; Chandran, D.; Liang, C.J. An experimental study on the erosion of stainless steel wire mesh sand screen using sand blasting technique. *J. Nat. Gas Sci. Eng.* **2019**, *65*, 267–274. [CrossRef]
16. Arifvianto, B.; Mahardika, M.; Salim, U.A. Comparison of Surface Characteristics of Medical-grade 316L Stainless Steel Processed by Sand-blasting, Slag Ball-blasting and Shot-blasting Treatments. *J. Eng. Technol. Sci.* **2020**, *52*, 1–13. [CrossRef]
17. Guo, C.Y.; Tang, A.T.H.; Tsoi, J.K.H.; Matinlinna, J.P. Effects of different blasting materials on charge generation and decay on titanium surface after sandblasting. *J. Mech. Behav. Biomed. Mater.* **2014**, *32*, 145–154. [CrossRef] [PubMed]
18. Caravaca, C.F.; Flamant, Q.; Anglada, M.; Gremillard, L.; Chevalier, J. Impact of sandblasting on the mechanical properties and aging resistance of alumina and zirconia based ceramics. *J. Eur. Ceram. Soc.* **2018**, *38*, 915–925. [CrossRef]
19. Li, D. *Urea-Formaldehyde Resin Adhesive*; Chemical Industry Press: Beijing, China, 2002; pp. 1–14.

Disclaimer/Publisher's Note: The statements, opinions and data contained in all publications are solely those of the individual author(s) and contributor(s) and not of MDPI and/or the editor(s). MDPI and/or the editor(s) disclaim responsibility for any injury to people or property resulting from any ideas, methods, instructions or products referred to in the content.

Article

Hierarchical Micro/Nanostructures with Anti-Reflection and Superhydrophobicity on the Silicon Surface Fabricated by Femtosecond Laser

Junyu Duan [1], Gui Long [1], Xu Xu [2,3,*], Weiming Liu [4], Chuankun Li [4], Liang Chen [5,6], Jianguo Zhang [1] and Junfeng Xiao [1,*]

1. State Key Laboratory of Intelligent Manufacturing Equipment and Technology, School of Mechanical Science and Engineering, Huazhong University of Science and Technology, Wuhan 430074, China; duanjunyu@hust.edu.cn (J.D.); longgui@hust.edu.cn (G.L.); zhangjg@hust.edu.cn (J.Z.)
2. Hubei Jiuzhiyang Infrared System Co., Ltd., Wuhan 430223, China
3. Wuhan National Laboratory for Optoelectronics, Huazhong Institute of Electro-Optics, Wuhan 430223, China
4. China Ship Development and Design Center, Wuhan 430064, China; 13667160564@163.com (W.L.); ck_lee3376@163.com (C.L.)
5. State Key Laboratory of High-End Heavy-Load Robots, Midea Group, Foshan 528300, China; chenliangwy22@163.com
6. Midea Corporate Research Center, Foshan 528311, China
* Correspondence: ioexuxu@163.com (X.X.); xiaojf@hust.edu.cn (J.X.)

Citation: Duan, J.; Long, G.; Xu, X.; Liu, W.; Li, C.; Chen, L.; Zhang, J.; Xiao, J. Hierarchical Micro/Nanostructures with Anti-Reflection and Superhydrophobicity on the Silicon Surface Fabricated by Femtosecond Laser. *Micromachines* **2024**, *15*, 1304. https://doi.org/10.3390/mi15111304

Academic Editor: Francesco Ruffino

Received: 19 September 2024
Revised: 17 October 2024
Accepted: 25 October 2024
Published: 27 October 2024

Copyright: © 2024 by the authors. Licensee MDPI, Basel, Switzerland. This article is an open access article distributed under the terms and conditions of the Creative Commons Attribution (CC BY) license (https://creativecommons.org/licenses/by/4.0/).

Abstract: In this paper, hierarchical micro/nano structures composed of periodic microstructures, laser-induced periodic surface structures (LIPSS), and nanoparticles were fabricated by femtosecond laser processing (LP). A layer of hydrophobic species was formed on the micro/nano structures through perfluorosilane modification (PM). The reflectivity and hydrophobicity's influence mechanisms of structural height, duty cycle, and size are experimentally elucidated. The average reflectivity of the silicon surface in the visible light band is reduced to 3.0% under the optimal parameters, and the surface exhibits a large contact angle of $172.3 \pm 0.8°$ and a low sliding angle of $4.2 \pm 1.4°$. Finally, the durability of the anti-reflection and superhydrophobicity is also confirmed. This study deepens our understanding of the principles of anti-reflection and superhydrophobicity and expands the design and preparation methods for self-cleaning and anti-reflective surfaces.

Keywords: micro/nano structures; anti-reflection; superhydrophobicity; femtosecond laser; durability

1. Introduction

As an abundant, environmentally friendly, and non-toxic semiconductor material, silicon is extensively utilized in various fields, including photodetection and photovoltaics [1–3]. However, its high reflectivity leads to substantial light loss [4]. Consequently, minimizing the reflection of incident light, enhancing the utilization rate of incident light, and increasing the photon absorption rate of silicon-based optical components have emerged as critical strategies for expanding and reinforcing the traditional applications of silicon devices. Traditional anti-reflective coatings, such as thin or multilayer films, have been widely used to achieve this goal [5–7]. However, most of these coatings are targeted at specific wavelength ranges and have poor durability. Therefore, the development of advanced micro/nano structures has emerged as a promising approach to enhance anti-reflective properties further [8–10]. It is possible to achieve significantly lower reflectance across a broad spectrum of wavelengths by leveraging the unique surface morphology and light-trapping mechanisms of micro/nano structures. These anti-reflective structures are often achieved by various methods, such as chemical etching [11], colloidal lithography and etching [12], and nanoimprinting [13]. It is worth noting that most reported methods require complex processes and typically carry a relatively high cost.

In addition, traditional anti-reflective surfaces are prone to the accumulation of dust, and the lack of self-cleaning functionality seriously affects their performance [14,15]. Superhydrophobic surfaces have garnered significant interest in recent years due to their unique functionalities and potential applications [16–18]. These surfaces exhibit remarkable water resistance, low adhesion of water droplets, and self-cleaning capabilities, making them ideal platforms for a wide range of practical applications [19]. The wettability of solid surfaces is determined by their structure morphology and chemical composition, while superhydrophobic surfaces are typically created by low surface energy substances and high roughness surfaces [16,20,21]. At present, the common methods for manufacturing superhydrophobic surfaces include chemical vapor deposition [22], chemical etching [23], laser processing [24], chemical reactions [25,26], etc.

The integration of superhydrophobic and anti-reflective features into a single surface presents a unique set of challenges and opportunities. The construction of micro/nano structures to achieve anti-reflection and superhydrophobicity simultaneously is a feasible method [27–29]. Laser processing has the advantages of high efficiency, low cost, and environmental friendliness among all manufacturing methods, and can easily fabricate micro/nano structures [30,31]. For example, L.B. Boinovich et al. fabricated a multimodal roughness with regular surface ripples on glass surface, and obtained a superhydrophobic surface with a contact angle of 160° and a rolling angle of less than 10° by adsorption of fluorosilane on the textured surface, but its optical performance changes have not been evaluated [32]. D.K. Chu et al. processed anti-reflection silicon with an excellent self-cleaning effect by combining an fs laser processing holes array with chemical modification [33]. The reflectance of the prepared surface was lower than 5%, and the surface exhibited a contact angle (CA) larger than 150° and a sliding angle (SA) lower than 3°. However, there is a lack of detailed research on the effects of different micro/nano structural morphologies, sizes, and chemical state changes on both the anti-reflective properties and hydrophobicity simultaneously [32–36]. It is possible to create structures that minimize light reflection while maximizing water resistance by carefully engineering the surface morphology and chemistry.

In this work, we fabricated hierarchical composite structures consisting of microstructure, LIPSS, and nanoparticles on silicon surfaces through simple single-step femtosecond laser processing. A layer of hydrophobic film formed on laser-ablated microstructures through perfluorosilane modification, obtaining superhydrophobic silicon surfaces with anti-reflection properties. By optimizing laser processing parameters, we explored and analyzed the effects of structural height, duty cycle, and size on reflectivity and superhydrophobicity. This work provides detailed analyses and insights into how laser processing parameters and chemical treatment affect these functionalities. Finally, after conducting a series of durability tests, it has been proven that the anti-reflective and self-cleaning surfaces have excellent stability to withstand external mechanical damage. The experimental results are significant for guiding researchers to efficiently and quickly prepare surfaces with durable superhydrophobicity and anti-reflection.

2. Experiments

2.1. Materials and Sample Preparation

Silicon wafer (30 × 30 × 5 mm^3, the resistivity of 1~10 Ω·cm, n-type, double-sided polished) were purchased as a shelf product (Harbin Tebo Technology, Harbin, China). The samples were consequently ultrasonically cleansed with acetone, anhydrous ethanol, and deionized water for 10 min, then dried and stored for later use.

2.2. Laser Processing and Perfluorosilane Modification

A GS-FIR20 femtosecond laser system with 400 fs pulses at a central wavelength of 1030 nm and a repetition rate of 600 kHz was utilized for the fabrication of micro/nanostructured surfaces. The laser beam was focused on the surface of the silicon sample and scanned in a pattern of crossed lines in an atmospheric environment. Considering the acquisition of regular surface morphology, the laser power and scanning speed were set to 6 W and 600 mm/s, respectively, and the laser spot size was about 30 µm. The energy density was 1.415 J/cm^2, and the number of pulses applied per spot was 30. The samples with different structural heights (H), duty cycles (D, defined as the ratio of the square of structural size to the square of the period size), and sizes (S) (Table 1; a detailed explanation of microstructure parameters can be found in Supplementary Figure S1a) were obtained by controlling the repeated number and scanning pitch. Then, all of the samples were washed with ethanol to remove processed dust and were slowly dried with pure nitrogen. At last, all the micro/nanostructured surfaces were immersed in the 1H, 1H, 2H, 2H-perfluorooctactyltrimethoxysilane alcohol solution with a concentration of 1% for 1 h, and then dried naturally. The laser processing schematic diagram, laser fabrication equipment, and laser processing parameters are shown in Supplementary Figures S1 and S2, and in Supplementary Table S1.

Table 1. Parameters of all samples.

S/N	1	2	3	4	5	6	7	8	9	10	11	12
Height (µm)	5	15	30	40			25			30		
Duty cycle			1		0.25	0.5	0.75	1			1	
Size (µm)			30				30		30	60	90	120

2.3. Characterization

The morphology of all the samples was examined by field-emission scanning electron microscopy (SEM, JSM-7600, JEOL, Tokyo, Japan) equipped with an energy-dispersive spectroscope (EDS, INCA-CH5, Oxford Instruments, Oxford, UK). The 3D topography measurement was characterized using a white-light interferometer (Zygo NewView 9000, ZYGO, CT, USA). The chemical composition of the samples was determined by X-ray photoelectron spectroscopy (XPS, AXIS-ULTRA DLD-600W, Shimadzu Kratos Corporation, kyoto, Japan).

A UV–Vis–NIR spectrophotometer (SolidSpec-3700, Shimadzu Corporation, kyoto, Japan) with an integrating sphere was utilized to investigate the light absorption behavior of every sample at 200–2500 nm. Contact angle (CA) and sliding angle (SA) measurements were carried out to evaluate the wettability of various surfaces by a video-based optical contact angle measuring device (OCA 20 from Data Physics Instruments). The droplets in this work are deionized water with a volume of 4 µL. The contact angles of every sample were examined at different randomly selected locations at least three times.

2.4. Durability Test

A scratch test was designed to evaluate the mechanical durability of the micro/nanostructured surfaces under severe abrasion conditions [37]. During the scratch test, a 1000-grit SiC sandpaper was selected as the abrasive surface. The micro/nanostructured surfaces were tested facing the abrasive surface with a fixed load of 200 g. The tape was stuck between the weight and the back of the sample with a linear speed of around 1 mm/s, which the sample and the weight can move together along with the dragging of the tape. The reflectivity and contact angle were recorded after 10 abrasion cycles. The schematic illustration of the experimental setup is shown in Supplementary Figure S3. The scotch tape test was used to evaluate the adherence performance of the as-fabricated superhydrophobic layer on the silicon substrate [38]. During the test, the scotch tape was pressed onto and peeled off from the sample surface, and the reflectivity and contact angle of the sample were tested after 10 peeling attempts. To verify the micro/nanostructured surfaces' long-term performance, the reflectivity and contact angle were recorded after 30 days of exposure to the natural environment.

3. Results and Discussion
3.1. Fabrication and Characterization of Micro/Nanostructured Surfaces

All micro/nanostructured surfaces were fabricated through single-step scanning by ultrafast laser processing. The typical machining time for a 30 × 30 mm sample, like the one in Figure 1b, was about 1.2 h. The height of different microstructures was controlled by the repeated number, and the duty cycle and size were determined by the scan pitch. The hierarchical morphology (Supplementary Figure S4) obtained by the white-light interferometer indicated that the designed micro/nanostructured surfaces had been prepared, and specific parameters are shown in Supplementary Table S2.

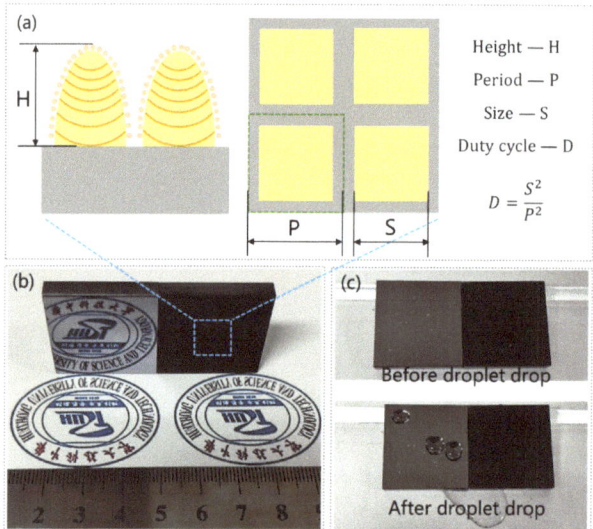

Figure 1. (a) Detailed explanation of microstructure parameters, (b) the anti-reflective effect of samples (The pattern is the emblem of Huazhong University of Science and Technology), and (c) the self-cleaning effect of samples before and after the droplet drop.

The surface morphology of the micro/nano structures was observed by SEM, as shown in Figure 2. All micro/nanostructured surfaces were hierarchical composite structures consisting of a periodic microstructure, laser-induced periodic surface structures (LIPSS), and nanoparticles. The microstructure presented a regular arrangement, and its period was related to the scan pitch (Figure 2d–f). The LIPSS was spontaneously formed on the surface of monocrystalline silicon by laser pulses, and its period was ~1 μm (Figure 2g–i). Nanoparticles with a diameter range of 100–500 nm were attached to the microstructures (Figure 2j–l). Scan pitch and repeated number would affect the surface laser heat affected area, which could significantly determine the structure morphology. As the number of repetitions increased, the surface LIPSS became regular, and more nanoparticles formed, as shown in Figure 2a,b. When the scan space was set to 60 μm, most of the nanoparticles accumulated at the top of the microstructure, and their size increased (Figure 2b,c).

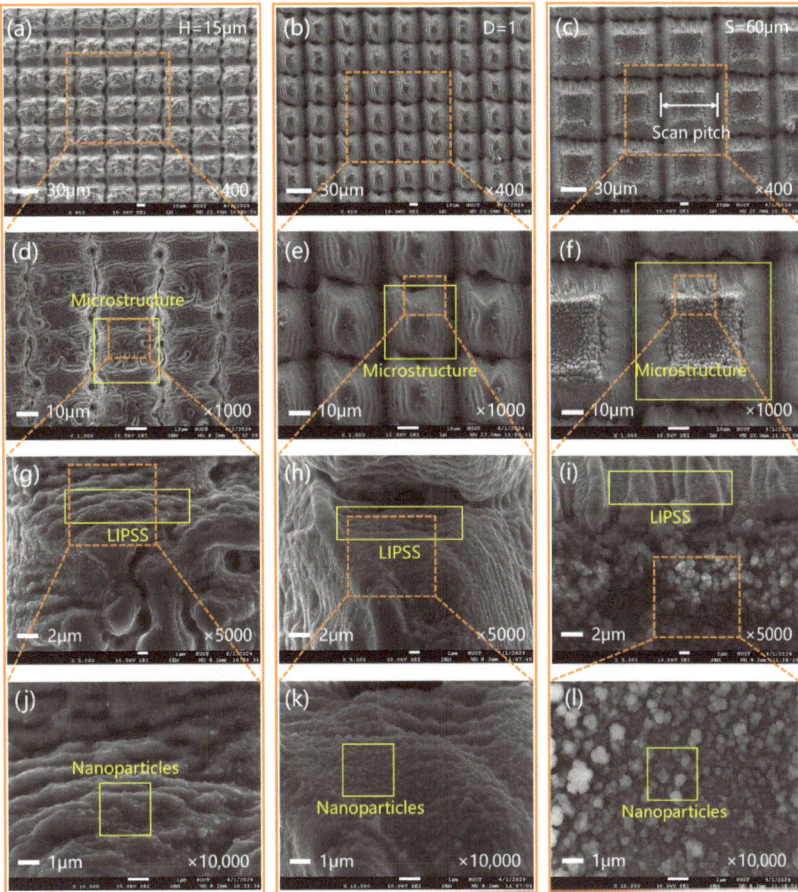

Figure 2. SEM images of micro/nanostructured surfaces. (**a**–**c**) Micro/nanostructures with different structural heights (H), duty cycles (D), and sizes (S). (**d**–**f**) Microstructures; (**g**–**i**) LIPSS; (**j**–**l**) nanoparticles.

3.2. Chemical Changes After Laser Processing and Perfluorosilane Modification

The elemental content of micro/nanostructured surfaces was analyzed qualitatively by EDS after laser processing (LP) and perfluorosilane modification (PM). After LP, the mass fraction of the surface oxygen element increased from 0.44% to 24.14% (Figure 3a,b). This indicates that laser processing caused oxidation reactions on the silicon surface. The mass fraction of fluorine element after LP was only 0.32%, but it increased to 18.15% after PM (Figure 3c). Due to the special structure and chemical properties of fluorosilane molecules, an extremely thin layer of fluorosilane molecular film was formed on the surface during the modification process, and the micro/nanostructure did not change. As shown in Figure 3d, the EDS surface scan obtained the element distribution information of the test area, and the fluorine element was uniformly distributed on the microstructure's surface.

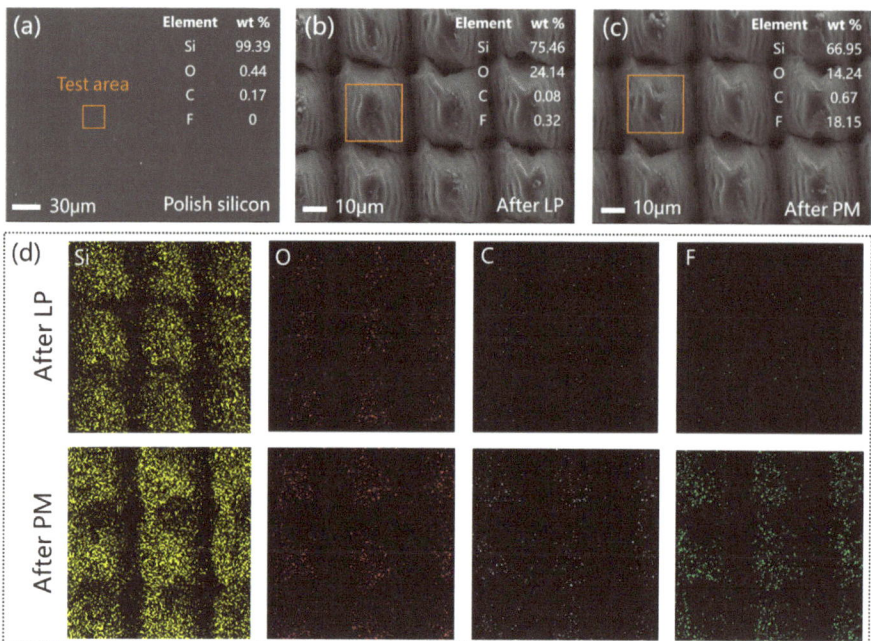

Figure 3. EDS results of micro/nanostructured surfaces. (**a**–**c**) Elemental content of the polished surface, after LP and after PM; (**d**) element distribution.

The chemical state of the surface element was analyzed qualitatively by XPS after LP and PM, as shown in Figure 4. The results of XPS show that the surface is mainly composed of Si, O, C, and F elements. After PM, the concentration of F atoms is 61.9%, which is 59.4% more than before. The composition of F increased remarkably, implying that the micro/nanostructured surface has been covered with fluoroalkyl molecular film. The high-resolution XPS spectra for C 1s (Figure 4b) show significant peaks at 291.3 eV and 293.7 eV, corresponding to the $-CF_2-$ peak and $-CF_3$ peak, respectively [39]. From the F 1s high-resolution XPS spectra shown in Figure 4c, the main peak can be found at approximately 688.6 eV, and it is attributed to the F–C covalent bond [40]. These results further prove that fluoroalkyl is grafted on the micro/nanostructured surfaces, and the outermost surface is mainly composed of low-surface energy $-CF_3$ and $-CF_2-$ components. As a result, the obtained surfaces were able to possess both micro/nano hierarchical structures and low surface energy at the same time, thus providing basic conditions for superhydrophobicity.

3.3. The Optical Property of the Micro/Nanostructured Surfaces

3.3.1. The Anti-Reflective Principle of Micro/Nanostructures

Micro/nanostructured surfaces can effectively improve surface light absorption and exhibit significant anti-reflective properties. As shown in Figure 2, a micro/nano hierarchical composite structure was obtained by laser one-step preparation. This micro/nanostructure can utilize the multiple internal reflections of micrometer structures to achieve the effect of geometric "light trapping" [10,41], while the presence of subwavelength structures enhances the absorption of electromagnetic waves, thereby achieving anti-reflection effects [42].

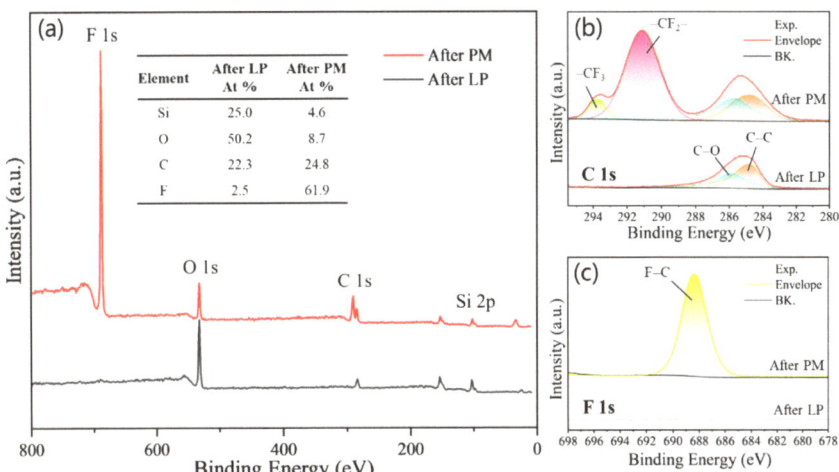

Figure 4. The XPS survey spectra of micro/nanostructured surfaces after LP and after PM. (**a**) XPS full spectrum. (**b**,**c**) High-resolution XPS spectra for C 1s and F 1s.

Figure 5 shows a schematic of the anti-reflective principle of light waves propagating through different characteristic structural surfaces. When the incident surface is almost flat, a portion of the light is absorbed by the substrate material and another portion of the light is reflected out of the surface, which is related to the inherent characteristics of substrate material, as shown in Figure 5a. The schematic diagram of the propagation of light waves incident on the surface of microstructures (Figure 5b) indicates that when the size of the spacing between microstructures is much larger than the wavelength of the incident light, microstructures exhibit excellent light capture performance. Electromagnetic waves are captured and reflected multiple times in the microstructure, and an increase in optical path length leads to an increase in energy absorption. In addition, the presence of submicron and nanostructures with structural dimensions similar to or smaller than the wavelength further reduces the reflection phenomenon caused by the sharp change in refractive index, thereby reducing surface light reflection (Figure 5c). Therefore, the surface with micro/nanostructures is conducive to enhancing its surface anti-reflective properties.

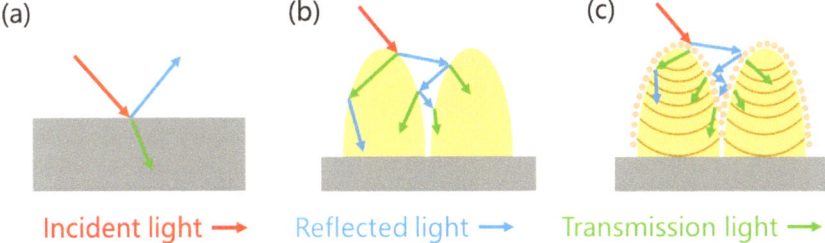

Figure 5. Schematic of the anti-reflective principle of light waves propagating through different characteristic structural surfaces. (**a**) Flat surface, (**b**) microstructures, and (**c**) micro/nanostructures.

3.3.2. The Anti-Reflective Performance of the Micro/Nanostructured Surfaces

To determine the optimal anti-reflection effect, the surface reflectance spectra of micro/nanostructures with different heights, duty cycles, and sizes were compared in the range of 200–2500 nm. Figure 5 shows the reflectance spectra of micro/nanostructured surfaces at 200–2500 nm. All surfaces have a reduced reflectivity compared to the original polished silicon surface. The results of reflectivity with different heights (Figure 6a) show that higher structures have better anti-reflective properties because the number of light wave oscillations increases and the effective refractive index changes more gradually with an increase in the height of the structures. Especially in the visible light band, the average reflectance at 400–800 nm and at a height of 5 μm is 11.6%. As the height of the structure increases, the average reflectivity decreases to 3.5% when reaching a height of 40 μm, which is more than 90% lower than the reflectivity of 36.5% on polished silicon surfaces.

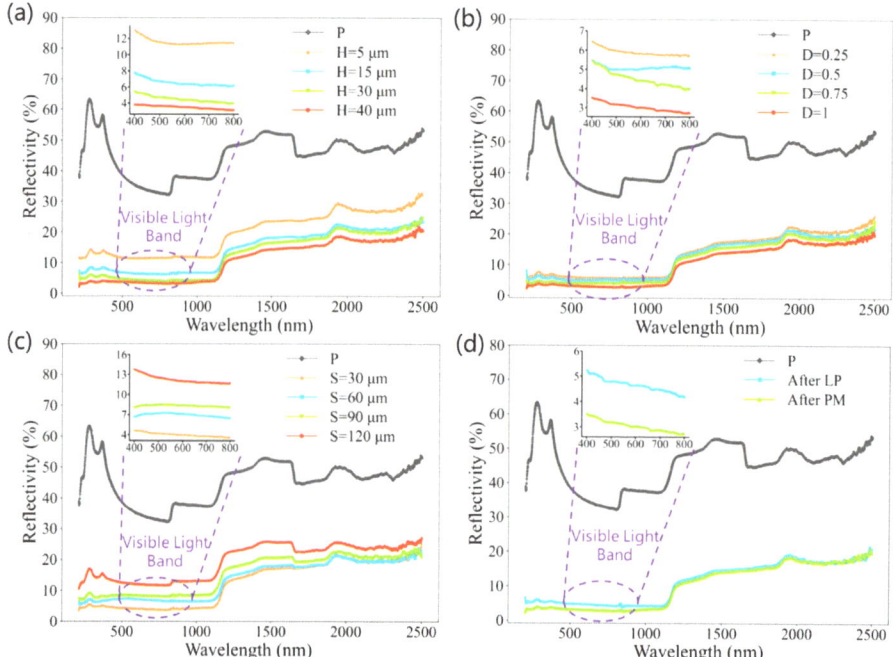

Figure 6. Reflectance spectra of micro/nanostructured surfaces at 200–2500 nm. (**a**) Different heights, (**b**) different duty cycles, and (**c**) different sizes. (**d**) After LP and after PM.

In Figure 6b, as the structural duty cycle increases, the surface reflectance shows a gradually decreasing trend. When the duty cycle increases from 0.25 to 1, the average reflectance of 400–800 nm decreases from 5.9% to 3.1%. An increase in the duty cycle means a decrease in the microstructure period, and there are more microstructure cavities per unit area to capture incident light, resulting in a decrease in surface reflectivity. The results in Figure 6c indicate that the size of the microstructure also has a significant impact on the surface reflectivity. The reflectivity of a structure with a size of 30 μm in the visible light range is only 4.0%, but as the structural size increases, the reflectivity continues to increase. Until the size increases to 120 μm, the average reflectivity increases to 12.2%, which is related to the fact that the increase in size causes more raw surfaces that have not been laser processed. Figure 6d shows the reflectivity after laser processing and perfluorosilane modification. After modification with perfluorosilane, the reflectivity of the sample decreased by 1.6% compared to the sample after laser processing. This is because fluoroalkyl silane has a very low refractive index (~1.3) at room temperature, much lower than the refractive index of the silicon substrate (~3.4), which can cause an obvious anti-reflective effect on the substrate surface.

Overall, the height, duty cycle, and size of microstructures are key factors affecting the surface reflectivity of micro/nanostructures, with the influence of height and size being more pronounced. The average lowest surface reflectivity in the visible light range is reduced to 3.0%, indicating that monocrystalline silicon has an excellent anti-reflection effect after preparing micro/nanostructures and fluorosilane molecular films.

3.4. Wettability of the Micro/Nanostructured Surfaces

The wettability is mainly dependent on the surface microstructure and chemical composition of the material surfaces. The superhydrophobic surface requires low surface energy and needs micro/nanostructures to preserve the air, which suspends the droplets [16].

To evaluate the wettability of the multi-scale micro/nanostructured silicon surface, water contact angle (WCA) measurements were carried out. The water contact angle measurement of $60.3 \pm 2.3°$ for the polished silicon surface is shown in Figure 7a. After laser processing, a large number of oxides are formed on the surface of silicon, and liquid droplets infiltrate into the microstructure according to the Wenzel formula [43], resulting in the surface of the micro/nanostructures exhibiting superhydrophilicity. To obtain hydrophobicity of the prepared surface, the typical fluoroalkyl silane modification is adopted here. After perfluorosilane modification, the surface treated with the laser changed from superhydrophilic to superhydrophobic. Rough structures can increase contact angles due to the decrease in the solid–liquid contact area according to the Cassie–Baxter formula [44], causing a superhydrophobic surface. Figure 7b shows a schematic diagram of the wettability transition. After the sample is immersed in the fluoroalkyl silane solution, hydrolysis and condensation reactions will occur, and the sample surface will self-assemble to form a single or multi-molecular film [29]. The outside of this film is covered by –CF3 and –CF2– groups, which have good hydrophobic properties [45]. In addition, the contact angle of the polished silicon surface after perfluorosilane modification only increased to $79.7 \pm 1.5°$, but the micro/nanostructured surface exhibits a large contact angle (CA) of $172.3 \pm 0.8°$ and a low sliding angle (SA) of $4.2 \pm 1.4°$ for water. This proves that both micro/nanostructures and low surface energy are indispensable conditions for the preparation of superhydrophobic surfaces.

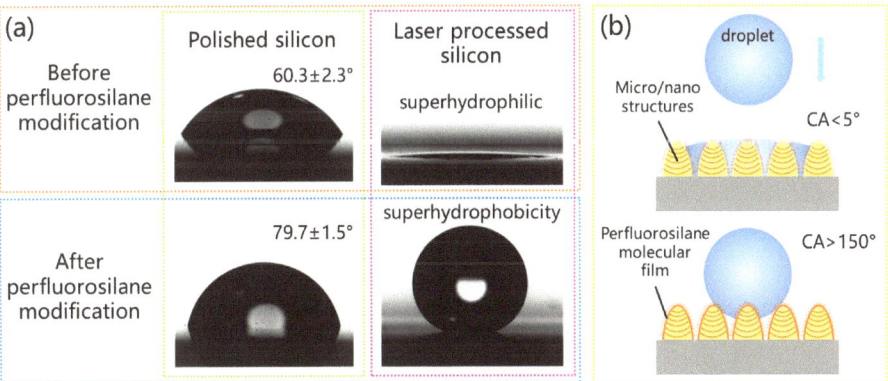

Figure 7. (a) The wettability of prepared surfaces before and after perfluorosilane modification. (b) Schematic diagram of wettability transition.

To obtain the trend of contact angle variances on micro/nanostructure surfaces with different heights, duty cycles, and sizes, the contact angles of all samples were measured, as shown in Figure 8. Figure 8a shows the contact angle measurement process of a 4 µL droplet. The results indicate that an increase in the height and duty cycle of the structure leads to the enhancement of surface hydrophobicity, and the smaller the structure size, the larger the contact angle. The change in contact angle can be explained by the Cassie model and the increased solid–liquid contact area [44]. This is consistent with the influence of structure on anti-reflection performance. Therefore, it can be concluded that a larger structural height, duty cycle, and smaller size corresponds to better anti-reflection and superhydrophobic performance.

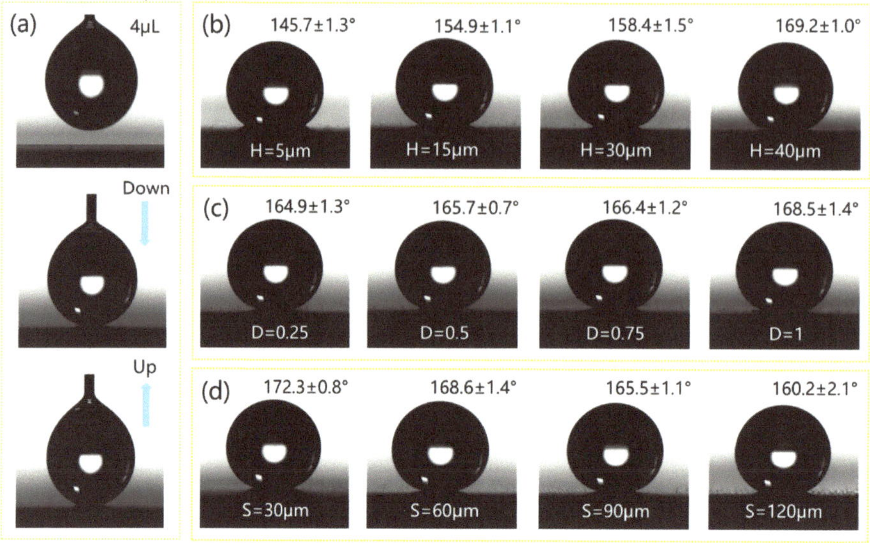

Figure 8. (**a**) The contact angle testing process of a 4 μL droplet. (**b**–**d**) The contact angle of micro/nanostructured surfaces with different heights, duty cycles, and sizes.

3.5. Durability of Anti-Reflection and Self-Cleaning Performance

Durability is a significant factor in practical engineering applications. In particular, physical and chemical damage can cause a serious loss of surface micro/nanostructure and chemical composition, which degrade the anti-reflection and super-hydrophobic properties. In this study, we used various methods to examine the durability of the micro/nanostructured surfaces, such as sandpaper abrasion, tape peeling, and long-term exposure to the natural environment. As shown in Figure 9, after various tests, the sample still maintains a low reflectivity and a contact angle exceeding 150°. After 30 days of exposure to the natural environment, SEM results (Figure 9e) showed that the micro/nanostructure did not change, but the surface oxidation increased the oxygen content, and the relative fluorine content decreased to 3.46%. The average reflectivity of the sample at 400–800 nm only increased from 5.4% to 6.0%, and the contact angle merely decreased by 1.4°. After ten cycles of powerful tape peeling, the fluoroalkyl molecular film was slightly damaged, and the fluorine content decreased to 2.97% (Figure 9d). However, the sample surface still exhibited excellent anti-reflection and superhydrophobic properties. Even after scratch tests, the surface structure and fluoroalkyl molecular film were damaged to a certain extent (Figure 9c). The average reflectivity of the sample at 400–800 nm was only 6.8%, and the contact angle was $156.7 \pm 2.5°$. These results demonstrate the mechanical stability and long-term anti-reflective superhydrophobic effect of the micro/nanostructured surfaces under various conditions.

The anti-reflection and self-cleaning effects of micro/nanostructured surfaces are shown in Figure 1b,c. In contrast to the specular reflection of the original polished surface, the prepared sample exhibits no reflective images, highlighting its excellent anti-reflective performance. In practice, contaminants and dust adhering to a superhydrophobic surface are easily washed off by water, so the surface exhibits self-cleaning properties (See Movie S1).

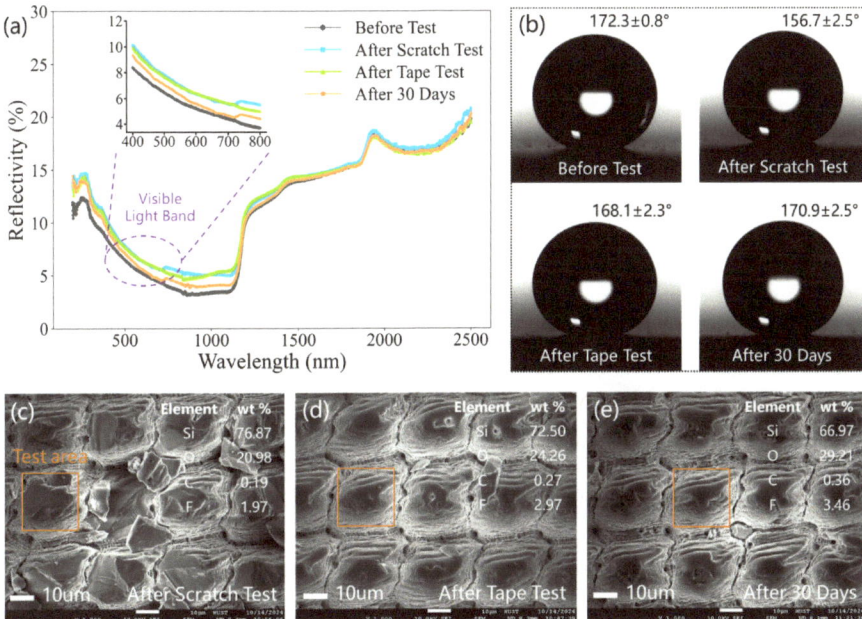

Figure 9. (**a**) Reflectance spectra of samples at 200–2500 nm after various tests. (**b**) The contact angle of samples after various tests. (**c**–**e**) Surface morphology and elemental content after scratch test, after tape test, and after 30 days.

4. Conclusions

In summary, we demonstrated a facile and simple approach to creating anti-reflection and self-cleaning silicon surfaces using laser processing and perfluorosilane modification. By adjusting the processing parameter, hierarchical micro/nanostructured surfaces with different heights, duty cycles, and sizes were obtained. The results indicate that surfaces with larger structural heights, duty cycles, and smaller structural dimensions have lower reflectivity and larger contact angles. The modification with perfluorosilane not only imparts low surface energy to the surface but also further reduces reflectivity. The average reflectivity of the silicon surface in the visible light band is reduced to 3.0% under the optimal parameters, and the surface exhibits a large CA of 172.3 ± 0.8 ° and a low SA of 4.2 ± 1.4°. Sandpaper abrasion, tape peeling, and exposure to natural environment tests show that the surface has excellent durability. This method could lead to applications in diverse areas including optical, detectors, and photovoltaics.

Supplementary Materials: The following supporting information can be downloaded at: https://www.mdpi.com/article/10.3390/mi15111304/s1.

Author Contributions: Conceptualization, J.D. and C.L.; Data curation, J.D.; Funding acquisition, X.X., L.C. and J.X.; Investigation, J.D., G.L., W.L. and C.L.; Methodology, L.C.; Resources, W.L.; Supervision, X.X., J.Z. and J.X.; Writing—original draft, J.D.; Writing—review and editing, G.L. and J.X. All authors have read and agreed to the published version of the manuscript.

Funding: This work was supported by the National Natural Science Foundation of China (52225506, 52188102) and the Program for HUST Academic Frontier Youth Team (No. 2019QYTD12).

Data Availability Statement: The original contributions presented in the study are included in the article/supplementary material, further inquiries can be directed to the corresponding authors.

Conflicts of Interest: Author Xu Xu was employed by the company Hubei Jiuzhiyang Infrared System Co., Ltd. The remaining authors declare that the research was conducted in the absence of any commercial or financial relationships that could be construed as a potential conflict of interest.

References

1. Brunner, R.; Sandfuchs, O.; Pacholski, C.; Morhard, C.; Spatz, J. Lessons from nature: Biomimetic subwavelength structures for high-performance optics. *Laser Photonics Rev.* **2011**, *6*, 641–659. [CrossRef]
2. Chee, K.W.A.; Tang, Z.; Lu, H.; Huang, F. Anti-reflective structures for photovoltaics: Numerical and experimental design. *Energy Rep.* **2018**, *4*, 266–273. [CrossRef]
3. Zhao, J.; Liu, L.; Wang, T.; Wang, X.; Du, X.; Hao, R.; Liu, J.; Zhang, J. Synchronous Phase-Shifting Interference for High Precision Phase Imaging of Objects Using Common Optics. *Sensors* **2023**, *23*, 4339. [CrossRef] [PubMed]
4. Yang, J.; Luo, F.; Kao, T.; Li, X.; Ho, G.; Teng, J.; Luo, X.; Hong, M. Design and fabrication of broadband ultralow reflectivity black Si surfaces by laser micro/nanoprocessing. *Light Sci. Appl.* **2014**, *3*, e185. [CrossRef]
5. Wang, J.; Zhang, J.; Liu, Z. Preparation of Anti-reflective Coatings on Solar Glass. *Asian J. Chem.* **2013**, *25*, 5787–5789. [CrossRef]
6. Papatzacos, P.H.; Akram, M.N.; Hector, O.; Lemarquis, F.; Moreau, A.; Lumeau, J.; Ohlckers, P. Temperature resistant anti-reflective coating on Si-wafer for long-wave infra-red imaging. *Heliyon* **2023**, *9*, e15888. [CrossRef]
7. Ali, K.; Khan, S.A.; Jafri, M.Z.M. Effect of Double Layer (SiO_2/TiO_2) Anti-reflective Coating on Silicon Solar Cells. *Int. J. Electrochem. Sci.* **2014**, *9*, 7865–7874. [CrossRef]
8. Raut, H.K.; Dinachali, S.S.; He, A.Y.; Ganesh, V.A.; Saifullah, M.S.M.; Law, J.; Ramakrishna, S. Robust and durable polyhedral oligomeric silsesquioxane-based anti-reflective nanostructures with broadband quasi-omnidirectional properties. *Energy Environ. Sci.* **2013**, *6*, 1929–1937. [CrossRef]
9. Busse, L.E.; Florea, C.M.; Frantz, J.A.; Shaw, L.B.; Aggarwal, I.D.; Poutous, M.K.; Joshi, R.; Sanghera, J.S. Anti-reflective surface structures for spinel ceramics and fused silica windows, lenses and optical fibers. *Opt. Mater. Express* **2014**, *4*, 2504–2515. [CrossRef]
10. Fan, P.; Bai, B.; Zhong, M.; Zhang, H.; Long, J.; Han, J.; Wang, W.; Jin, G. General Strategy toward Dual-Scale-Controlled Metallic Micro-Nano Hybrid Structures with Ultralow Reflectance. *ACS Nano* **2017**, *11*, 7401–7408. [CrossRef]
11. Zhang, D.; Jiang, S.; Tao, K.; Jia, R.; Ge, H.; Li, X.; Wang, B.; Li, M.; Ji, Z.; Gao, Z.; et al. Fabrication of inverted pyramid structure for high-efficiency silicon solar cells using metal assisted chemical etching method with $CuSO_4$ etchant. *Solar Energy Mater. Sol. Cells* **2021**, *230*, 111200. [CrossRef]
12. Chan, L.W.; Morse, D.E.; Gordon, M.J. Moth eye-inspired anti-reflective surfaces for improved IR optical systems & visible LEDs fabricated with colloidal lithography and etching. *Bioinspiration Biomim.* **2018**, *13*, 041001. [CrossRef]
13. Yanagishita, T.; Nishio, K.; Masuda, H. Anti-Reflection Structures on Lenses by Nanoimprinting Using Ordered Anodic Porous Alumina. *Appl. Phys. Express* **2009**, *2*, 022001. [CrossRef]
14. Said, S.A.M.; Walwil, H.M. Fundamental studies on dust fouling effects on PV module performance. *Sol. Energy* **2014**, *107*, 328–337. [CrossRef]
15. Said, S.A.M.; Al-Aqeeli, N.; Walwil, H.M. The potential of using textured and anti-reflective coated glasses in minimizing dust fouling. *Sol. Energy* **2015**, *113*, 295–302. [CrossRef]
16. Zhang, X.; Shi, F.; Niu, J.; Jiang, Y.; Wang, Z. Superhydrophobic surfaces: From structural control to functional application. *J. Mater. Chem.* **2008**, *18*, 621–633. [CrossRef]
17. Lu, Y.; Sathasivam, S.; Song, J.; Crick, C.; Carmalt, C.; Parkin, I. Robust self-cleaning surfaces that function when exposed to either air or oil. *Science* **2015**, *347*, 1132–1135. [CrossRef]
18. Weng, W.; Zheng, X.; Tenjimbayashi, M.; Watanabe, I.; Naito, M. De-icing performance evolution with increasing hydrophobicity by regulating surface topography. *Sci. Technol. Adv. Mater.* **2024**, *25*, 2334199. [CrossRef]
19. Bhushan, B.; Jung, Y.C. Natural and biomimetic artificial surfaces for superhydrophobicity, self-cleaning, low adhesion, and drag reduction. *Prog. Mater. Sci.* **2011**, *56*, 1–108. [CrossRef]
20. Saira, I. Fabrication and characterization of durable superhydrophobic and superoleophobic surfaces on stainless steel mesh substrates. *Mater. Res. Express* **2024**, *11*, 036401. [CrossRef]
21. Xiao, J.; Luo, Y.; Niu, M.; Wang, Q.; Wu, J.; Liu, X.; Xu, J. Study of imbibition in various geometries using phase field method. *Capillarity* **2019**, *2*, 57–65. [CrossRef]
22. Wang, L.; Liu, K.; Yin, M.; Yin, B.; Liu, X.; Tang, S. Anti-reflection silica coating simultaneously achieving superhydrophobicity and robustness. *J. Sol-Gel Sci. Technol.* **2024**, *109*, 835–848. [CrossRef]
23. Zhang, F.; Wu, R.; Zhang, H.; Ye, Y.; Chen, Z.; Zhang, A. Novel Superhydrophobic Copper Mesh-Based Centrifugal Device for Edible Oil-Water Separation. *ACS Omega* **2024**, *9*, 16303–16310. [CrossRef] [PubMed]
24. Wang, H.; Deng, D.; Zhai, Z.; Yao, Y. Laser-processed functional surface structures for multi-functional applications—A review. *J. Manuf. Process.* **2024**, *116*, 247–283. [CrossRef]
25. Yasuda, K.; Hayashi, Y.; Homma, T. Fabrication of Superhydrophobic Nanostructures on Glass Surfaces Using Hydrogen Fluoride Gas. *ACS Omega* **2024**, *9*, 12204–12210. [CrossRef] [PubMed]
26. Minakov, A.V.; Pryazhnikov, M.I.; Neverov, A.L.; Sukhodaev, P.O.; Zhigarev, V.A. Wettability, interfacial tension, and capillary imbibition of nanomaterial-modified cross-linked gels for hydraulic fracturing. *Capillarity* **2024**, *12*, 27–40. [CrossRef]

27. Choi, S.J.; Huh, S.Y. Direct Structuring of a Biomimetic Anti-Reflective, Self-Cleaning Surface for Light Harvesting in Organic Solar Cells. *Macromol. Rapid Commun.* **2010**, *31*, 539–544. [CrossRef] [PubMed]
28. Zhao, S.; Du, H.; Ma, Z.; Li, W.; Zhao, H.; Wen, C.; Ren, L. Hierarchical composite structure to simultaneously realize superior superhydrophobicity and anti-reflection. *Appl. Surf. Sci.* **2023**, *611*, 155652. [CrossRef]
29. Zhang, J.; Zhu, Y.; Li, Z.; Zhou, W.; Zheng, J.; Yang, D. Preparation of anti-reflection glass surface with self-cleaning and anti-dust by ammonium hydroxide hydrothermal method. *Mater. Express* **2015**, *5*, 280–290. [CrossRef]
30. Hu, Y.; Duan, J.; Yang, X.; Zhang, C.; Fu, W. Wettability and biological responses of titanium surface's biomimetic hexagonal microstructure. *J. Biomater. Appl.* **2022**, *37*, 1112–1123. [CrossRef]
31. Du, M.; Sun, Q.; Jiao, W.; Shen, L.; Chen, X.; Xiao, J.; Xu, J. Fabrication of Antireflection Micro/Nanostructures on the Surface of Aluminum Alloy by Femtosecond Laser. *Micromachines* **2021**, *12*, 1406. [CrossRef] [PubMed]
32. Boinovich, L.B.; Domantovskiy, A.G.; Emelyanenko, A.M.; Pashinin, A.S.; Ionin, A.A.; Kudryashov, S.I.; Saltuganov, P.N. Femtosecond laser treatment for the design of electro-insulating superhydrophobic coatings with enhanced wear resistance on glass. *ACS Appl. Mater. Interfaces* **2014**, *6*, 2080–2085. [CrossRef] [PubMed]
33. Chu, D.; Yao, P.; Huang, C. Anti-reflection silicon with self-cleaning processed by femtosecond laser. *Opt. Laser Technol.* **2021**, *136*, 106790. [CrossRef]
34. Oh, S.; Cho, J.W.; Lee, J.; Han, J.; Kim, S.K.; Nam, Y. A Scalable Haze-Free Antireflective Hierarchical Surface with Self-Cleaning Capability. *Adv. Sci.* **2022**, *9*, 2202781. [CrossRef] [PubMed]
35. Verma, L.K.; Sakhuja, M.; Son, J.; Danner, A.J.; Yang, H.; Zeng, H.C.; Bhatia, C.S. Self-cleaning and antireflective packaging glass for solar modules. *Renew. Energy* **2011**, *36*, 2489–2493. [CrossRef]
36. Wang, K.; Zhang, Y.; Chen, J.; Li, Q.; Tang, F.; Ye, X.; Zheng, W. Wide-Spectrum Antireflective Properties of Germanium by Femtosecond Laser Raster-Type In Situ Repetitive Direct Writing Technique. *Coatings* **2024**, *14*, 262. [CrossRef]
37. Wang, H.; Zhuang, J.; Yu, J.; Qi, H.; Ma, Y.; Wang, H.; Guo, Z. Fabrication of Anti-Reflective Surface with Superhydrophobicity/High Oleophobicity and Enhanced Mechanical Durability via Nanosecond Laser Surface Texturing. *Materials* **2020**, *13*, 5691. [CrossRef]
38. Barthwal, S.; Kim, Y.S.; Lim, S.-H. Mechanically Robust Superamphiphobic Aluminum Surface with Nanopore-Embedded Microtexture. *Langmuir* **2013**, *29*, 11966–11974. [CrossRef]
39. Hoque, E.; DeRose, J.A.; Hoffmann, P.; Mathieu, H.J.; Bhushan, B.; Cichomski, M. Phosphonate self-assembled monolayers on aluminum surfaces. *J. Chem. Phys.* **2006**, *124*, 174710. [CrossRef]
40. Lee, J.-M.; Kim, S.J.; Kim, J.W.; Kang, P.H.; Nho, Y.C.; Lee, Y.-S. A high resolution XPS study of sidewall functionalized MWCNTs by fluorination. *J. Ind. Eng. Chem.* **2009**, *15*, 66–71. [CrossRef]
41. Zheng, B.; Wang, W.; Jiang, G.; Mei, X. Fabrication of broadband antireflective black metal surfaces with ultra-light-trapping structures by picosecond laser texturing and chemical fluorination. *Appl. Phys. B* **2016**, *122*, 180. [CrossRef]
42. Duan, M.; Wu, J.; Zhang, Y.; Zhang, N.; Chen, J.; Lei, Z.; Yi, Z.; Ye, X. Ultra-Low-Reflective, Self-Cleaning Surface by Fabrication Dual-Scale Hierarchical Optical Structures on Silicon. *Coatings* **2021**, *11*, 1541. [CrossRef]
43. Wenzel, R.N. Resistance of solid surfaces to wetting by water. *Ind. Eng. Chem.* **1936**, *28*, 988–994. [CrossRef]
44. Cassie, A.B.D.; Baxter, S. Wettability of porous surfaces. *Trans. Faraday Soc.* **1944**, *40*, 546–551. [CrossRef]
45. Lu, Z.; Wang, P.; Zhang, D. Super-hydrophobic film fabricated on aluminium surface as a barrier to atmospheric corrosion in a marine environment. *Corros. Sci.* **2015**, *91*, 287–296. [CrossRef]

Disclaimer/Publisher's Note: The statements, opinions and data contained in all publications are solely those of the individual author(s) and contributor(s) and not of MDPI and/or the editor(s). MDPI and/or the editor(s) disclaim responsibility for any injury to people or property resulting from any ideas, methods, instructions or products referred to in the content.

Article

A Normal Displacement Model and Compensation Method of Polishing Tool for Precision CNC Polishing of Aspheric Surface

Yongjie Shi [1,*], Min Su [2], Qianqian Cao [3] and Di Zheng [4]

1   School of Intelligent Manufacturing, Jiaxing Vocational Technical College, Jiaxing 314036, China
2   Ministry of Basic Education, Jiaxing Vocational Technical College, Jiaxing 314036, China; sum103@126.com
3   College of Information Science and Engineering, Jiaxing University, Jiaxing 314001, China; qqcao@zjxu.edu.cn
4   School of Mechatronics and Energy Engineering, Ningbo Tech University, Ningbo 315100, China
*   Correspondence: jiesy2007@126.com

Abstract: The position accuracy of the polishing tool affects the surface quality of the polished aspheric surface. The contact deformation among the polishing tool, abrasives, and aspheric part can cause a displacement, which, in turn, will cause a position error of the polishing tool, which will lead to a significant change in the polishing force. In order to resolve this error, this paper proposed a method of normal displacement compensation for a computer numerical controlled (CNC) polishing system by controlling the polishing force. Firstly, the coupling principle between the polishing force and the position of the polishing tool is expounded, and the relationship between normal displacement and deformation is analyzed. Based on Hertz's theory, a model of normal displacement is established. Then, on the basis of the decoupled polishing system developed, a normal displacement compensation method was proposed. Finally, a group of comparative experiments was carried out to verify the effectiveness of the proposed method. Compared with no displacement compensation, when the part was polished with the normal displacement compensation method, the value of roughness decreased from 0.4 μm to 0.21 μm, and the unevenness coefficient of surface roughness decreased from 112.5% to 19%. The experimental results show that the polishing quality is improved greatly, and the aspheric surfaces can be polished more uniformly with the method proposed in this paper.

Keywords: displacement compensation; CNC polishing; aspheric surface; deformation

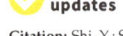

Citation: Shi, Y.; Su, M.; Cao, Q.; Zheng, D. A Normal Displacement Model and Compensation Method of Polishing Tool for Precision CNC Polishing of Aspheric Surface. *Micromachines* 2024, 15, 1300. https://doi.org/10.3390/mi15111300

Academic Editor: Reza Teimouri

Received: 26 September 2024
Revised: 23 October 2024
Accepted: 23 October 2024
Published: 25 October 2024

Copyright: © 2024 by the authors. Licensee MDPI, Basel, Switzerland. This article is an open access article distributed under the terms and conditions of the Creative Commons Attribution (CC BY) license (https:// creativecommons.org/licenses/by/ 4.0/).

1. Introduction

Aspheric parts are widely used as basic components in aerospace, electronics, defense, and other fields due to their excellent optical properties [1]. At present, the industry mainly adopts grinding and polishing as its final finishing processing. With the rapid increase in the need for high-quality aspheric parts, the development of aspheric ultra-precision machining technology and equipment is becoming much more important [2], and many automatic polishing techniques have been proposed [3].

However, in the common automatic polishing system, the polishing force is generally generated and controlled by the motion of the polishing tool and the contact deformation between the polishing tool and the aspheric surface [4]. The contact deformation will lead to a normal displacement error of the polishing tool, and then the polishing force will change greatly. This means that there is a close coupling relationship between the polishing force and the tool displacement. It is difficult to ensure the uniformity of material removal, profile accuracy, and surface quality. Thus, it is very important to control and compensate for this normal displacement change so as to ensure that the polishing force does not change abruptly.

In order to solve this problem, many researchers have put forward many useful approaches [5,6]. Some scholars have studied the key technology of deterministic polishing.

Zhang et al. [7] studied the process planning of the automatic polishing of the curved surface using a five-axis machining tool. Some scholars have studied the wheel polishing technology [8]. Yao et al. [9] designed a pneumatic floating structure used to compensate for the Z motion error of the robot to realize the stable control of polishing pressure based on industrial robots. Some scholars have studied ballonet polishing [10,11]. In those studies, many polishing force control devices were proposed.

On the other hand, some scholars have studied the polishing path of the polishing tool. Zhao et al. [12] proposed a revised Archimedes spiral polishing path, which is generated based on the modified tool–workpiece contact model and the pointwise searching algorithm. Han [13] proposed an adaptive polishing path optimization method based on footprint evolution, which considers the influence of curvature on footprint evolution. Qu et al. [14] proposed an optimized Archimedes spiral path to ensure uniform material removal depth in aspheric polishing. In those studies, the effect of the surface curvature variations on material removal was eliminated.

As an overview of the above research, those approaches mainly focus on force control strategies and tool path planning, but it does not take into account the effect of the normal displacement on the position and posture of the polishing tool, and there is no study taken on the normal displacement compensation. Therefore, it is necessary to study the relationship between displacement and deformation in depth and propose a displacement compensation method.

This study focused on the model and method of the normal displacement compensation based on a computer numerical controlled (CNC) machine tool. The material removal principle was introduced, and the coupling principle of polishing force and position of the polishing tool was analyzed. Then, the models of normal displacement were established, considering the contact condition among the polishing tool, part, and abrasives. Further, based on the polishing system, the normal displacement compensation method was proposed. Subsequently, the validity of the model and method is examined experimentally.

The main structure of this paper as shown in Figure 1 is organized as follows: Section 2 derives the coupling principle of polishing force and position; Section 3 analyzes the normal displacement and establishes a normal displacement model; Section 4 proposes a polishing system and a normal displacement compensation method; Section 5 shows experiments study and gives the experimental results; Section 6 presented discussions; finally, Section 7 makes some conclusions.

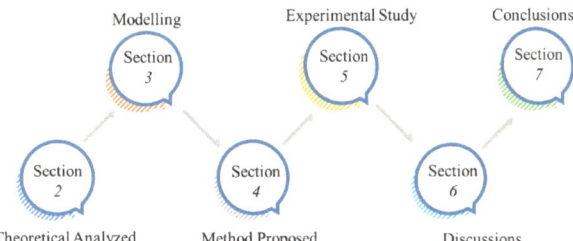

Figure 1. Structure of the paper.

2. Coupling Principle of Polishing Force and Position

2.1. Toolhead and Abrasive

The toolhead, which contacts with the aspheric surface, is a carrier for coating or embedding abrasives. Its material hardness is generally lower than that of the part, and the structure is uniform and dense. At the same time, it also has a certain abrasive embedment and immersion. The surface accuracy and retention of the toolhead affect the polishing quality. The shape of the abrasive particles is mostly irregular. In this paper, it is assumed that the particles are spherical and their shape and size are uniform.

2.2. Material Removal Principle

It is the result of the interaction among the toolhead, the abrasive particles, and the part that leads to tiny material removal. In this paper, soft toolhead and free abrasive polishing methods are mainly used. During the process of polishing, the abrasive is coated on the surface of the soft toolhead. The particles are free between the toolhead and the part, which means the particles have greater freedom of movement. They can be fixed and semi-fixed on the toolhead and can also slide and roll between the toolhead and the part surface.

The material removal mechanism is shown in Figure 2. The whole mechanical action process is a process of extrusion, sliding, plowing, and cutting. The aspheric surface undergoes elastic deformation, plastic deformation, and micro-cutting. This is the main reason for the normal displacement of the toolhead. Compared with the wear and cutting caused by particles, the wear of the part caused by the toolhead can be neglected.

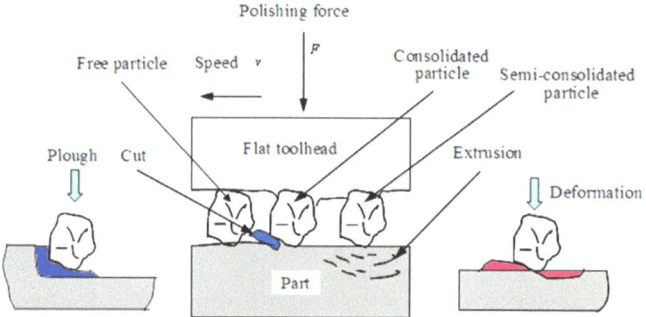

Figure 2. Material removal principle.

2.3. Force–Position Coupling Process

Figure 3 shows the coupling process between the polishing force and the position of the toolhead. When the aspheric part is polished, in order to ensure machining accuracy, it is hoped that the part will be polished with the ideal polishing force F_A, and the toolhead moves along the ideal polishing point A according to the pre-set trajectory, the ideal posture of toolhead can be express by θ_1. However, due to the elastic–plastic deformation among the toolhead, the abrasive, and the part, there is a normal displacement δ of the toolhead, which will cause an error in the position of the toolhead.

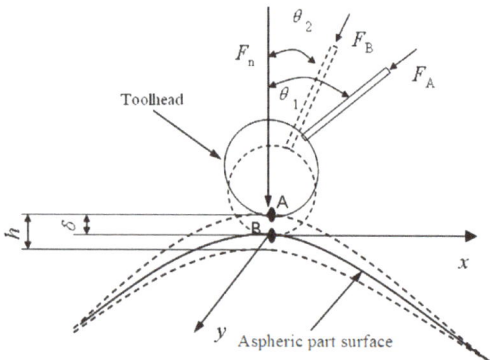

Figure 3. Process of force–position coupling.

It can be seen from Figure 3 that the actual position of the toolhead changes from the ideal position A to the actual position B, and the actual posture of the toolhead changes from θ_1 to θ_2. At this time, the ideal polishing force F_A and the normal polishing force F_{An} applied to the polishing point change into the actual values F_B and F_{Bn}, respectively. It can be seen that due to the effect of normal displacement, the polishing force and the position of the toolhead all are changed, which leads to the polishing force and position coupling closely.

Therefore, in order to eliminate the error caused by normal displacement to ensure the quality and efficiency of the polishing, in the actual polishing, the normal displacement needs to be compensated to ensure that the actual polishing force does not change.

3. Modeling of Normal Displacement

During the polishing process, the normal displacement among the toolhead, the aspheric surface, and the abrasive mainly comes from contact deformation. When the toolhead and the aspheric surface contact each other, the elastic deformation occurs first. As the deformation increases, the deformation changes from elastic deformation to plastic deformation. As the residence time increases, under the action of the abrasive, micro-cutting occurs so that the aspheric surface material can be removed, and the toolhead produces a certain amount of wear.

3.1. Analysis of Normal Displacement Change

The change process of the normal displacement after the contact between the toolhead and the aspheric part is shown in Figure 4. On the contact interface composed of toolhead, abrasive, and aspheric part, the hardness of the toolhead is the lowest of the three. At the beginning of contact, the large elastic–plastic deformation and wear of the toolhead will occur. Figure 4b shows the elastic deformation δ_{e0} between the toolhead and the aspheric surface, and Figure 4c shows the elastic deformation δ_{e1} between the toolhead and the abrasive.

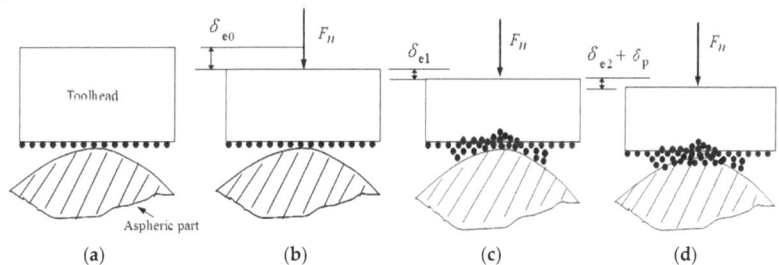

Figure 4. The change in normal displacement among toolhead, aspheric part, and abrasive. (**a**) Initial contact; (**b**) deformation δ_{e0}; (**c**) deformation δ_{e1}; (**d**) elastic–plastic deformation.

With the continuation of polishing, under the action of the polishing force, the depth of the abrasive cut into the aspheric surface is very small. There is only elastic deformation δ_{e2} occurring on the aspheric surface. Then, with the contact stress applied by the abrasive on the aspheric surface gradually increasing, elastic deformation of the aspheric surface changes into plastic deformation δ_p, as shown in Figure 4d. Finally, as the depth of the abrasive cut into the surface increases, small cutting occurs, and the material is removed.

In summary, the normal displacement δ in the polishing process mainly includes the elastic deformation δ_0 between the toolhead and the aspheric surface, the elastic deformation δ_{e1} between the toolhead and the abrasive, the elastic deformation δ_{e2} and plastic deformation δ_p between the abrasive and the aspheric surface. δ can be expressed as follows:

$$\delta = \delta_{e0} + \delta_{e1} + \delta_{e2} + \delta_p \tag{1}$$

3.2. Abrasive Distribution Model

Generally, it is believed that the contact between the toolhead and the part can be equivalent to the contact between a rough surface and a smooth surface. During the polishing process, the microscopic interaction among the toolhead, aspheric surface, and abrasive particle is shown in Figure 5.

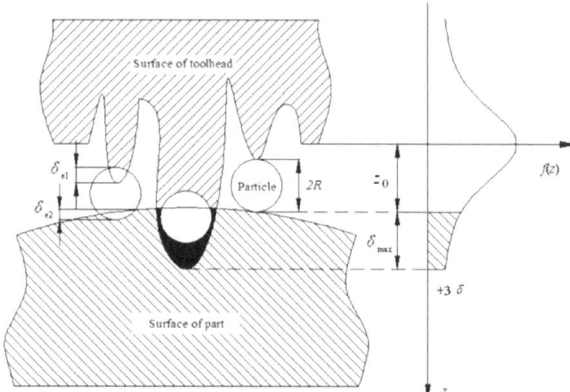

Figure 5. Micro-contact condition of polishing area.

It can be seen that not all particles interact with the part surface. Some particles plow or cut the part to produce the plastic deformation δ_p, some particles squeeze and slip on the part surface to produce elastic deformation δ_e, and some particles only roll on the part surface without any force and deformation. The number of particles per unit area N_0 can be calculated as follows [15]:

$$N_0 = \left(\frac{3V_g}{4\pi R^3}\right)^{2/3} \quad (2)$$

Here, V_g is the particle rate. According to the statistical model, in the small polishing contact area, the profile of the particle attached to the surface of the toolhead is approximated as follows:

$$f(z) = \frac{1}{\sqrt{2\pi}\sigma} e^{-\frac{(z)^2}{2\sigma^2}} \quad (3)$$

where z is the protrusion height of the abrasive grain, σ is the standard deviation, and $f(z)$ is the distribution function of particle height on the surface of the toolhead. Let z_0 be the distance between the part surface and the reference plane. When $z > z_0$, elastic and plastic deformation occurs among the toolhead, the abrasive particles, and the aspheric surface. Let δ_{max} be the maximum deformation in the interval $(-3\sigma, 3\sigma)$. Because of $\int_{-3\sigma}^{3\sigma} f(z)dz = 0.999$, the distribution function of the surface profile can be expressed as follows:

$$f(z) = \begin{cases} 0 & -\infty < z < -3\sigma \\ \frac{1}{\sqrt{2\pi}\sigma} e^{-\frac{z^2}{2\sigma^2}} & -3\sigma < z < 3\sigma \\ 0 & 3\sigma < z < \infty \end{cases} \quad (4)$$

3.3. Normal Displacement Modeling

Based on the force balance among the toolhead, part, and abrasive in the polishing contact area, the deformation and the resulting displacement can be analyzed. Several assumptions are made as follows:

① The particle hardness H_{Bm} is higher than that of the part and toolhead, and elastic-plastic deformation between them only occurs when they are in contact with each other;

② Each particle only interacts with the part surface once at the same polishing point;
③ The contact area is very small, and all contact points within the area have the same curvature radius;
④ All particles are spherical and have the same average radius.

3.3.1. Elastic Deformation

The micro-contact among the toolhead, particles, and aspheric surface is shown in Figure 6.

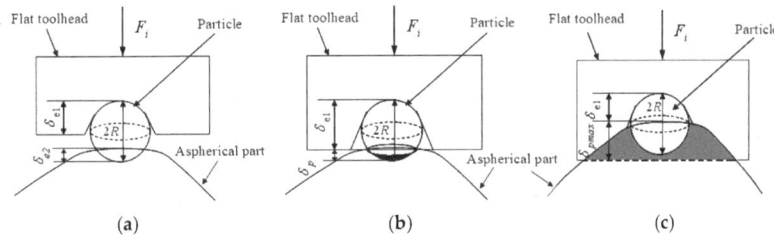

Figure 6. The microaction process of a single particle. (**a**) Elastic deformation; (**b**) elastic–plastic deformation; (**c**) fully plastic deformation.

In the initial stage of polishing, there is only elastic deformation. According to the Hertz theory [16], the elastic deformation δ_{e0} between the toolhead and the part can be given as follows:

$$\delta_{e0} = \left(\frac{9F_{e0}^2}{16R_{\rho 0}E_{0*}^2} \right)^{1/3} \quad (5)$$

where F_{e0} is the polishing force applied perpendicularly to the part surface at the polishing point; E_0^* is the relative elastic modulus of toolhead and part, given by $1/E_0^* = (1 - \nu_1^2)/E_1 + (1 - \nu_2^2)/E_2$, where E_1 and E_2, and ν_1 and ν_2 are the elastic modulus and the Poisson's rates of the toolhead and part, respectively; $R_{\rho 0}$ is the equivalent curvature radius, because of the curvature radius of plane toolhead tends to ∞, so $R_{\rho 0} = R_e$, where R_e is the curvature radius of the part.

Figure 6a illustrates the elastic deformation among the single particle, the part, and the toolhead. The deformation δ_{e1} and the average pressure P_{m1} between the toolhead and the particle can be obtained as follows:

$$\delta_{e1} = \left(\frac{9F_i^2}{16R_{\rho 1}E_1*^2} \right)^{1/3} \quad (6)$$

$$P_{m1} = \frac{1}{\pi} \left(\frac{16E_1 *^2 F_i}{9R_{\rho 1}^2} \right)^{1/3} \quad (7)$$

Here, F_i is the force applied to a single abrasive particle, E_1^* and $R_{\rho 1}$ are the relative elastic modulus and equivalent curvature radius of toolhead and particle, respectively, given by $1/E_1^* = (1 - \nu_1^2)/E_1 + (1 - \nu_3^2)/E_3$, $R_{\rho 1} = R$, where E_3 and ν_3 are the elastic modulus and Poisson's rates of toolhead and particle, respectively; R is the radius of the particle.

Similarly, the deformation δ_{e2} and the average pressure P_{m2} between the particle and the part can be obtained as follows:

$$\delta_{e2} = \left(\frac{9F_i^2}{16R_{\rho 2}E_2*^2} \right)^{1/3} \quad (8)$$

$$P_{m2} = \frac{1}{\pi}\left(\frac{16E_2*^2 F_i}{9R_{\rho 2}^2}\right)^{1/3} \tag{9}$$

where E_2* and $R_{\rho 2}$ are the relative elastic modulus and equivalent curvature radius of the particle and part, respectively, given by $1/E_2* = (1 - \nu_2^2)/E_2 + (1 - \nu_3^2)/E_3$, $1/R_{\rho 2} = 1/R + 1/R_e$.

Existing studies show that when the average contact pressure $P_{m2} \leq H_{Bf}/3$ (H_{Bf} is the Brinell hardness of the part) [17], only elastic deformation occurs between the part and the particle. When $P_{m2} = H_{Bf}/3$, the plastic flow begins in the surface layer of the part, and plastic deformation occurs. Therefore, the maximum elastic deformation δ_{e2max} between the part and the particle can be obtained from Equations (8) and (9):

$$\delta_{e2max} = \frac{\pi^2 R_{\rho 2} H_{Bf}^2}{16 E_2*^2} \tag{10}$$

From Equation (9), F_i can be written as follows:

$$F_i = \frac{\pi^3 R_{\rho 2}^2 H_{Bf}^3}{48 E_2*^2} \tag{11}$$

Because the elastic contact force between the particle and the toolhead is equal to the plastic contact force between the particle and the part. Substituting Equation (11) into Equation (6), the maximum elastic deformation δ_{e1max} can be obtained as follows:

$$\delta_{e1max} = \left(\frac{R_{\rho 2}^4 H_{Bf}^4}{16 R_{\rho 1} E_1 *^2 E_2*^4}\right)^{1/3} \tag{12}$$

Particularly, when the number of particles is large enough, there is no contact between the toolhead and the part, and the maximum normal displacement δ_{max} can be obtained as follows:

$$\delta_{max} = \delta_{e0} + \delta_{e1max} + \delta_{e2max} \tag{13}$$

3.3.2. Plastic Deformation

When $P_{m2} \geq H_{Bf}/3$, δ_{e2} exceeds the δ_{e2max}, the plastic deformation occurs, and the particle will plow or cut the part surface. The micro-contact is shown in Figure 6b,c.

(1) $\delta_{e1} + \delta_p \leq 2R$

Assuming that a_p is the radius of the contact zone between the particle and the part, we can obtain

$$a_p = \sqrt{\delta_p(2R - \delta_p)} = \sqrt{(z - z_0)(2R + z_0 - z)} \tag{14}$$

When $\delta_{e1} + \delta_p \leq 2R$, in the range of $[z_0 + \delta_{e2max}, z_0 + 2R]$, the force F_{p1} that causes plastic deformation of part can be given as follows:

$$F_{p1} = N_0 F_i \int_{z_0 + \delta_{e2max}}^{z_0 + 2R} f(z) dz \tag{15}$$

In this case, the average pressure $P_{m2} = H_{Bf}/2$, F_i can be expressed as follows:

$$F_i = \pi a_p^2 P_{m2} = \frac{\pi}{2}(z - z_0)(2R + z_0 - z) \times H_{Bf} \tag{16}$$

From Equations (15) and (16), F_{p1} can be obtained:

$$F_{p1} = \frac{\sqrt{\pi} H_{Bf} N_0}{2\sqrt{2}\sigma} \int_{z_0+\delta_{e2max}}^{z_0+2R} (z-z_0)(2R+z_0-z) \times e^{-\frac{z^2}{2\sigma^2}} dz \qquad (17)$$

Since plastic deformation δ_p is very small, F_i can also be expressed as follows:

$$F_i = H_{Bf} \times \pi \times R \times \delta_p \qquad (18)$$

On the other hand, there is elastic contact between the toolhead and the particle. The contact force can be given as follows:

$$F_{ie} = \frac{4}{3} E_1 * R^{1/2} \delta_{e1}^{3/2} \qquad (19)$$

$F_{ie} = F_i$, based on Equations (18) and (19), we can obtain

$$\delta_p = \frac{4 E_1 * R^{1/2} \delta_{e1}^{3/2}}{3\pi H_{Bf} R} \qquad (20)$$

Especially when $\delta_{e1} + \delta_p = 2R$, based on Equation (20), we can obtain

$$\delta_p^3 + R(9\pi^2 H_{Bf}^2/16 E_1*^2 - 6)\delta_p^2 + 12R^2 \delta_p - 8R^3 = 0 \qquad (21)$$

(2) $\delta_{e1} + \delta_p \geq 2R$

When $\delta_{e1} + \delta_p \geq 2R$, it can be seen from Figure 6c, $a_p = R$; thus, F_i can be expressed as follows:

$$F_i = \pi a_p^2 P_{m2} = \frac{\pi}{2} R^2 \times H_{Bf} \qquad (22)$$

When $\delta_{e1} + \delta_p = 3\sigma$, from Equation (20), δ_p can be written as follows:

$$\delta_p^3 + (9\pi^2 H_{Bf}^2 R/16 E_1*^2 - 9\sigma)\delta_p^2 + 27\sigma^2 \delta_p - 27\sigma^3 = 0 \qquad (23)$$

3.3.3. Normal Displacement Model

Based on the above analysis, it can be seen that during the polishing process, when the materials were removed by particles, complete plastic deformation occurs between the particles and the part, then $\delta_{e2} = \delta_{e2max}$, $\delta_p = \delta_{pmax} = 3\sigma$. Furthermore, the normal displacement δ can be obtained as follows:

$$\delta = \delta_{e0} + \delta_{e2max} + \delta_{pmax} = \left(\frac{9 F_{e0}^2}{16 R_{\rho 0} E_0*^2}\right)^{1/3} + \frac{\pi^2 R_{\rho 2} H_{Bf}^2}{16 E_{2*}^2} + 3\sigma \qquad (24)$$

4. Polishing System and Displacement Compensation Method

4.1. Description of the Polishing System

A polishing system developed for NC polishing of aspheric surface is shown in Figure 7. The system mainly includes three subsystems: (1) the polishing force control subsystem based on magnetorheological torque servo device (MRT) is mainly composed of magnetorheological torque servo device, controllable current source, torque detection, and control system. During the polishing process, the polishing force is provided and controlled by the MRT, and the constant force polishing is achieved; (2) a position and posture control subsystem, mainly composed of a CNC system, polishing toolhead, transmission, tool holder, and so on. The polishing toolhead designed in this paper can adapt to the curvature variations of the part surface. During the polishing process, the CNC system controls the trajectory of the polishing tool system, and MRT and the transmission device ensure that the toolhead is always perpendicular to the aspheric surface feed; therefore, it is possible to compensate for the normal displacement of the toolhead, and the ideal polishing trajectory will not be changed; (3) the main motion subsystem, which was provided by the NC lathe to control the rotation speed of the aspheric part.

Figure 7. Polishing system.

4.2. Magnetorheological Torque Servo Device (MRT)

The magnetorheological torque servo device (MRT) [18] developed by the authors is a force control device based on the magnetorheological (MR) effect. According to this device, the servo control of torque can be achieved. During the polishing process, the torque is converted into polishing force by means of a polishing tool system, and the servo control of polishing force can be achieved. Figure 8 shows the structure of MRT, which works in shear mode. The MR fluid becomes solidified in milliseconds when a magnetic field is applied, resulting in a shear yield stress τ_b. When the working disc is driven by the input shaft, MR fluid is sheared, thereby generating an output torque. By changing the current applied to the coil of the MRT, the magnetic field strength passing through the MR fluid can be changed, and then the shear yield stress and output torque can be changed. Through experimental tests, the torque model of the MRT designed in this paper can be obtained [18]:

$$T = 1.1398I - 0.3812 \tag{25}$$

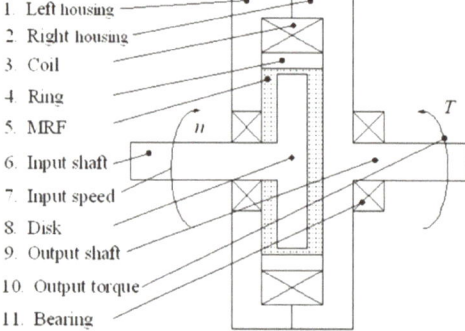

Figure 8. MRT structure.

4.3. Displacement Compensation Method

Based on the polishing system developed by the author, during the polishing process, it is required that the tool axis vector direction of the toolhead at the polishing point is consistent with the normal direction of the aspheric surface, that is, perpendicular to the tangent direction of the aspheric surface, so as to ensure that the polishing force can be applied vertically to the aspheric surface without changing the posture of toolhead. At this time, the posture angle of the toolhead is $\theta = 0$. The trajectory of the polishing tool system is shown in Figure 9.

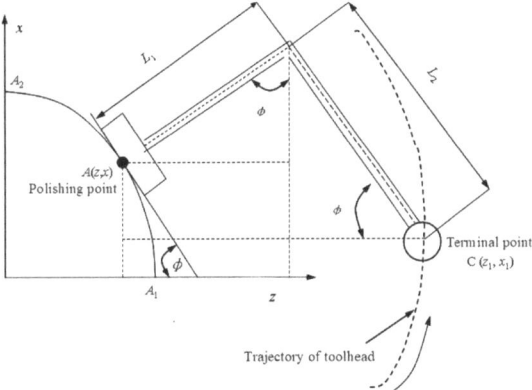

Figure 9. Trajectory of polishing tool.

A coordinate system of part is established with the center O of the aspheric end face as the origin, and the generatrix equation of part in the xoz plane can be expressed as $x = f(z)$ (A_1A_2, as shown in Figure 9). The following supposition can be made: $A(z, x)$ is the coordinate of the polishing point, and φ (acute angle) is the angle between the tangent plane of the part at the polishing point and the z-axis, then $\tan \varphi = dx/dz$. The terminal point $C(z_1, x_1)$ of the tool holder can be seen as the cutter location point of the polishing tool. Its trajectory, which is controlled by the programming of the CNC lathe, can be described as follows:

$$\begin{cases} z_1 = z + (L_1) \sin \varphi + L_2 \cos \varphi \\ x_1 = x + (L_1) \cos \varphi - L_2 \sin \varphi \\ x = f(z) \\ \tan \varphi = f'(z) \end{cases} \quad (26)$$

where L_1 is the length from the tool holder to the polishing point.

As analyzed in Section 3.3, during the polishing process, the normal displacement δ will be generated among the surface of the polishing head, abrasive particles, and aspheric part. This will cause the actual position of the toolhead on the aspheric surface to change. The position changes in the z and x directions are $\Delta z = \delta \sin\varphi$, $\Delta x = \delta \cos\varphi$. In order to ensure that the actual position of the toolhead does not change, it is necessary to compensate for the normal displacement δ in the trajectory equation. Then, the trajectory of terminal point C can be written as follows:

$$\begin{cases} z_1 = z + (L_1 + \delta) \sin \varphi + L_2 \cos \varphi \\ x_1 = x + (L_1 + \delta) \cos \varphi - L_2 \sin \varphi \\ x = f(z) \\ \tan \varphi = f'(z) \end{cases} \quad (27)$$

Equation (27) is the normal displacement compensation model of the toolhead in the motion space. When the tool holder of the NC lathe is fed according to this trajectory, it can not only ensure that the toolhead feeds perpendicularly to the surface of the aspheric part but also compensate for the position change caused by the normal displacement.

5. Experimental Study

5.1. Experimental Setup

To verify the validity of the model and method proposed in this paper, polishing experiments were carried out based on the polishing system developed. In order to ensure the initial conditions such as geometry, accuracy, and surface roughness are consistent, two aluminum ellipsoid parts with the same material and the same finish turning were prepared for comparison, which is shown in Figure 10a. The long semi-axis and short semi-axis of the ellipsoid were 23 mm and 15 mm, respectively. Diamond paste with a particle size of 0.25 µm was used. A disc-shaped wool felt was used in the polishing tool with a diameter of 25 mm and a thickness of 8 mm. The surface roughness is measured using the SRM-1 (D) surface roughness measuring instrument. The specific parameters during the polishing process are shown in Table 1.

Table 1. Polishing parameter.

Classification	Diamond Grits	Part	Wool Felt
Elastic modulus E (GPa)	1050	70	0.015
Poisson's ratio ν	0.2	0.34	0.079
Brinell hardness H_B (GPa)	102	0.06	-

Two sets of experiments were conducted to verify the above analysis. In the first set, during the polishing process, one part was polished without displacement compensation; that is, the position trajectory of the polishing toolhead adopts Equation (27). In the second experiment, another part was polished using the displacement compensation method. That is, the position trajectory of the polishing toolhead adopts Equation (28). During the two polishing processes, the other polishing parameters, such as the polishing force controlled by MRT, the rotation speed, and the feed rate of the ellipsoid part controlled by the NC lathe, are all the same. The actual polishing force is obtained indirectly by measuring the actual torque output by MRT in the polishing process in real time.

5.2. Experimental Results

The experimental results are shown in Figure 10, where Figure 10a shows the aluminum ellipsoid part surface before polishing, Figure 10b shows the part surface after polishing without the normal displacement compensation method, and Figure 10c shows the part surface after polishing with normal displacement compensation method. The surface roughness was measured at four different areas on each part surface. Each different area was measured five times, and the average values were obtained as the roughness value of the measured areas. The measurement results are listed in Table 2.

Table 2. Measurement results of surface roughness.

Areas	Roughness Before Polishing Ra [µm]	Roughness After Polishing Ra [µm]		Polishing Time t [min]
		Without Displacement Compensation	With Displacement Compensation	
④	1.82	0.066	0.023	
③	1.62	0.051	0.021	45
②	1.58	0.021	0.019	
①	1.56	0.022	0.020	

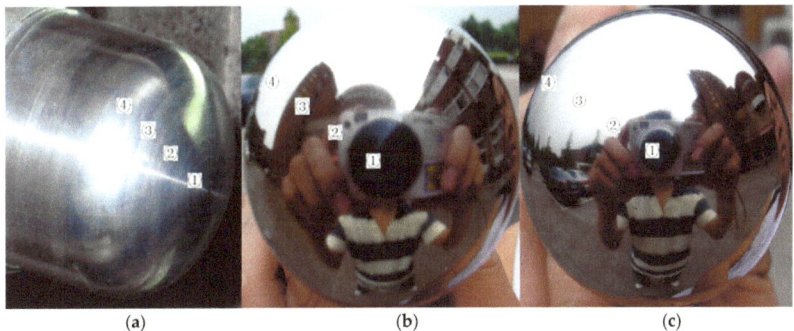

(a)   (b)   (c)

Figure 10. Surface condition of part before and after polished. (**a**) Before polishing; (**b**) polishing with no compensation; (**c**) polishing with compensation.

6. Discussion

6.1. Surface Roughness

As can be seen from Figure 10b,c. that the mirror-like surfaces are obtained using the polishing experimental system developed in Section 4.1. In the ordinary workshop environment, within 45min, the polishing quality of the aspheric surface can be greatly improved, whether or not the displacement compensation model and method are used. It can be calculated from Table 2 that the average value of surface roughness of parts decreases from 1.645 µm before being polished to 0.04 µm and 0.021 µm after being polished without and with displacement compensation, respectively. The above experimental results provide strong proof of the effectiveness of the developed polishing system. It also can be seen that the value of surface roughness of areas ③ and ④ with displacement compensation is significantly lower than that of no displacement compensation.

At the same time, it should be noted from Table 2 that the roughness values in the area ① of the two parts are quite close for both treatment modes. This is mainly because in area ①, the gyratory radius on the aspheric surface approaches zero, and the rotational speed approaches zero. Although the normal displacement compensation method was used, the material removal rate in this area was also small, resulting in a close roughness value obtained by the two methods.

6.2. Unevenness Coefficients

It can be calculated from Table 2 that when the part was polished with the normal displacement compensation method, the unevenness coefficient of surface roughness was 19%, and the other part was 112.5%, which was polished without normal displacement compensation. The unevenness coefficient ∇R_a, is expressed as follows:

$$\nabla R_a = \frac{R_{a,max} - R_{a,min}}{\overline{R}_a} \tag{28}$$

where $R_{a,max}$, $R_{a,min}$, and $\overline{R}_a$ are the maximum, minimum, and average values of surface roughness in the four detection areas, respectively.

It is noted that the values of the unevenness coefficient obtained with the normal displacement compensation method are much smaller than those obtained without normal displacement compensation. It is easy to find the reason for using the model and control method proposed in this article; the normal displacement change of each polishing point can be compensated by the motion trajectory of the toolhead. Furthermore, the position error of the polishing point is reduced, and the consistency of surface roughness is improved. Therefore, it can be concluded that more uniform roughness and better surface quality can be obtained using the normal displacement compensation method.

6.3. Effect of Particle Shape

In the paper, the normal displacement model considers ideal spherical particles, and a higher surface quality is obtained using a polishing system developed. However, it is well known that the shape and size distribution of particles affect the quality and efficiency of the polishing. Therefore, irregular particles should be discussed.

(1) Material area removal rate. Spherical particles tend to roll more on the part surface, resulting in less material removal. In contrast, irregularly shaped particles, due to their sharp edges or protrusions, are more likely to cut or plow the surface, thus removing material more effectively. It is easier to achieve material removal by using irregular particles than regular abrasives.

(2) Deformation of the part surface. According to Section 3, whether it is spherical particles or irregular particles, the contact area is very small, and the local deformation is minimal. The difference is that spherical particles produce a more uniform stress distribution on the part surface, while irregular particles can generate higher stress concentrations in the contact area, which can easily cause scratches or cracks on the part surface, thus affecting surface quality.

In this paper, the same type of particles produced by the same company were used in the experiment. When the size of the particles is much smaller than the curvature radius of the aspherical surface, it can be considered that the motion behavior of irregular abrasive particles is consistent with that of spherical particles. Of course, the impact of particle shape on the experimental system developed in this paper still needs further experimental verification, which is also one of our future research directions.

7. Conclusions

This study has established a model of normal displacement and proposed the compensation method based on a decoupling polishing system. The deformation among the polishing toolhead, abrasive particle, and part has been studied using the Hertz contact theory. The normal displacement model is established. Then, the trajectory of the polishing tool is pre-planned, and the compensation method of normal displacement is proposed based on the polishing system. Experiments were carried out to examine our model and method. The major results of this study are summarized as follows:

(1) When the part was polished without the displacement compensation method, the values of roughness varied greatly with the variation in the curvature radius. The reason is that the normal displacement is mainly affected by the curvature radius of the aspheric surface when the plane toolhead is used. The normal displacement decreases sharply with the increasing curvature radius, and the values of roughness decrease accordingly. This illustrates that displacement compensation is necessary.

(2) The polishing quality can be improved significantly, and the average value of surface roughness of parts is 0.021 μm after being polished with the displacement compensation model and method.

(3) The values of the unevenness coefficient obtained with the normal displacement compensation method are much smaller than those obtained without normal displace-

ment compensation. The results show that the displacement compensation method proposed and the experimental system developed in this paper can achieve a more uniform surface quality and reduce the position error of the toolhead. This further proves that the method can effectively solve the problem of position error control in polishing, which has certain novelty and effectiveness.

(4) The model and method proposed in this paper can effectively achieve the normal displacement compensation, the quality of polishing can be improved greatly, and a more uniform surface quality can be obtained.

Our experimental study confirms that the normal displacement between the toolhead and the part should be compensated during the polishing process when the radius of curvature of the part surface varies. The model and method proposed in this paper can compensate for the normal displacement, and it is a new method for position error control.

Author Contributions: Methodology, Y.S.; Validation, D.Z.; Formal analysis, M.S.; Data curation, M.S. and Q.C. All authors have read and agreed to the published version of the manuscript.

Funding: This research was funded by the Tianjin Key Laboratory of Civil Aircraft Airworthiness and Maintenance Open Fund of Tianjin of China, grant number TJCAAM202102.

Data Availability Statement: The original contributions presented in the study are included in the article, further inquiries can be directed to the corresponding author.

Acknowledgments: Thanks very much for the support of the School of Intelligent Manufacturing of Jiaxing Vocational Technical College and the Ministry of Basic Education of Jiaxing Vocational Technical College.

Conflicts of Interest: The authors declare no conflicts of interest.

References

1. Jiang, L.; Zheng, J.X.; Peng, W.M.; Li, W.H.; Han, Y.J.; Zhang, S.H.; Zhou, N.N.; Qin, T.; Qian, L.M. Research Progress of Ultra-precision Polishing Technologies for Basic Components of Spacecraft. *Surf. Technol.* **2022**, *51*, 1–19.
2. Peng, Y.F.; Shen, B.Y.; Wang, Z.Z.; Yang, P.; Bi, G. Review on polishing technology of small-scale aspheric optics. *Int. J. Adv. Manuf. Technol.* **2021**, *115*, 965–987. [CrossRef]
3. Kumar, M.; Alok, A.; Das, M. Experimental and Simulation Study of Magnetorheological Miniature Gear-Profile Polishing (MRMGPP) Method Using Flow Restrictor. *J. Mech. Sci. Technol.* **2021**, *35*, 5151–5159. [CrossRef]
4. Shi, Y.J.; Zheng, D.; Hu, L.Y.; Wang, L.S. Modeling and analysis of force-position-posture decoupling for NC polishing of aspheric surface with constant material removal rates. *J. Mech. Sci. Technol.* **2012**, *58*, 1061–1073.
5. Jacobs, S.D.; Golini, D.; Hsu, Y.; Puchebner, B.E.; Strafford, D.; Kordonski, W.I.; Prokhorov, I.V.; Fess, E.M.; Pietrowski, D.; Kordonski, V.W. Magnetorheological finishing: A deterministic process for optics manufacturing. *SPIE* **1995**, *2576*, 372–382.
6. Kuriyagawa, T.; Saeki, M.; Syoji, K. Electrorheological fluid-assisted ultraprecision polishing for small three-dimensional parts. *Precis. Eng.* **2002**, *26*, 370–380. [CrossRef]
7. Zhang, L.; Ding, C.; Fan, C.; Wang, Q.; Wang, K.J. Process planning of the automatic polishing of the curved surface using a five-axis machining tool. *Int. J. Adv. Manuf. Technol.* **2022**, *120*, 7205–7218. [CrossRef]
8. Zhang, S.; Xie, B. *Non-Spherical Surface Simulation Processing Based on WheelPolishing Technology*; SPIE: Bellingham, WA, USA, 2020; Volume 11567.
9. Yao, Y.S.; Li, Q.X.; Ding, J.; Wang, Y.; Ma, Z.; Fan, X. Investigation of an Influence Function Model as a Self-Rotating Wheel Polishing Tool and Its Application in High-Precision Optical Fabrication. *Appl. Sci.* **2022**, *12*, 3296. [CrossRef]
10. Ji, S.M.; Yuan, Q.L.; Zhang, L. Study of the Removing Depth of the Polishing Surface Based on an ovel Spinning-Inflated-Ballonet Polishing Tool. *Mater. Sci. Forum* **2006**, *532*, 452–455. [CrossRef]
11. Zeng, S.; Blunt, L. An experimental study on the correlation f polishing force and material removal or bonnet polishing of cobalt chrome alloy. *Int. J. Adv. Manuf. Technol.* **2014**, *73*, 185–193. [CrossRef]
12. Zhao, Q.Z.; Zhang, L.; Han, Y.J.; Fan, C. Polishing path generation for physical uniform coverage of the aspheric surface based on the Archimedes spiral in bonnet polishing. *Proc. Inst. Mech. Eng. Part B J. Eng. Manuf.* **2019**, *233*, 2251–2263. [CrossRef]
13. Han, Y.J.; Wang, C.; Zhang, H.Y.; Yu, M.H.; Chang, X.C.; Dong, J.; Zhang, Y.F. Adaptive polishing path optimization for free-form uniform polishing based on footprint evolution. *Int. J. Adv. Manuf. Technol.* **2024**, *130*, 4311–4324. [CrossRef]
14. Qu, X.T.; Liu, Q.L.; Wang, H.Y.; Liu, H.Z.; Sun, H.C. A spiral path generation method for achieving uniform material removal depth in aspheric surface polishing. *Int. J. Adv. Manuf. Technol.* **2022**, *119*, 3247–3263. [CrossRef]
15. Zhao, Y.W.; Chang, L. A micro-contact and wear model for chemical-mechanical polishing of silicon wafers. *Wear* **2002**, *252*, 220–226. [CrossRef]

16. Johnson, K.L. *Contact Mechanics*; Cambridge University Press: Cambridge, UK, 1985.
17. Majid, R.M. A review of elasto-plastic shakedown analysis with limited plastic deformations and displacements. *Period. Polytech. Civ. Eng.* **2018**, *62*, 812–817.
18. Shi, Y.J.; Zheng, D.; Zhan, J.M.; Wang, L.S. Design, modeling and testing of torque servo driver based on Magnetorheology. In Proceedings of the 2010 International Conference on Measuring Technology and Mechatronics Automation, Changsha, China, 13–14 March 2010; pp. 1081–1084.

Disclaimer/Publisher's Note: The statements, opinions and data contained in all publications are solely those of the individual author(s) and contributor(s) and not of MDPI and/or the editor(s). MDPI and/or the editor(s) disclaim responsibility for any injury to people or property resulting from any ideas, methods, instructions or products referred to in the content.

Article

Integration of Metrology in Grinding and Polishing Processes for Rotationally Symmetrical Aspherical Surfaces with Optimized Material Removal Functions

Ravi Pratap Singh * and Yaolong Chen *

Department of Mechanical Engineering, Xi'an Jiaotong University; 28 Xianning West Road, Xi'an 710049, China
* Correspondence: staticravi78@gmail.com (R.P.S.); chenzwei@mail.xjtu.edu.cn (Y.C.)

Abstract: Aspherical surfaces, with their varying curvature, minimize aberrations and enhance clarity, making them essential in optics, aerospace, medical devices, and telecommunications. However, manufacturing these surfaces is challenging because of systematic errors in CNC equipment, tool wear, measurement inaccuracies, and environmental disturbances. These issues necessitate precise error compensation to achieve the desired surface shape. Traditional methods for spherical optics are inadequate for aspherical components, making accurate surface shape error detection and compensation crucial. This study integrates advanced metrology with optimized material removal functions in the grinding and polishing processes. By combining numerical control technology, computer technology, and data analysis, we developed CAM software (version 1) tailored for aspherical surfaces. This software uses a compensation correction algorithm to process error data and generate NC programs for machining. Our approach automates and digitizes the grinding and polishing process, improving efficiency and surface accuracy. This advancement enables high-precision mass production of rotationally symmetrical aspherical optical components, addressing existing manufacturing challenges and enhancing optical system performance.

Keywords: metrology; quality control; material removal function; error compensation; CAM software

Citation: Singh, R.P.; Chen, Y. Integration of Metrology in Grinding and Polishing Processes for Rotationally Symmetrical Aspherical Surfaces with Optimized Material Removal Functions. *Micromachines* 2024, 15, 1276. https://doi.org/10.3390/mi15101276

Academic Editors: Yao Liu and Jinjie Zhou

Received: 13 September 2024
Revised: 18 October 2024
Accepted: 20 October 2024
Published: 21 October 2024

Copyright: © 2024 by the authors. Licensee MDPI, Basel, Switzerland. This article is an open access article distributed under the terms and conditions of the Creative Commons Attribution (CC BY) license (https://creativecommons.org/licenses/by/4.0/).

1. Introduction

In modern society, optical components are integral to a wide range of industries, including optics, aerospace, medical devices, and telecommunications [1,2]. These components typically encompass planes, spheres, and, more recently, aspherical surfaces. The refractive index and direction of incident light differ across these types, leading to variations in imaging effects. Planar and spherical optical elements, with their straightforward geometries, can achieve high surface accuracy through traditional machining techniques, such as grinding, polishing, lapping, centering, diamond turning, and sometimes chemical etching [3]. However, despite their high yield and ease of mass production, spherical optical systems inherently suffer from optical aberrations such as low definition and peripheral distortion due to the differing focal points for off-axis rays. To mitigate these aberrations, traditional optical designs often employ multiple spherical mirrors with varying radii of curvature [4]. While this can improve imaging quality, it also results in more complex and costly optical systems.

In contrast, aspheric lenses, characterized by their non-spherical surfaces, represent a significant advancement in optical technology [5]. The inclusion of higher-order curvature allows for independent correction of spherical aberration, leading to more efficient and simpler optical systems [6]. Aspherical surfaces have the ability to focus light more accurately, improving the performance of telescopes and satellite imaging systems to achieve high-resolution images of space and the Earth's surface. These properties make them indispensable for space exploration and Earth observation missions [7]. The automotive industry also benefits significantly from aspherical optics. Modern vehicles utilize aspherical lenses in headlights and rear-view mirrors to improve driver visibility and safety. Aspherical headlights provide better illumination patterns, reducing glare for oncoming traffic and enhancing night-time driving conditions. In rear-view mirrors, aspherical designs reduce blind spots, offering a wider field of view and improving overall vehicle safety [8]. In the medical field, aspherical lenses and mirrors are used in a variety of diagnostic and therapeutic devices. For instance, in endoscopy, aspherical lenses provide clearer and more detailed images of internal organs, aiding in accurate diagnosis and treatment. Ophthalmic devices, such as corrective eyeglass lenses and contact lenses, also benefit from aspherical designs to improve vision correction by minimizing distortions [9]. The telecommunications industry employs aspherical lenses in fiber optics to improve signal clarity and transmission efficiency. Aspherical components help focus light precisely on optical fibers, reducing signal loss and enhancing the performance of communication networks. This precision is critical for high-speed data transmission and reliable internet connectivity [10].

Despite the numerous advantages and broad applications of aspherical surfaces, their manufacturing presents significant challenges. The complex geometry of aspherical surfaces, with their varying curvature, demands exceptional precision during production [11]. Several key difficulties arise in achieving the high precision required for these surfaces [12]. The production process is highly susceptible to systematic errors in CNC (Computer Numerical Control) equipment. These errors can result from inaccuracies in machine calibration, misalignment of components, and limitations in the control algorithms. Such inaccuracies are amplified when dealing with the intricate shapes of aspherical surfaces, leading to deviations from the desired surface profile [13]. Tool wear is a critical factor affecting precision. During the grinding and polishing processes, tools undergo wear and tear, altering their shapes and effectiveness. This wear is not uniform and can lead to uneven material removal, introducing errors in the surface geometry. Continuous monitoring and compensation for tool wear are essential to maintain precision, yet this adds complexity to the manufacturing process [14]. To monitor tool wear effectively and ensure the desired surface quality, various methods and devices are employed. Advanced microscopy techniques play a pivotal role in this regard. For example, 3D focus variation microscopes utilize multiple focal planes to construct a detailed 3D profile of the surface, enabling the identification of wear patterns that may not be visible through traditional methods. Interferometric microscopy offers high-resolution surface measurements by analyzing interference patterns created by the interaction of light waves, allowing for the detection of minute changes in surface geometry due to wear [15]. Confocal microscopy is another valuable technique, providing optical sectioning capabilities to generate high-resolution images. This method is particularly useful for examining surface roughness and detecting wear at various depths, thereby contributing to a more comprehensive understanding of tool performance over time [16]. In addition to optical methods, elastomeric tactile sensors are gaining traction in wear measurement. These sensors can conform to the surface profile and directly measure wear by assessing changes in surface texture and roughness. Their flexibility allows for real-time monitoring during the manufacturing process, providing immediate feedback on tool performance and wear characteristics [17].

Measurement inaccuracies of aspherical surfaces also pose significant challenges. Accurate measurement of aspherical surfaces is difficult because of their non-uniform curvature [18]. Traditional measurement techniques used for spherical and planar surfaces are inadequate for aspherical geometries [19]. Advanced metrology tools are required to capture the surface profile precisely, but these tools themselves are susceptible to calibration errors and environmental influences such as temperature fluctuations and vibrations [20]. Environmental disturbances further complicate the production process. Variations in temperature, humidity, and vibrations can affect both the manufacturing equipment and the material being processed [21]. Such disturbances can lead to thermal expansion or contraction of the materials, misalignment of equipment, and other issues that compromise the precision of the finished aspherical surface [22].

To address these challenges, precise error compensation strategies are necessary. Traditional methods used for spherical optics do not suffice for aspherical components, necessitating the development of specialized approaches. This study aims to tackle these issues by integrating advanced metrology with optimized material removal functions in the grinding and polishing processes. Through the combination of numerical control technology, computer technology, and data analysis, we developed CAM software specifically designed for aspherical surfaces. This software employs a compensation correction algorithm to process error data and generate accurate NC programs for machining, thereby automating and digitizing the grinding and polishing process.

2. Analysis of Material Removal Mechanism and Its Coordinate System

2.1. Material Removal Mechanism in Grinding

Grinding is a crucial process in the production of aspherical surfaces, characterized by high precision and material removal rates. The fundamental principle involves oblique cutting, where the contact path between the workpiece and the diamond cutting tool forms a circular trajectory during machining. This circular plane is angled relative to the workpiece axis, creating what is known as a truncated circle. In this method, variations in error have minimal impact on the surface quality, allowing for high-precision surfaces to be achieved with fewer coordinate variables. The schematic diagram of the generating process is shown in Figure 1. The formula for calculating the processing angle is given as

$$\sin \alpha = \frac{D_m}{2(R \pm r)} \quad (1)$$

where D_m = diamond grinding wheel pitch diameter, R = machined spherical radius, r = diamond grinding wheel end arc, and α = axis swing angle.

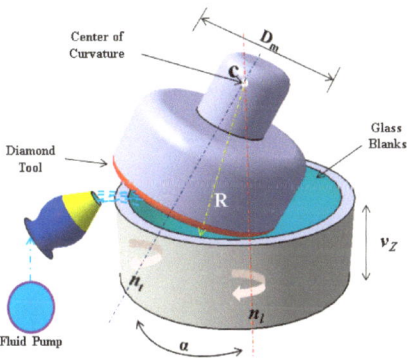

Figure 1. Schematic diagram of the grinding process.

2.2. Material Removal Mechanism in Polishing

The polishing process of an aspherical lens begins with grinding to shape the lens surface into a preform close to the desired aspherical profile. This preform still contains irregularities and inaccuracies that must be eliminated by polishing to achieve the final precision optics. Figure 2 illustrates a schematic diagram of the polishing process.

Modeling the entire polishing process mathematically and physically is challenging because of its complexity. Factors like temperature, slurry concentration, particle size, pH value, workpiece speed, and feed rate all influence material removal rates. Consequently, creating an accurate mathematical relationship between all polishing parameters is nearly impossible. To address these challenges, researchers have developed mathematical models with certain assumptions. One of the most notable is the Preston equation, proposed by F.W. Preston in 1927. The Preston equation describes the CNC polishing process as a linear relationship over a broad numerical range, suggesting that the material removal rate is proportional to the pressure applied by the polishing pad and the relative velocity between the pad and the workpiece. The mathematical model for the amount of material removed during polishing per unit time is given by [23].

$$\frac{dZ}{dx} = KV(x,y,t)P(x,y,t) \tag{2}$$

where dZ/dx—the amount of material removed per unit time; K—Preston constant, with polishing die material, workpiece material, polishing liquid concentration, temperature, and other factors; $V(x, y, t)$—the relative speed of the polishing head and the workpiece at the point of contact degree; and $P(x, y, t)$—the instantaneous pressure between the polishing head and the workpiece at the point of contact.

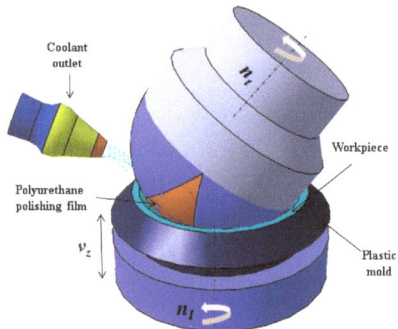

Figure 2. Schematic diagram of the polishing process.

2.3. Coordinate System for Describing Rotationally Symmetrical Aspheric Surfaces

Manufacturing rotationally symmetrical aspheric surfaces poses significant challenges because of their complex shapes. Currently, the standard right-hand Cartesian coordinate system is used internationally to describe aspheric surfaces, as shown in Figure 3a. The coordinate system is configured with the vertex of the aspheric surface positioned at the origin. The z-axis serves as the optical axis, oriented from left to right.

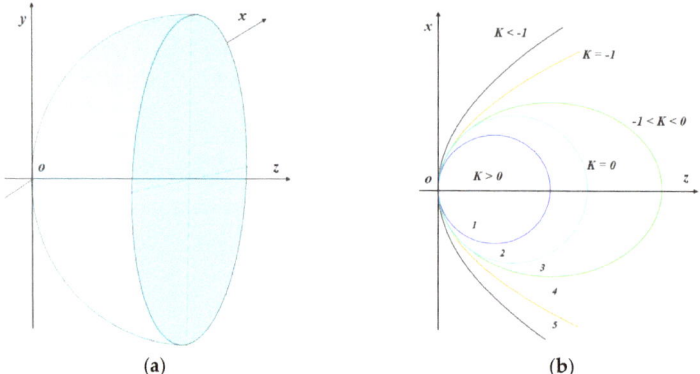

Figure 3. Cartesian coordinate system of aspheric surfaces. (**a**) Aspheric surface. (**b**) Quadratic curve for with K.

Aspheric surfaces are typically described using meridian cross-section curves. In this setup, the *xoz* coordinate plane represents the plane of the meridian section. For any point on an aspheric surface, the radius of curvature is determined by the position of the center of curvature relative to the vertex. Specifically, if the center of curvature is to the right of the vertex, the radius of curvature is positive. If the center of curvature is to the left of the vertex, the radius of curvature is negative. This convention ensures a consistent and precise description of aspheric surfaces, facilitating accurate design and analysis in optical applications.

The equation for the meridian section curve of an aspheric surface is typically composed of the following parts: a datum quadric surface and an additional polynomial. The additional polynomials are generally expressed as polynomials of even power series. The equation can be written as

$$z(x) = \frac{x^2}{R_0 \left[1 + \sqrt{1 - (1+K)(x/R_0)^2}\right]} + \sum_{i=1}^{n} A_i x^{2i+2}, x \in \left[0, \frac{\Phi_0}{2}\right] \qquad (3)$$

where $z(x)$ represents the height of the surface along the z-axis, R_0 is the radius of curvature at the vertex of the aspheric curve, defined as $R_0 = 1/c$, where c is the curvature at the apex of the aspheric curve, Φ_0 is the clear aperture of the workpiece to be processed, K is the quadratic coefficient (with $K = -e^2$), e is the eccentricity of the aspheric surface, and A_i (where $i = 1, 2, \ldots n$) denotes the surface coefficients of the higher order terms of the spheric surface.

Because of the nature of rotational symmetry, the range of values for x is desirable as non-negative fractions. The quadratic coefficient K varies based on the quadratic curve types, as shown in Figure 3b. The radius of curvature at each point on the rotationally symmetrical aspheric surface is different, and it can be calculated as

$$R(x) = \frac{\left[1 + \left(\frac{dz}{dx}\right)^2\right]^{\frac{3}{2}}}{\left|\frac{d^2z}{dx^2}\right|} \qquad (4)$$

where dz/dx refers to the first derivative of the meridian section equation at any point on the surface of the workpiece, and d^2z/dx^2 refers to the second derivative of the surface equation at any point on the workpiece.

In the material removal process of aspheric components, the spherical profile is first machined, which differs from the aspherical profile. The difference between the corresponding points in the x-direction of the surface profile in the z-direction is the asphericity, and the maximum difference is the maximum asphericity A_{max}^0. The sphere with the minimum value of maximum asphericity is called the best reference sphere. A schematic diagram of the best reference sphere is given in Figure 4.

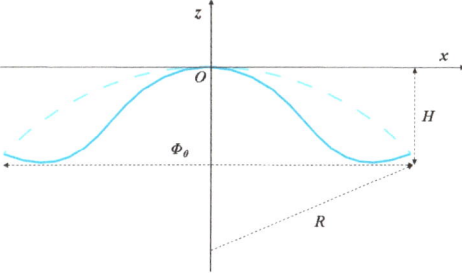

Figure 4. Schematic diagram of the best reference sphere.

The maximum asphericity A_{max}^0 can be calculated as

$$R = \frac{\Phi_0^2 + 4H^2}{8H}$$
$$A_{\max}^0 = -\frac{K\Phi_0^4}{512R^3} \tag{5}$$

where Φ_0 is the diameter of the workpiece and H is the height. From Equation (5), the maximum degree of asphericity can be achieved when $x = \Phi_0/2\sqrt{2}$.

3. Establishment of the Optimized Material Removal Model

3.1. Mathematical Model of Material Removal

In the process of material removal, the finished workpiece surface is typically divided into numerous small areas to determine the material removal amount on the surface. The material removal amount for each differential area is then calculated, and the overall removal distribution is accumulated. This method requires extensive calculations, so an idealized assumption is proposed to simplify the process. Firstly, considering the physical model, material removal on the workpiece surface occurs during both grinding and polishing. During grinding, large amounts of material are removed to form the rough shape of the workpiece. In polishing, the feed speed of the polishing head is significantly lower than its rotation speed. Therefore, the material removal by the polishing head per unit time can be analyzed based solely on the movement of the polishing head, neglecting the minor error influence caused by its feed.

Secondly, from a mathematical perspective, maintaining a constant material removal rate by the polishing head is challenging. Generally, physical and chemical reactions occur during polishing, causing surface distortion of the workpiece. Without using fluid non-contact polishing or a flexible polymer polishing head, it is impossible to ensure constant pressure at each point of contact. Additionally, considering the edge effect, when the polishing head moves to the edge of the workpiece, the contact width between the polishing head and the workpiece changes, affecting the relative contact pressure and thus altering the material removal rate. However, if the polishing head is not exposed, the relative contact pressure remains constant, and the rotation speed is relatively low, making it feasible to consider the material removal rate constant.

The mathematical model of polishing material removal is illustrated in Figure 5. In the equal compression polishing process, the contact width between the polishing head and the workpiece remains constant, denoted as L, and the workpiece diameter is $2x_0$. The polishing process aims to reduce material continually, ideally polishing all materials to the lowest point of error based on the given surface error data. Considering residence time $D(x, y)$ as an independent variable, residence time in the polishing process is given as

$$T = \int_{-x_0}^{x_0} \int_0^L D(x,y)\mathrm{d}x\mathrm{d}y \tag{6}$$

where $D(x, y)$ is the residence time of the polishing head, and T is the total residence time of the polishing process. When the polishing head does not move, the average removal amount per unit time $R(x, y)$ is given as

$$R(x,y) = \lim_{T \to \infty} \left[\frac{1}{T} \int_{-x_0}^{x_0} \Delta h(x,y)\mathrm{d}t \right] \tag{7}$$

If the removal function applies to any polishing area, the total amount of material removed from the workpiece surface is the sum of removal amounts at each point. Thus, the removal function of the polishing head can be denoted as $\delta\alpha$ and the residence time function of the workpiece surface as $\delta\beta$. By superimposing numerous elements over the polishing head path, the material removal expression can be derived as follows:

$$\Delta h(x,y) = \lim_{\alpha \to 0, \beta \to 0} \sum_\alpha \sum_\beta R(x-\alpha, y-\beta)D(\alpha,\beta)\delta\alpha\delta\beta \tag{8}$$

When $\delta\alpha\delta\beta$ approaches 0, $\delta\alpha\delta\beta$ can be infinitely reduced to the area element $d\alpha d\beta$, and then there are

$$\Delta h(x,y) = \int_\alpha \int_\beta [R(x-\alpha, y-\beta)D(\alpha,\beta)d\alpha d\beta] \tag{9}$$

The total material removed from the workpiece is expressed as a two-dimensional convolution of the removal function $R(x, y)$ and the residence time $D(x, y)$ as follows:

$$H(x,y) = R(x,y) * D(x,y) \tag{10}$$

After polishing, the residual error $E(x, y)$ can be calculated as

$$E(x,y) = H(x,y) - R(x,y) * D(x,y) \tag{11}$$

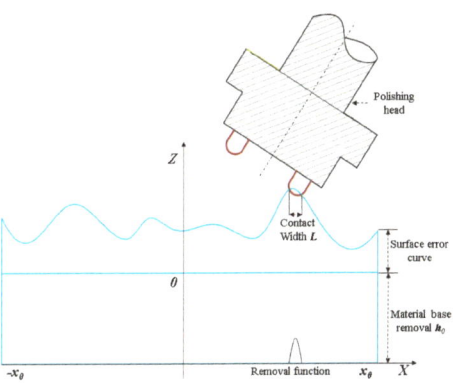

Figure 5. Mathematical model of material removal.

3.2. Equivalent Material Removal Model of Aspheric Polishing

A schematic diagram of equivalent material removal model of aspheric polishing is given in Figure 6. xoz is the workpiece coordinate system, the origin of the coordinate system coincides with the vertex of the workpiece surface, and $p(x, z)$ is any point on the workpiece.

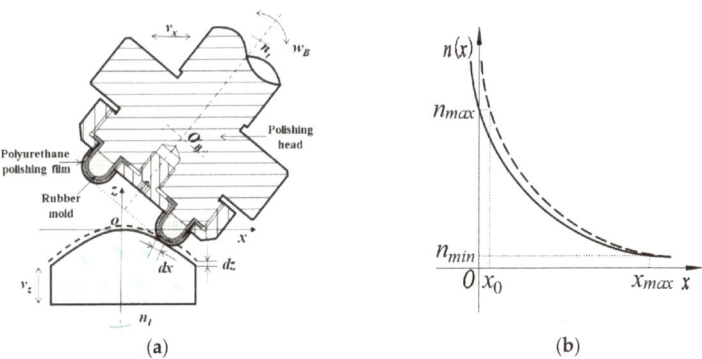

Figure 6. Schematic diagram of the equivalent model of material removal. (**a**) Equivalent material removal. (**b**) Relation between workpiece rotation speed and machining position.

During a small unit of time, the polishing head moves a tiny distance dx in the x-direction and the workpiece shifts by dz in the z-direction. As the workpiece rotates at speed $n(x)$, the polishing head effectively removes material in a cylindrical shape on the workpiece surface. This cylindrical volume has a length of $2\pi x n(x)$, a width of dz, and a height of dx. This process can be described as

$$dV = 2\pi x n(x) dx dz \tag{12}$$

Consider two points $p_1(x_1, y_1)$ and $p_2(x_2, y_2)$ at different positions on the workpiece. To ensure that the material removal per unit time is the same, it is necessary to satisfy,

$$2\pi x_1 n(x_1) dx dz = 2\pi x_2 n(x_2) dh dx \tag{13}$$

Equation (13) can be reduced to

$$x n(x) = Const \tag{14}$$

From the relationship between the workpiece rotation speed and machining position in Figure 6b, when the polishing head reaches position $x \to 0$, the theoretical rotation speed $n(x) \to \infty$. However, in practical machining, the workpiece rotation speed cannot actually reach infinity. Therefore, the maximum rotation speed of the workpiece at the coordinate origin, $x = x_0$, represents a practical limit. To find the actual rotation speed curve, we shift the theoretical rotation speed curve leftward by x_0. Here, the solid line depicts the actual rotation speed curve, while the dotted line represents the theoretical one. From here, Equation (14) can be written as

$$(x + x_0) n(x + x_0) = Const \tag{15}$$

During numerical control polishing, the workpiece rotational speed reaches its maximum n_{max} at the center $x = 0$ and its minimum n_{min} at the edge $x = \phi_0/2$. Based on these two points, we obtain

$$\begin{cases} x_0 n_{max} = x_{max} n_{min} = Const \\ x_{max} - x_0 = \frac{\phi}{2} \end{cases} \tag{16}$$

From Equation (16), we obtain

$$x_0 = \frac{\phi n_{min}}{2(n_{max} - n_{min})} \tag{17}$$

Combining Equations (15)–(17), the rotational speed at each machining position on the surface of the workpiece is given as

$$n(x) = \frac{\phi n_{max} n_{min}}{\phi n_{min} + 2x(n_{max} - n_{min})} \tag{18}$$

From Equation (18), we observe that once the max and min speeds of the workpiece are established, the rotational speed at any point $p(x, z)$ on the workpiece surface can be uniquely determined. During polishing, since there is less material at the center and more at the edges of the workpiece, ensuring consistent material removal across all points requires varying processing times at different points. To synchronize the amount of material removed at each point, the feed speed $F(x)$ of the polishing head must be adjusted relative to the rotational speed $n(x)$ of the workpiece. This relationship is expressed as follows when the polishing head advances by a step Δx:

$$F(x) = n(x) \cdot \Delta x \tag{19}$$

When the processing feed step is set, the feed speed of the polishing head is directly proportional to the workpiece speed. This means the head feed rate depends on the machining position. The feed speed at each processing point on the workpiece surface can be obtained by combining Equations (18) and (19)

$$F(x) = \frac{\phi n_{\max} n_{\min} \Delta x}{\phi n_{\min} + 2x(n_{\max} - n_{\min})} \tag{20}$$

3.3. Calculation of the Feature Removal Amount

A schematic diagram of the material removal during aspheric motion polishing is shown in Figure 7. In the coordinate system OXY, the contact area is wide. If the degree is a, then $\rho_{2max} = \rho_1 + a$, $\rho_{2min} = \rho_{1-a}$, and the angle range corresponding to the arc l_1l_2, is $[\theta_1, \theta_2]$. Here, θ_1, θ_2 can be calculated as

$$\theta_2 = \arccos\left(\frac{\rho_1^2 + \rho_2^2 - a^2}{2\rho_1\rho_2}\right), \theta_1 = -\arccos\left(\frac{\rho_1^2 + \rho_2^2 - a^2}{2\rho_1\rho_2}\right) \tag{21}$$

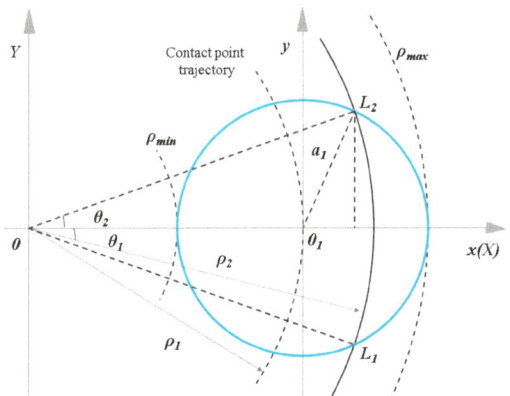

Figure 7. Schematic diagram of material removal during aspheric motion polishing.

The residence time dt of the polishing head on the workpiece surface is proportional to the length of the polishing track dl, while the feed of the polishing head is inversely proportional to the length of the polishing track dl. This relationship can be expressed as

$$dt = \frac{dl}{F} = \frac{\rho d\theta}{F} \tag{22}$$

Combining Equations (2) and (22), we obtain

$$\frac{dh(x,y)}{dl} = \frac{KP(x,y)V(x,y)}{F} \tag{23}$$

When converted to polar coordinates, we obtain

$$\frac{dh(\rho,\theta)}{dl} = \frac{KP(\rho,\theta)V(\rho,\theta)\rho}{F} \tag{24}$$

When the polishing head is fed in the negative direction of the x-axis, the feed speed in the contact area also moves negatively along the x-axis, increasing from ρ_{2min} to ρ_{2max}. This indicates there is

$$F = \frac{\rho}{\rho_1} F_0 \tag{25}$$

Combining Equations (24) and (25), we obtain

$$h(\rho,\theta) = \frac{K\rho_1}{F_0} \int_{\theta_1}^{\theta_2} P(\rho,\theta)V(\rho,\theta)d\theta \tag{26}$$

Using polar coordinates to indicate the pressure distribution in the contact area

$$P(\rho,\theta) = P_0\sqrt{1 - \frac{(\rho_2\cos\theta - \rho_1)^2 + (\rho_2\sin\theta)^2}{a^2}} \quad (27)$$

By substituting Equations (18) and (27) in Equation (27), we obtain the expression of feature material removal amount

$$h(\rho,\theta) = \frac{3K\rho_1 P}{F_0 a^3} \int_{\theta_1}^{\theta_2} \sqrt{(a^2 + \rho_1^2 + \rho_2^2 - 2\rho_1\rho_2\cos\theta)} * \sqrt{\left[\begin{array}{c}(n_T\rho_1)^2 + \left(n_T + \frac{\phi n_{\max}n_{\min}}{\phi n_{\min} + 2x(n_{\max} - n_{\min})}\right)^2 \rho_2^2 - \\ 2\rho_1\rho_2 n_T\left(n_T + \frac{\phi n_{\max}n_{\min}}{\phi n_{\min} + 2x(n_{\max} - n_{\min})}\right)\cos\theta\end{array}\right]} d\theta \quad (28)$$

3.4. Pressure Distribution Model in the Contact Area

In aspheric polishing, the polishing film is typically softer than the aspheric components, allowing the polishing head and workpiece surface to establish adaptive contact. This adaptive contact ensures effective engagement as the polishing head feeds across different curvature points of the workpiece surface. During the polishing process, the polishing head applies a normal force to the workpiece surface, creating a small contact area where material removal occurs because of the elastic deformation of both the polishing head and the workpiece surface. The pressure distribution in this contact area is often analyzed using Hertz contact theory, which describes how compressive stress is distributed when two elastic bodies are in contact.

Given that the radius of the polishing head edge is much smaller than the radius of curvature of the non-spherical surface, their contact can be approximated as surface-to-surface contact. This results in the formation of a small elliptical contact region when the surfaces come into contact and compress each other. The size and shape of this contact area, as well as the pressure distribution, vary with changes in the polishing head size and the workpiece curvature radius. The ellipse contact area can be expressed as

$$\frac{x^2}{a^2} + \frac{y^2}{b^2} = 1 \quad (29)$$

where a is the long half-axis of the ellipse and b is the short of the ellipse. a and b are expressed as

$$\begin{cases} a = m^3\sqrt{\frac{3F}{4A}\left(\frac{1-\mu_1^2}{E_1} + \frac{1-\mu_2^2}{E_2}\right)} \\ b = n^3\sqrt{\frac{3F}{4A}\left(\frac{1-\mu_1^2}{E_1} + \frac{1-\mu_2^2}{E_2}\right)} \end{cases} \quad (30)$$

where F is the normal force of the polishing head on the workpiece; μ_1 and μ_2 are the Poisson's ratio of the polishing film and workpiece material, and E_1 and E_2 are the modulus of elasticity of the polishing film and workpiece material, respectively. Now, we denote p_0 as the maximum compressive stress in the contact area and P as the maximum pressure

$$P = \iint p\,dA = \frac{2}{3}\pi ab p_0 \quad (31)$$

From Equation (30), we obtain

$$p_0 = \frac{3P}{2\pi ab} \quad (32)$$

Therefore, the pressure distribution in the elliptical contact region can be expressed as

$$p(x,y) = p_0\sqrt{1 - \frac{x^2}{a^2} - \frac{y^2}{b^2}} \quad (33)$$

By substituting the value of a and b in Equation (33), the theoretical pressure distribution in the elliptical contact region can be obtained as

$$P = \frac{3F}{2\pi mn \left(\frac{3F}{4A}\left(\frac{1-\mu_1^2}{E_1}+\frac{1-\mu_2^2}{E_2}\right)\right)^{\frac{2}{3}}} * \sqrt{1 - \frac{x^2}{m^2\left(\frac{3F}{4A}\left(\frac{1-\mu_1^2}{E_1}+\frac{1-\mu_2^2}{E_2}\right)\right)^{\frac{2}{3}}} - \frac{y^2}{n^2\left(\frac{3F}{4A}\left(\frac{1-\mu_1^2}{E_1}+\frac{1-\mu_2^2}{E_2}\right)\right)^{\frac{2}{3}}}} \quad (34)$$

When the radius of the polishing head wheel end is significantly smaller than the radius of curvature of the workpiece, the contact area can be approximated as a circle, simplifying practical engineering issues. As depicted in Figure 8, which illustrates the contact between the polishing head and the aspheric workpiece, the geometric relationship between the two determines the contact width

$$r - \sqrt{r^2 - \left(\frac{L}{2}\right)^2} + R - \sqrt{R^2 - \left(\frac{L}{2}\right)^2} = \delta \quad (35)$$

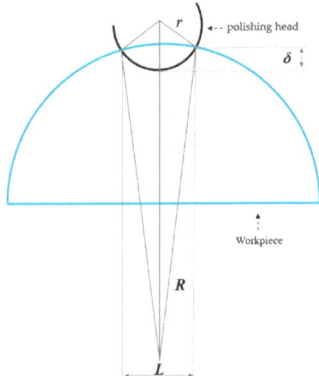

Figure 8. Schematic diagram of the contact between the polishing head and workpiece.

The width of contact area L is reduced to

$$L = 2\sqrt{R^2 - \frac{\left[R^2 - r^2 + (r+R-\delta)^2\right]^2}{4(r+R-\delta)^2}} \quad (36)$$

where r is the radius of the arc of the wheel end of the polishing head; R is the radius of curvature of the workpiece at the contact point; δ is the compression of the polishing head; and L represents the width of the contact between the polishing head and the workpiece.

3.5. Relative Linear Velocity Distribution Model for the Contact Area

According to Preston's equation, the material removal rate between the polishing head and workpiece is influenced by their relative linear velocity. Adjusting this velocity can enhance polishing efficiency. Figure 9 illustrates the polishing trajectory from the top view, where the polishing head rotates at speed n_T, the workpiece at speed n_L, and their feed speed in the x-direction is significantly smaller, minimally affecting the process.

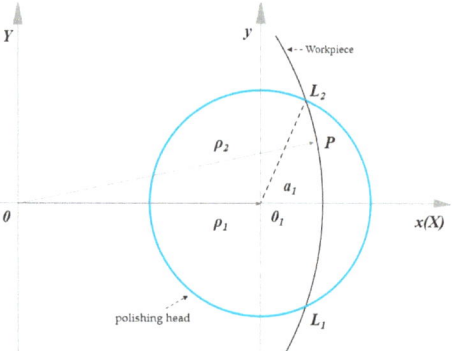

Figure 9. Schematic diagram of the contact area of the polishing trajectory.

Assume the center of the workpiece is O, the center of the contact area between the polishing head and the workpiece is O_1, and point P is any point in the contact area. As the workpiece rotates, point P travels an arc length of l_1l_2, with a distance from the radius of the contact area at O_1 to point P denoted as a_1. Polar coordinates are more suitable for describing curved motion compared with Cartesian coordinates. Point O serves as the pole, O_x as the polar axis, and the polar radius of point O_1 is ρ_1. Furthermore, point O_1 acts as the pole for its own coordinate system, where its polar radius is ρ_1, and for point P, its polar radius is ρ_2 with a polar angle θ. The distance from point P to point O_1 is given by

$$a_1 = \sqrt{(\rho_2 \cos\theta - \rho)^2 + (\rho_2 \sin\theta)^2} \tag{37}$$

The linear velocity of the polishing head at the fixed-point P is given as

$$V_T = 2\pi n_T.a_1 = 2\pi n_T \sqrt{\rho_1^2 + \rho_2^2 - 2\rho_1\rho_2 \cos\theta} \tag{38}$$

The linear velocity of the workpiece at point P is given as,

$$V_L = 2\pi n_L.|OP| = 2\pi n_L \rho_2 \tag{39}$$

By combining Equations (37)–(39), we obtain the relative linear velocity of the polishing head and the workpiece at the contact point P

$$\begin{aligned} V &= \sqrt{V_T^2 + V_L^2 - 2V_TV_L \frac{\rho_2^2 + (\rho_2 \cos\theta - \rho_1)^2 - \rho_1^2}{2\rho_2(\rho_2 \cos\theta - \rho_1)}} \\ &= 2\pi \sqrt{(n_T\rho_1)^2 + (n_T + n_L)^2 \rho_2^2 - 2\rho_1\rho_2 n_T(n_T + n_L) \cos\theta} \end{aligned} \tag{40}$$

4. CAM Software Development for Rotational Symmetrical Aspheric Surfaces

4.1. Overall Architecture of CAM Software

The aspheric CNC grinding and polishing CAM software is mainly tailored for Chenna Automation CNC LGS200 grinding and LPS200 polishing processes. Based on software requirements, its various functions are subdivided into several functional modules, forming the foundational units of the system. The overall structure, shown in Figure 10, includes modules such as parameter control, chart visualization, NC program generation and verification, surface shape error correction, and detection data processing.

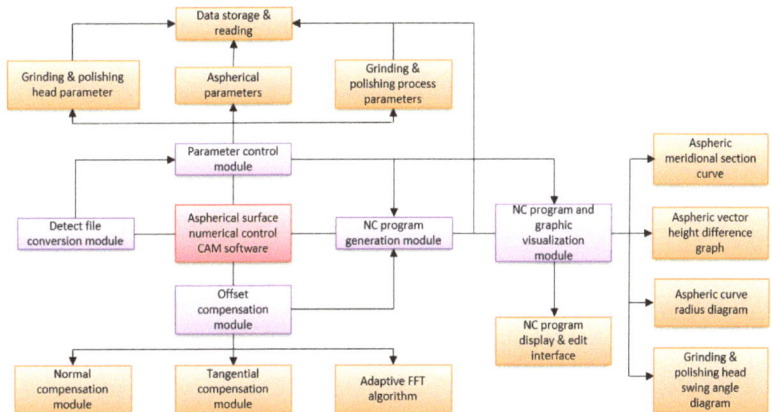

Figure 10. Schematic diagram of the contact area of the polishing trajectory.

4.2. CAM Software Function Module Design

The parameter control module consists of the following main parts: (a) tool parameter setting and (b) workpiece processing coordinate setting. In Figure 11a, the tool parameter setting involves handling various parameters. These include polishing tool details like the B-axis offset angle, polishing head length L_T, polishing head middle diameter D_m, wheel end arc radius r, distance from rotary center to hydraulic jaw surface L_B, basic machine tool parameters X_{basic} and Z_{basic}. The workpiece processing coordinate setting involves the curvature radius, aperture size, and quadratic coefficient K, and higher-order surface coefficients are also managed in this section. It also includes the program name, compensation, concave and convex surface, model, caliber, direction for processing, etc.

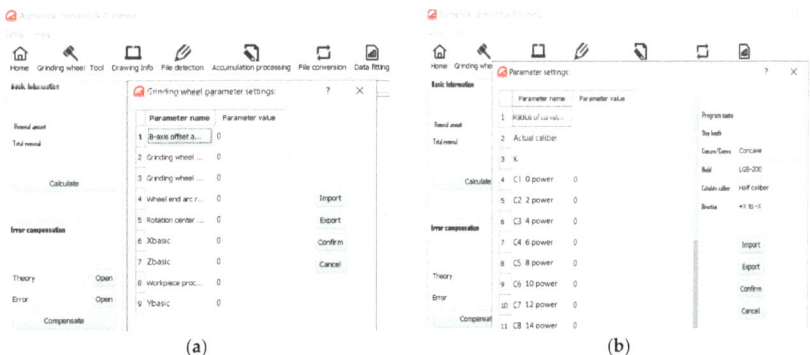

Figure 11. CAM software function module. (**a**) Tool parameter setting. (**b**) Workpiece parameter setting.

5. Experimental Verification of CAM Software

To verify the rotational symmetry of the developed CAM software, a closed-loop error compensation study was conducted. The experiment focused on a rotationally symmetrical aspheric optical element with a diameter of 15 mm, made of H-K9L glass. UV glue was applied to the contact area between the workpiece and the tooling, and a dial gauge was used to ensure coaxial alignment. To reduce the thermal effect during grinding and polishing, a water-soluble coolant was used. The detailed setup of the closed-loop experiment is shown in Figure 12, and the parameters for the grinding wheel, polishing tool, and the grinding and polishing processes are provided in Tables 1–4.

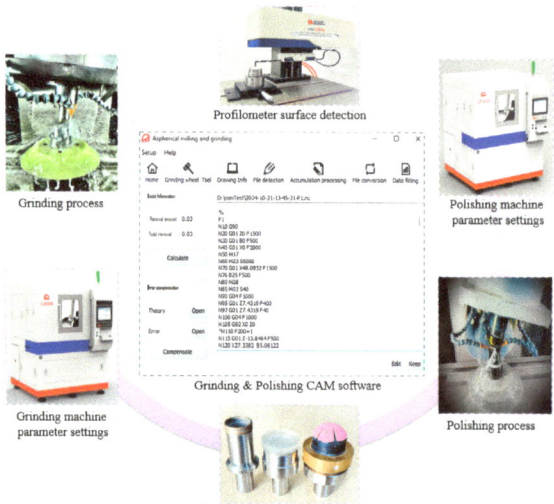

Figure 12. Closed-loop error compensation with CAM software.

Table 1. Basic parameter of the cylindrical diamond grinding wheel.

Grinding Wheel Parameter	Parameter Value
Grinding wheel length/mm	68.26
Grinding wheel diameter/mm	32.05
Radius of wheel end arc/mm	1.08
Diamond grain size	D20
Diamond concentration/%	80
Binders	Bronze, graphite powder

Table 2. Basic parameter of the polishing head.

Polishing Head Parameter	Parameter Value
Polishing head length/mm	70
Polishing head middle diameter/mm	27
Radius of the wheel end arc/mm	23
Polishing film material	Polyurethane polishing film

Table 3. Aspheric grinding process parameters.

Process Parameter	Parameter Value
Grinding wheel speed/rpm	3500
Workpiece speed/rpm	40
Grinding wheel feed speed/(m/min)	20
Amount of material removal/mm	0.01

Table 4. Aspheric polishing process parameters.

Process Parameter	Parameter Value
Polishing head speed/rpm	3000
Workpiece speed/rpm	20–1000
Feed speed of the polishing head/(m/min)	2–10

Before conducting the aspheric polishing experiment, the aspheric elements were first ground to achieve the desired aspheric profile using the LGS 200, a three-axis CNC machine.

The polishing process was then carried out using the LPS 200, an ultra-precision optical CNC polishing machine. Both machines were made in China, independently developed by Suzhou Chenna Automation Technology Co., Ltd. (Suzhou, China) After each grinding and polishing stage, surface errors were detected using a Taylor-Hobson profiler from the United Kingdom to measure the surface shape error of the aspheric elements.

To verify the closed-loop error compensation process experimentally, we input the basic parameters of the grinding wheel and the aspheric workpiece into the CAM software to generate the initial NC code for grinding. Similarly, the polishing head and workpiece parameters were used to generate the NC code for polishing. The CAM software produced theoretical and actual data files, along with the NC code for both processes. After each grinding operation, surface errors were measured using a profilometer, which generated a .mod file containing the measured error data. The CAM software then converted the .mod file into an Excel format, saving it as the actual measured surface error data. Both the theoretical data from the initial input and the actual measured data were loaded into the CAM software for analysis. The software compensated for the errors and generated new NC codes for the next iteration.

This iterative feedback loop is key to maintaining precision without requiring active control of environmental factors. Even when temperature or vibration introduces small deviations, the CAM software detects and compensates for these errors at each step. Thus, any discrepancies are continuously corrected by generating updated NC codes, ensuring precision over multiple iterations. After completing the grinding phase to remove scratches and irregularities, the same compensation process was applied during polishing to achieve the desired surface finish. This dual-step approach ensured both dimensional accuracy and surface quality, with results from both processes presented in Figure 13.

As shown in Figure 13a, the initial fine grinding of the aspheric surface produced a surface shape error with a PV value of 21.71 μm and an RMS value of 6.99 μm. These relatively high error values indicated the need for surface shape compensation. Using the developed CAM software, a new NC code was generated for this compensation. After applying the correction, the surface shape error was significantly reduced, achieving a PV value of 3.01 μm and an RMS value of 0.59 μm. The error curve flattened considerably, showing that the surface was much closer to the desired shape, with the PV and RMS convergence rates reaching 86.1% and 91.5%, respectively. These results clearly demonstrate effective error convergence, meeting the processing requirements for the aspheric surface and allowing the experiment to proceed to the polishing phase.

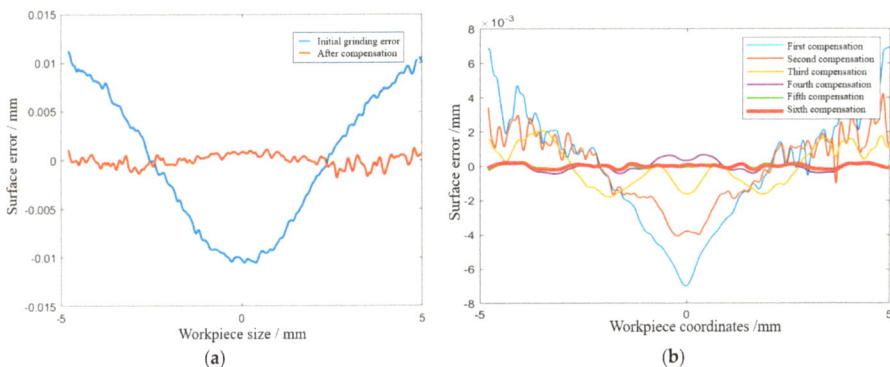

Figure 13. Surface error during grinding and polishing. (**a**) Surface error during grinding. (**b**) Surface error during polishing.

In the subsequent polishing process, as shown in Figure 13b, by the sixth error compensation, the surface shape error was further reduced to a PV value of 0.4004 μm and an

RMS value of 0.1107 μm. The error curve at this stage was nearly a flat straight line, with the PV convergence rate reaching 97.1%. This indicates a highly effective error convergence, confirming that the polishing process successfully refined the surface to meet stringent optical quality standards.

6. Conclusions

This paper presents a novel closed-loop framework for processing rotationally symmetrical aspherical surfaces using optimized material removal functions. Developed on the Qt platform, the CAM software integrates parameter control, chart visualization, sequential NC program generation, surface shape error correction, and detection data processing. The software facilitates seamless feedback between theoretical and measured data, refining NC programs through the entry of polishing parameters and data correction using an Adaptive Fourier Transform Algorithm.

The designed CAM software generates both theoretical predictions and pre-machining NC code based on the input parameters of the cutting tool and workpiece. After machining, the workpiece is measured using a profilometer, and the measured data are integrated back into the software for comparison with theoretical models. This iterative error compensation approach allows the system to generate an optimized NC code for precise surface finishing. Experimental results from grinding and polishing processes confirm that this method effectively controls surface shape errors within the sub-micron accuracy range, demonstrating the software's reliability and precision.

Unlike previous studies that focus on individual aspects of surface finishing, this research offers a fully integrated closed-loop solution, combining real-time error compensation and adaptive data correction. The iterative feedback mechanism eliminates the need for external tools, enhancing precision while reducing processing time. The use of the Adaptive Fourier Transform Algorithm for data refinement further strengthens the software's ability to handle complex geometries. This comprehensive approach advances existing surface finishing techniques, offering a streamlined workflow with sub-micron precision. The framework has significant potential for improving the manufacturing of high-precision aspherical optical elements and other advanced optical components.

Author Contributions: Conceptualization, R.P.S. and Y.C.; methodology, R.P.S. and Y.C.; software development, R.P.S.; validation, R.P.S. and Y.C.; project administration, Y.C.; writing—original draft, R.P.S.; writing—review and editing, R.P.S. and Y.C.; funding acquisition, Y.C. All authors have read and agreed to the published version of the manuscript.

Funding: This research received no external funding.

Data Availability Statement: The original contributions presented in this study are included in this article. Further inquiries can be directed to the corresponding author.

Acknowledgments: The authors would like to thank the anonymous reviewers for their constructive comments and suggestions, which greatly improved the quality of this manuscript.

Conflicts of Interest: The authors declare no conflicts of interest.

References

1. Qu, X.; Liu, Q.; Wang, H.; Liu, H.; Sun, H. A spiral path generation method for achieving uniform material removal depth in aspheric surface polishing. *Int. J. Adv. Manuf. Technol.* **2022**, *119*, 3247–3263. [CrossRef]
2. Sun, G.; Wang, S.; Zhao, Q.; Ji, X.; Ding, J. Material removal and surface generation mechanisms in rotary ultrasonic vibration–assisted aspheric grinding of glass ceramics. *Int. J. Adv. Manuf. Technol.* **2024**, *130*, 3721–3740. [CrossRef]
3. Yin, S.; Jia, H.; Zhang, G.; Chen, F.; Zhu, K. *Review of Small Aspheric Glass Lens Molding Technologies*; Higher Education Press: Beijing, China, 2017. [CrossRef]
4. Shank, C.V.; Nooradian, A. *Harnessing Light*; National Academies Press: Washington, DC, USA, 1998. [CrossRef]
5. Visser, D.; Gijsbers, T.G.; Jorna, R.A.M. Molds and measurements for replicated aspheric lenses for optical recording. *Appl. Opt.* **1985**, *24*, 1848–1852. [CrossRef] [PubMed]
6. Peng, Y.; Shen, B.; Wang, Z.; Yang, P.; Yang, W.; Bi, G. Review on polishing technology of small-scale aspheric optics. *Int. J. Adv. Manuf. Technol.* **2021**, *115*, 965–987. [CrossRef]

7. Martin, H.M. Design and manufacture of 8.4m primary mirror segments and supports for the GMT. In *Optomechanical Technologies for Astronomy*; SPIE: Bellingham, WA, USA, 2006; p. 62730E. [CrossRef]
8. Nkrumah, J.K.; Cai, Y.; Jafaripournimchahi, A. *A Review of Automotive Intelligent and Adaptive Headlight Beams Intensity Control Approaches*; SAGE Publications Inc.: Thousand Oaks, CA, USA, 2024. [CrossRef]
9. Chen, J. Spectacle lenses with slightly aspherical lenslets for myopia control: Clinical trial design and baseline data. *BMC Ophthalmol.* **2022**, *22*, 345. [CrossRef] [PubMed]
10. Zhong, S.; Xu, C.; Duan, L.; Zhang, F.; Duan, J.A. Optimizing the efficiency of a laser diode and single-mode fiber coupling using multi-aspherical lenses. *Opt. Fiber Technol.* **2022**, *68*, 102781. [CrossRef]
11. Hinn, M.; Pisarski, A. Efficient grinding and polishing processes for asphere manufacturing. In *Optifab*; SPIE: Bellingham, WA, USA, 2013; p. 88840I. [CrossRef]
12. Singh, R.; Xue, C.; Chen, Y. Integration of metrology in the manufacturing processes with smart control systems. In *Thirteenth International Conference on Information Optics and Photonics (CIOP 2022)*; SPIE: Xi'an, China, 2022; p. 107. [CrossRef]
13. Singh, R.P.; Chen, Y. An innovative approach to measuring radius of curvature and form error of spherical optics with an interferometer. In *Digital Optical Technologies 2023*; SPIE: Munich, Germany, 2023; p. 60. [CrossRef]
14. Walker, D.; Brooks, D. The precessions tooling for polishing and figuring flat spheric asurface. *Opt. Express* **2003**, *11*, 958. [CrossRef] [PubMed]
15. Yang, J.; Duan, J.; Li, T.; Hu, C.; Liang, J.; Shi, T. Tool wear monitoring in milling based on fine-grained image classification of machined surface images. *Sensors* **2022**, *22*, 8416. [CrossRef] [PubMed]
16. Zucker, R.M.; Price, O.T. Practical confocal microscopy and the evaluation of system performance. *Methods* **1999**, *18*, 447–458. [CrossRef] [PubMed]
17. Peta, K.; Stemp, W.J.; Chen, R.; Love, G.; Brown, C.A. Multiscale characterizations of topographic measurements on lithic materials and microwear using a GelSight Max: Investigating potential archaeological applications. *J. Archaeol. Sci. Rep.* **2024**, *57*, 104637. [CrossRef]
18. Sun, W.; McBride, J.W.; Hill, M. A new approach to characterising aspheric surfaces. *Precis. Eng.* **2010**, *34*, 171–179. [CrossRef]
19. Gao, W.; McBride, J.W.; Hill, M. On-machine and in-process surface metrology for precision manufacturing. *CIRP Ann.* **2019**, *68*, 843–866. [CrossRef]
20. Karodkar, D.; Gardner, N.; Bergner, B.C. Traceable radius of curvature measurements on a Micro-Interferometer. In *Optical Manufacturing and Testing V*; SPIE: Bellingham, WA, USA, 2003; Volume 5180, pp. 261–273. [CrossRef]
21. Honghong, G.; Singh, R.P. Influence of Ultrasonic Vibration-Assisted Ball-End Milling on the Cutting Performance of AZ31B Magnesium Alloy. *Int. J. Recent Technol. Eng.* **2020**, *9*, 271–276. [CrossRef]
22. Singh, R.P.; Chen, Y. Comprehensive advancements in automatic digital manufacturing of spherical and aspherical optics. In *Optical Metrology and Inspection for Industrial Applications X*; SPIE: Beijing, China, 2023; p. 11. [CrossRef]
23. Jones, R.A. Optimization of computer controlled polishing. *Appl. Opt.* **1977**, *16*, 218–224. [CrossRef] [PubMed]

Disclaimer/Publisher's Note: The statements, opinions and data contained in all publications are solely those of the individual author(s) and contributor(s) and not of MDPI and/or the editor(s). MDPI and/or the editor(s) disclaim responsibility for any injury to people or property resulting from any ideas, methods, instructions or products referred to in the content.

Article

Investigation on the Machinability of Polycrystalline ZnS by Micro-Laser-Assisted Diamond Cutting

Haoqi Luo [1], Xue Wang [2], Lin Qin [3], Hongxin Zhao [2], Deqing Zhu [2], Shanyi Ma [1], Jianguo Zhang [1] and Junfeng Xiao [1,*]

[1] State Key Laboratory of Intelligent Manufacturing Equipment and Technology, School of Mechanical Science and Engineering, Huazhong University of Science and Technology, Wuhan 430074, China; m202370528@hust.edu.cn (H.L.)
[2] Beijing Precision Engineering Institute for Aircraft Industry, Aviation Industry Corporation of China, Beijing 100076, China
[3] Shanghai Aerospace Control Technology Institute, Shanghai 201109, China
* Correspondence: xiaojf@hust.edu.cn

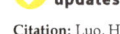

Citation: Luo, H.; Wang, X.; Qin, L.; Zhao, H.; Zhu, D.; Ma, S.; Zhang, J.; Xiao, J. Investigation on the Machinability of Polycrystalline ZnS by Micro-Laser-Assisted Diamond Cutting. *Micromachines* **2024**, *15*, 1275. https://doi.org/10.3390/mi15101275

Academic Editors: Francesco Ruffino and Reza Teimouri

Received: 9 September 2024
Revised: 17 October 2024
Accepted: 18 October 2024
Published: 21 October 2024

Copyright: © 2024 by the authors. Licensee MDPI, Basel, Switzerland. This article is an open access article distributed under the terms and conditions of the Creative Commons Attribution (CC BY) license (https://creativecommons.org/licenses/by/4.0/).

Abstract: Polycrystalline ZnS is a typical infrared optical material. It is widely used in advanced optical systems due to its excellent optical properties. The machining accuracy of polycrystalline ZnS optical elements must satisfy the requirements of high-performance system development. However, the soft and brittle nature of the material poses a challenge for high-quality and efficient machining. In recent years, in situ laser-assisted diamond cutting has been proven to be an effective method for ultra-precision cutting of brittle materials. In this study, the mechanism of in situ laser-assisted cutting on ultra-precision cutting machinability enhancement of ZnS was investigated. Firstly, the physical properties of ZnS were characterized by high-temperature nanoindentation experiments. The result revealed an increase in ductile machinability of ZnS due to plastic deformation and a decrease in microhardness and Young's modulus at high temperatures. It provided a fundamental theory for the ductile–brittle transition of ZnS. Subsequently, a series of diamond-cutting experiments were carried out to study the removal mechanism of ZnS during in situ laser-assisted cutting. It was found that the mass damage initiation depth groove generated by in situ laser-assisted cutting increased by 57.99% compared to the groove generated by ordinary cutting. It was found that micron-sized pits were suppressed under in situ laser-assisted cutting. The main damage form of HIP-ZnS was changed from flake spalling and pits to radial cleavage cracks. Additionally, the laser can suppress the removal mode difference of different grain crystallographic and ensure the ductile region processing. Finally, planning cutting experiments were carried out to verify that a smooth and uniform surface with Sa of 3.607 nm was achieved at a laser power of 20 W, which was 73.58% better than normal cutting. The main components of roughness were grain boundary steps and submicron pit. This study provides a promising method for ultra-precision cutting of ZnS.

Keywords: polycrystalline ZnS; in-process-heat laser-assisted diamond machining; brittle-to-ductile transition; soft brittle material

1. Introduction

Zinc sulfide (ZnS) is an important infrared optical material with excellent optical transmittance and mechanical and thermal properties. Hot isostatically pressed ZnS (HIP-ZnS) is also known as multispectral ZnS due to its high transmittance in the band of 400 nm–1200 nm. HIP-ZnS has great application prospects in infrared optics and aerospace. At present, it is widely used in infrared imagers, measuring instruments, fairing and focusing lenses for laser systems, etc. For infrared optical elements, the lower the surface roughness and number of surface defects, the smaller the total scattering value of the workpiece [1,2] and the greater the optical properties of the elements. Additionally, surface and subsurface damage layers reduce the laser-damaged threshold and induce component failure. HIP-ZnS is a typical soft–brittle material [3,4], and conventional grinding could induce great subsurface damage. Diamond-cutting technology is an ultra-precision machining method with the capability of processing high-precision and low-damage surfaces. However, it is prone to brittle fracture during the ordinary cutting process due to the low fracture toughness of HIP-ZnS. As a superior infrared material [5], how to improve the machining quality and efficiency of ZnS has become a critical challenge.

Extensive research has been conducted on the cutting mechanisms of polycrystalline ZnS, particularly regarding its behavior in brittle and ductile regimes. Li et al. [6] integrated the material properties of ZnS with the machining mechanics of hard and brittle materials, investigating its turning characteristics via micro-morphological analysis and XRD spectra. Their findings indicated that the likelihood of ductile cutting increases when the crystal axis is aligned with the cutting direction. Zhang et al. [7] proposed a modified stress-assisted nanocutting model based on the virtual boundary between the workpiece and tool, exploring the nanocutting process under various external stresses. Arif et al. [8] introduced a cutting energy-related model to predict the brittle-to-ductile transition in diamond turning of polycrystalline brittle materials, accounting for workpiece properties, tool geometry, and cutting parameters. Yang et al. [9] developed an ultra-precision diamond turning process with trapezoidal modulation, maintaining constant chip thickness and cutting direction, offering new strategies to enhance the brittle-to-ductile transition and machining performance. Similarly, Yao et al. [10] calculated energy consumption in both ductile and brittle cutting modes during ultra-precision machining of ZnS, establishing the critical uncut chip thickness for brittle-to-ductile transition and outlining the necessary machining conditions. However, despite these advancements, surfaces produced through the diamond turning of ZnS still exhibit crater-like defects due to particle fragmentation, limiting the precision fabrication of optical components.

To better understand the ZnS damage mechanisms during machining and address the associated challenges, extensive research has been conducted. Salzman et al. [11] utilized magnetorheological finishing with chemically modified fluids to investigate the material-removal behavior of single-crystal ZnS, aiming to address the challenges posed by its anisotropy. While this technique proved effective, its lower machining efficiency compared to turning and limitations in processing optical components with complex geometries—such as aspherical, freeform, and diffractive surfaces—present challenges. In contrast, Guo et al. [12] explored the influence of cutting parameters on ZnS surface quality through an orthogonal experimental approach, optimizing parameters to achieve a surface roughness of 2.3 nm. Zong et al. [13] proposed an innovative oblique-cutting model to enhance surface quality by facilitating a brittle–ductile transition during machining. The effectiveness of this method was verified through turning experiments and finite element simulations, offering new insights into the plastic machining of ZnS. Building on this, Navare et al. [14] employed micro-laser-assisted single-point diamond turning to study the effects of diamond tool crystal orientation on machining outcomes, presenting new strategies for improving ZnS surface quality. Despite these advances, the microscopic deformation mechanisms of polycrystalline ZnS during cutting remain insufficiently explored. Chen et al. [15] categorized the brittle fracture of ceramic matrix composites into macroscopic and macroscopic scales based on distinct material units. Zheng et al. [2] developed a coarse-grained molecular dynamics approach to systematically investigate the effects of material properties and cutting depth on removal patterns. Their transitional depth of cut (TDoC) model accurately predicted the material removal behavior, offering a robust framework for analyzing the cutting dynamics of polycrystalline ZnS. Unlike single-crystal materials, polycrystalline materials experience non-uniform fracture during machining due to the effects of crystal orientation, grain boundaries, twins, and grain size. Consequently, optimizing diamond-cutting technology is crucial to mitigate these non-uniform fractures and enhance the machinability of polycrystalline materials.

Lasers are widely used in the ultra-precision machining of various materials as an assistance due to their high energy density, easy control, and small heat-affected zone.

In situ laser-assisted machining improves the ductile cutting properties of materials by focusing a laser beam through the diamond tool on the cutting zone. In situ laser-assisted machining has been applied in the manufacturing of brittle materials in recent years. It has been demonstrated that silicon [16], germanium [17], calcium fluoride [18], zinc selenide [5], tungsten [19,20], tungsten carbide [21], silicon carbide [22], microcrystalline glass [23], and fused silica [24] can be processed with smooth surface by in situ laser-assisted cutting. He et al. [25] carried out molecular dynamics simulations of Ti alloy laser-assisted machining. They discussed the effect of laser power on the micro-deformation and damage of the material and verified the laser-assisted suppression of the subsurface damage layer. Ma et al. [26] investigated the effect of different machining parameters on the surface roughness of alumina ceramics. They characterized the surface morphology after processing and found that the crystal breakage was compacted to form a dense layer heated by a laser. Cracks were significantly inhibited, and the integrity of the processed alumina ceramic surface was improved. Pu et al. [27] characterized and compared the 3D morphology of laser-assisted cutting Si_3N_4 surfaces under different material removal modes and concluded that the relationship between different material removal modes and 3D morphology parameters. Surface roughness decreases with increasing laser power. Kong et al. [28] analyzed and compared the differences in chip morphology of titanium alloys in OC and LAC. They revealed that chips from in situ laser-assisted cutting are excessive from the continuous type to the segmented type, with less stress in the shear region and faster formation of shear bands. Yang et al. [29–31] first investigated the deformation mechanism, nano-wear mechanism, and material removal mechanism of KDP crystals using the molecular dynamics method. They established scratching maps to minimize the subsurface damage of KDP crystals. They carried out simulations of nano-scratches at different depths under different crystal orientations. It was found that the anisotropy of the material has a great influence on the deformation mechanism of KDPs under nano-wear. Based on this, yang established scratching maps as a guideline for quantitatively defining the "depth/radius removal mechanism" on different lattice surfaces of KDPs, which contributed to the study of the failure mechanism of KDPs. This contributes to the study of the failure mechanism of KDPs. Liu et al. [32] investigated the influence of temperature on the removal mechanism of materials by nano-scratch experiment. They revealed that the hardness and elastic modulus of the material change with temperature and explained the high ductility of KDP crystals at high temperatures, which provides a basis for the low-damage machining of soft and brittle materials under a thermal field. Geng et al. [5] verified the feasibility of laser-assisted processing techniques for improving the processing efficiency and surface integrity of ZnSe ceramics through comparative and orthogonal experiments. Unlike common single-crystal hard brittle materials (silicon, germanium, silicon carbide, etc.), ZnS is a typical polycrystalline soft brittle material. With lower hardness and fracture toughness, it is susceptible to fracture during diamond cutting, reducing surface quality. ZnS is more similar to ZnSe, but ZnS has higher fracture strength, hardness, and flexural strength. The grain size of ZnS is about 150 μm, which is larger than ZnSe with a grain size range of 30–50 μm.

In summary, although some scholars have studied the cutting mechanism and process of ZnS, the problems of low efficiency and low surface quality of ZnS cutting in machining have not been solved. In situ lasers can improve the material properties of the cutting area, reducing surface and subsurface damage and facilitating the enhancement of ductile machinability of soft and brittle materials. However, the mechanism of in situ laser-assisted cutting of ZnS has not been investigated. This study explores the main damage forms and material removal mechanisms of HIP-ZnS crystals in in situ laser-assisted cutting. Firstly, high-temperature nanoindentation studies were carried out to reveal the influence of temperature on the hardness, elastic modulus, and plastic deformation capacity of ZnS. Subsequently, in situ laser-assisted cutting experiments on HIP-ZnS were carried out. Optical microscopy and a white light interferometer (WLI) were utilized to characterize and analyze the material removal mode and main damage forms during the cutting process.

The results revealed that the property differences of different grains were suppressed to a certain extent with laser irradiation, and, thus, the material cutting performance and machined surface quality were improved. The large critical undeformed chip thickness can be applied to the HIP-ZnS turning process, which is conducive to the improvement of the material removal rate. It provides a promising method for ultra-precision manufacturing of HIP-ZnS.

2. Materials and Methods

2.1. Materials

In this study, multispectral ZnS crystals were used, which were supplied by the GRINM Guojing Advanced Materials Co., Ltd. (Beijing, China). The multispectral ZnS crystals were prepared by applying hot isostatic pressure treatment (HIP) to standard ZnS crystals made by chemical vapor deposition (CVD). HIP-ZnS is polycrystalline, and the grains in HIP-ZnS are mainly present as face-centered cubic crystals (FCC). The FCC ZnS undergoes cleavage in the form of a dodecahedron with six cleavage directions and a perfect cleavage surface of (011). Twin crystals are mostly present in aggregates parallel to <111> and <211>. A white light interferometer (Newview 9000, Zygo Corporation, Middlefield, CT, USA) was used to observe the polished surface quality of the samples. The polished surface of the multispectral ZnS crystals is shown in Figure 1a,b, with a surface roughness Sa (arithmetic mean height of the plane) of less than 2 nm. The phase composition of the material was characterized and analyzed by X-ray diffraction (XRD 7000, Shimadzu corporation, Kyoto, Japan). As shown in Figure 1c, the diffraction intensity of the sample is mainly from α-phase ZnS (sphalerite crystal type). The grains in the multispectral HIP-ZnS crystals are all in the α-phase, and <111> is their major crystallographic orientation. The absorbance, transmittance, and reflectance of the material were tested by a UV–visible near-infrared spectrophotometer (SolidSpec-3700, Shimadzu corporation, Kyoto, Japan), and, as shown in Figure 1d, the absorbance of the HIP-ZnS crystals for the light of 1064 nm wavelength was 5.9015%.

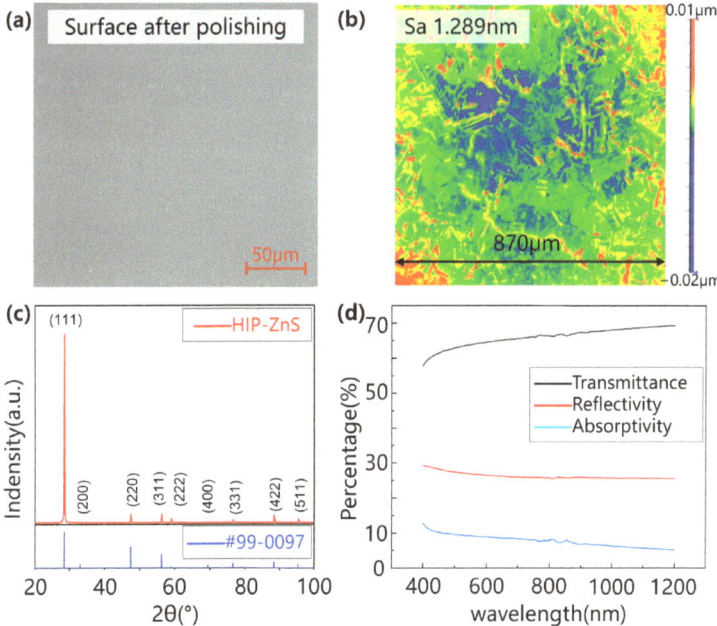

Figure 1. Characterization of ZnS samples. (**a**) Polished surface by optical microscope; (**b**) polished surface by WLI; (**c**) XRD pattern; (**d**) absorption spectra of the used workpiece.

2.2. High-Temperature Nanoindentation

The high-temperature nanoindentation experiment generates the load-displacement curve of the material to obtain the mechanical properties such as microhardness and Young's modulus of the material. The plastic deformation capacity of the material can be known by analyzing the curve. The maximum undeformed chip thickness of the material is correlated with the microhardness, Young's modulus, and fracture toughness of the material. Conducting high-temperature nanoindentation experiments can provide theoretical support for ZnS groove experiments. High-temperature nanoindentation experiments were carried out on polished ZnS samples (15 mm × 15 mm × 1 mm) using a high-temperature nanoindenter (Nano Indenter® G200, KLA foundation, Ann Arbor, MI, USA), as shown in Figure 2. The tip radius of the diamond Berkovich probe was about 100 nm, and the fusion angle was 142.30°. The diamond indenter is susceptible to oxidation at high temperatures, and the thermal drift of the indenter will increase dramatically. This leads to a large shift in the displacement-load curve and a decrease in the reliability of the experimental results. The laser heating temperature of the experimental platform and the limited maximum platform temperature determined that the ambient temperature used in the nanoindentation test was 25 °C. High-temperature conditions are set at 100 °C, 200 °C, 300 °C, and 400 °C. In this study, five nanoindentations were performed at five temperature conditions to collect load depth data in order to remove random error effects and ensure more credible results. The nanoindentation interval was set to 350 μm based on the grain size and grain distribution of the ZnS crystals. The triboindenter was operated in the load-controlled mode, with constant loading and unloading rates for every test cycle. The maximum load was 20 mN, the loading and unloading times were 10 s, and the load holding time was 2 s. A Bruker in situ SPM imaging system was used to obtain the indentation morphology. Prior to the indentation experiment, the sample stage was heated by a laser beam with the heating and cooling rate set to 10 °C/min, which heated the sample to a set temperature using the principle of heat conduction. Subsequently, the probe tip was moved into the heated test chamber to ensure isothermal contact conditions between the indenter and the sample. We automatically adjusted the thermal drift of the probe within 3 nm/s using the program to achieve thermal equilibrium between the sample and the probe.

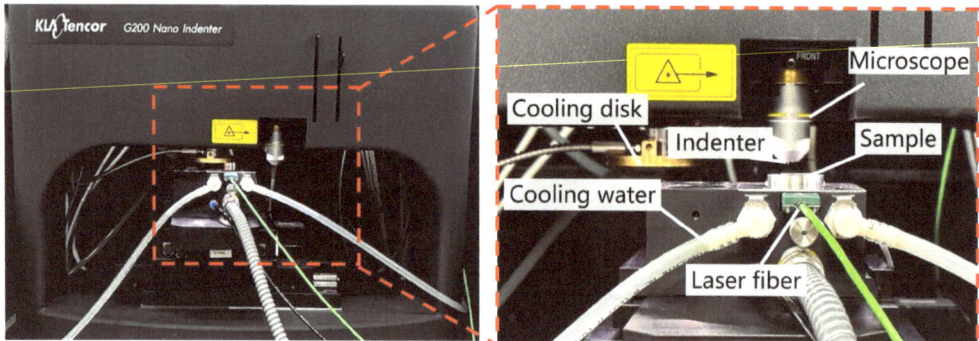

Figure 2. High-temperature nanoindentation equipment.

2.3. Grooving Experiments

The experiment was conducted on an ultra-precision machine tool (Nanoform X, Precitech, Keene, NH, USA) with a self-developed in-LADM system, as shown in Figure 3a. Groove experiments and planning experiments were performed to study the machinability improvement of HIP-ZnS materials with laser assistance, as shown in Figure 3b. Figure 3b shows the schematic diagram of groove experimental machining, and the experiment was carried out on a four-axis machine tool. Multiple grooves are machined by rotating the spindle. C represents the angle of each groove spacing and, in order to reduce the influence of different grain directions on the experiment and reduce the systematic error of the experiment, the value of C is taken to be 5°. L represents the length dimension of the workpiece used, and H represents the thickness dimension of the workpiece. The workpiece used in this experiment is a 15 mm × 15 mm × 3 mm rectangular sample. The laser generator utilizes Nd:YAG as the active medium, providing the continuous-wave laser with a wavelength of 1064 nm, which has a Gaussian profile. The laser beam was guided through a specially designed optical beam shaper module, focusing the laser beam from 3 mm to about 80 μm in diameter. Laser applied to near-transparent diamond tool edges, heating the workpiece by adjusting the laser power in the range of 0–25 W. Based on the simulation results of COMSOL software version 6.1, the material temperature at the center of the laser spot reaches 1200 °C when the laser power comes to 30 W. The center temperature is higher than the sublimation temperature of ZnS, and the material is prone to high-temperature oxidation and ablation. A power of 25 W was selected as the highest laser power. Since the laser beam exhibits a Gaussian distribution, there was a large temperature gradient around the laser irradiation area given by the short groove cutting distance and large feed rate. To attenuate its effect, the laser spot position is precisely adjusted to coincide with the center of the diamond tip. No cutting fluid was used in the groove-cutting experiments to minimize thermal stress. We used a multispectral absorbing power meter (30 (150) A-BB-18, Ophir, Jerusalem, Israel) to ascertain that the cutting interface received the proper amount of laser power and that the laser emitted through the right region of the tool. A single-crystal diamond (SCD) tool was used to feed horizontally at a constant cutting speed along the X direction, while the depth of cut was continuously varied along the Z direction from 0 to 4 μm during the groove cutting process. The SCD tool used in the experiment had a tip radius of 0.479 mm, a rake angle of −35°, and a clearance angle of 10°. The shift between the removal modes of HIP-ZnS crystals under OC and LAC was analyzed by groove experiments. The detailed machining conditions are shown in Table 1. Furthermore, planning experiments were carried out to verify the effect of laser on the machinability of the material during the continuous cutting process. The detailed machining conditions are shown in Table 2.

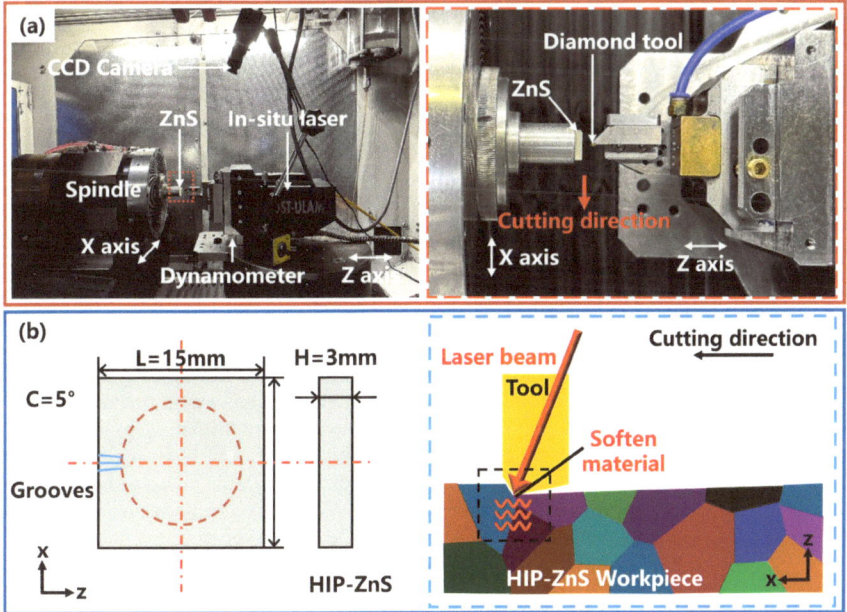

Figure 3. Cutting experiment: (**a**) equipment and (**b**) machining schematic diagram.

Table 1. Conditions of grooving experiments.

Experimental Parameters	Data
Workpiece material	HIP-ZnS
Tool material	Single-crystal diamond
Tool nose radius	0.5 mm
Rake angle	$-35°$
Clearance angle	$10°$
Cutting speeds	500 mm/min, 200 mm/min, 100 mm/min, 50 mm/min
Slope ratio	1:1000
Laser power	5 W, 10 W, 15 W, 20 W, 25 W

Table 2. Parameters of planning experiments.

Experimental Parameters	Data
Workpiece material	HIP-ZnS
Tool material	Single-crystal diamond
Tool nose radius	0.5 mm
Rake angle	$-35°$
Clearance angle	$10°$
Cutting speeds	500 mm/min
Feed rate	4 μm/rev
Depth of cutting	4 μm
Laser power	0 W, 20 W

2.4. Surface Characterization

The morphology of the grooves was observed by an optical microscope (Carl Zeiss Axiolab 5, Oberkochen, Germany), and the main forms of defects in the material and the influence of twinning and grain orientation were analyzed. After the planning experiments, a white light interferometer (WLI, Newview 9000, Zygo Corporation, Middlefield, CT, USA) was used to observe the cutting planes to verify the effect of material grain orientation, grain boundaries, and twins on the material removal mode.

3. Results and Discussions

3.1. Temperature Response of Material Properties

Figure 4a depicts the load-displacement curves of the material at temperatures of 25 °C, 100 °C, 200 °C, 300 °C, and 400 °C. The results show that the load-displacement curves changed significantly with increasing temperature and the indentation became deeper under the same load condition. The average Young's modulus of HIP-ZnS was 97.73 GPa at room temperature and 55.01 GPa and 34.95 GPa at 200 °C and 400 °C, respectively, as shown in Figure 4c. The average microhardness of ZnS was 2.12 GPa at room temperature and 1.55 GPa and 1.11 GPa, as shown in Figure 4d. Both Young's modulus and hardness of HIP-ZnS decreased significantly with increasing temperature: Young's modulus decreased by 64.25% and hardness decreased by 47.69%. Dislocation activity is enhanced at high temperatures and the efficiency of cleavage increases with temperature. By transferring energy to the material by the laser, the defects in the crystal are more likely to be displaced, making the material more susceptible to plastic deformation. The stiffness of the material decreases during the softening process at high temperatures, which makes it easy to produce large elastic deformation, and its ductile deformation ability is enhanced, as shown in Figure 4b, and the depth of ductile deformation increases from 395.53287 nm at 25 °C to 624.8939 nm at 400 °C, which is conducive to obtaining a good surface quality and a desirable tool life.

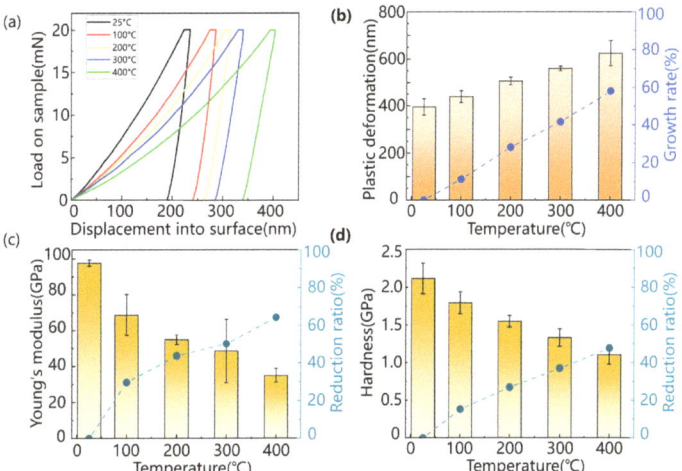

Figure 4. High-temperature nanoindentation results: (**a**) load–displacement curves under different temperatures; variation in (**b**) plastic deformation, (**c**) Young's modulus, and (**d**) materials' hardness with increasing temperatures.

3.2. Material Cutting Removal Mechanism

Grooving experiments were carried out to investigate the influence of in situ lasers on the machinability of HIP-ZnS. Investigate the changing pattern of the material-removal mechanism with the change of the in situ laser power and cutting speed. Firstly, the grooving experiments were performed at laser powers of 0 W, 5 W, 10 W, 15 W, 20 W, and 25 W, and the experimental results are shown in Figure 5. The area comprising the yellow dashed line in the groove is the ductile region, and the dark blue dashed line is demarcated as the cleavage cracks. When grooving under OC at a small depth of cut, the material is removed in ductile form, and a smooth surface is presented in the groove, as shown in Figure 5a. As the depth of cutting increases, the material first undergoes violent brittle fracture, forming large crater defects on the surface. Subsequently, a large crack is presented in the groove, and the percentage of ductile removal of the material remains at a high level When the depth of the cut reaches around 160 nm, the crack density increases, and the material is removed in a brittle mode. The onset depth of extensive chipping is defined as the depth at which the percentage of damage in the groove exceeds 60%, as shown by the black line in Figure 5. Accompanying the increase in power, the black lines in Figure 5a–f gradually move towards the back end of the grooves, indicating that the depth of large-area chipping initiation is increased. The depth of ZnS initiation of the large-area brittle fracture is increased in the presence of a laser, which improves the depth of the cut in planning and is beneficial for obtaining high-quality ZnS surfaces with high effectiveness. Under 25 W laser power, the material is removed in ductile form at small depths of cut. As the depth of cut increases, the cutting force increases, the material undergoes cleavage, and the cracks appear in the grooves, during which the crack density and crack size increase and gradually evolve into craters, as shown in Figure 5f. By analyzing the machined grooves, it was found that most of the material at the front of the grooves was removed plastically in the grooves under different experimental conditions. However, brittle fractures still occurred in some grains, producing craters and cleavage cracks. Compared with the material removal process in OC, the localized damage stage of grooves in LAC has a larger area of ductile area and reduced cratering and spalling. Cleavage cracks are produced in polycrystalline materials due to the occurrence of cleavage during mechanical processing. They are a common brittle processing defect in polycrystalline materials. Compared with the point-like and flake-like craters produced by grain crushing and pulling out, the cleavage cracks are regarded as brittle fractures of a lower degree. It is also seen that, with the reduction in crater defects in the grooves with increasing power in the grooves under different experimental conditions, the brittle defects are more in the form of cleavage cracks, which are considered to be an improvement in processing properties. Thus, the laser energy field can improve the material cutting properties and enhance the machining quality. In the interest of further investigation of the effect of thermal density on the material, a set of rate comparison tests at the same laser power is carried out.

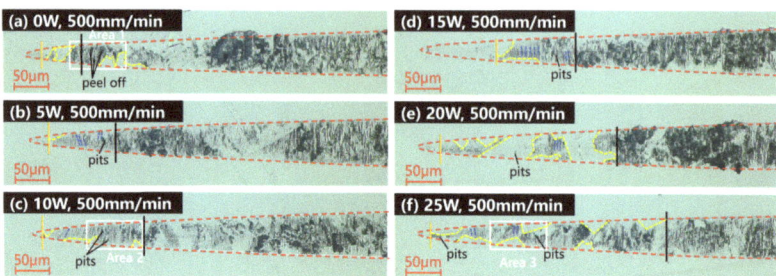

Figure 5. Power comparison grooving experimental results under (**a**) laser power of 0 W, cutting speed of 500 mm/min, (**b**) laser power of 5 W, cutting speed of 500 mm/min, (**c**) laser power of 10 W, cutting speed of 500 mm/min, (**d**) laser power of 0 W, cutting speed of 500 mm/min, (**e**) laser power of 5 W, cutting speed of 500 mm/min, (**f**) laser power of 10 W, cutting speed of 500 mm/min.

Under the laser power of 25 W, the grooving experiments were carried out at cutting speeds of 200 mm/min, 100 mm/min, and 50 mm/min, respectively, and the experimental results are shown in Figure 6. The area bounded by the yellow dashed line in the groove is the ductile region, and the dark blue dashed line labels the cleavage cracks. The onset of the brittle–ductile transition phase is marked with an orange solid line, and the onset of large-scale collapse is marked with a black solid line. Compared with the experimental results of the 500 mm/min cutting speed shown in Figure 5f, the area of the ductile removal region at the initial of the groove increases with the increasing cutting speed, but there are still a few defects. This is due to localized brittle chipping caused by individual grains being pulled out due to their hardness. The location where cracks and pits commence in the grooves to the point where large-scale collapse initiates is defined as the brittle–ductile mixing removal stage, as indicated by the orange line in Figures 5 and 6. In the ductile–brittle mixing removal stage, the density and size of the craters in the grooves decrease with the reduction of the cutting speed. This is consistent with the reduced microhardness, Young's modulus, and ductile deformability of the material at high temperatures. The increase in laser energy density improves the cutting performance of HIP-ZnS.

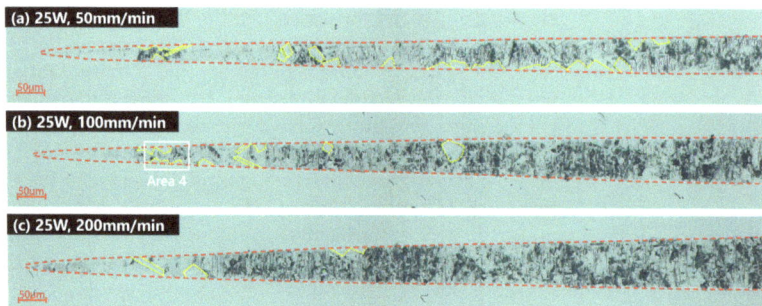

Figure 6. Cutting speed comparison grooving experimental results under (**a**) laser power of 25 W, cutting speed of 50 mm/min, (**b**) laser power of 25 W, cutting speed of 100 mm/min, (**c**) laser power of 25 W, cutting speed of 200 mm/min.

Unlike single-crystal materials, the form of material removal at the initial of the grooves under the same processing conditions is influenced by the grain orientation, with the ductile removal region remaining initially locally brittle and the transition region of the brittle–ductile transition being large in size. It was found that, at the beginning of the groove, both brittle and ductile removal modes existed. This is due to the different orientations of the grain slip systems in polycrystalline materials, and the microscopic removal characteristics of the material are closely related to grain orientation and twinning. Although the groove initiation section has a small depth of cut, the cutting force is small, and the cutting energy is not sufficient to sustain crack expansion and consolidation at small depths of cut. However, the hardness of ZnS crystal particles is affected by the crystal direction, and particles with high hardness are pulled out by the cutting force, forming large pits and chipping on the cutting surface. The traditional evaluation index of cutting performance is the brittle–ductile transition depth, which is difficult to use to accurately describe the changes in the processing performance of the material. As the depth of cut increases further, the cutting force increases, more cracks are generated and undergo expansion and merging, and the crack density increases rapidly and is eventually distributed over the entire groove surface. At this point, the material is removed in a brittle mode, so the depth of large chipping initiation is selected as the cutting performance evaluation index in this study. The onset depth of extensive chipping is defined in this study as the depth at which the percentage of damage in the groove exceeds when 60%, as shown by the black line in Figure 5. 'Damage' is defined as defects such as craters, cleavage cracks, etc., formed by brittle fracture. It is reflected in the optical microscope picture of the groove as a dark part with a different color from the polished surface. In this paper, an image recognition method was used to write a program to identify the density of defects in the grooves and to calculate the percentage of damage in the grooves. At the same time, in order to quantitatively study the changes in the brittle–ductile mixing region under the laser-assisted action and to reduce the influence of special individual cases, the defect density in the groove in the interval of 150 nm–300 nm was selected as an auxiliary evaluation index. As shown in Figure 7, the onset depth of large-area chipping of ZnS grooves increases with the increase in laser power, and the defect density of grooves in the 150 nm–300 nm interval decreases with the increase in laser power.

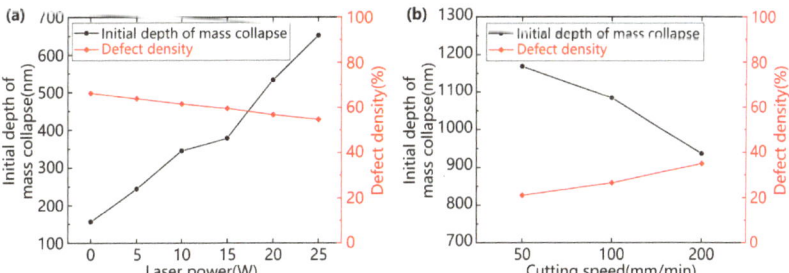

Figure 7. Initial depth of mass collapse chipping and damage density curves: (a) power comparison experiment; (b) cutting speed comparison experiment.

In pursuit of investigating whether the laser can inhibit the cutting variability of different grain orientations and sizes, the grooves were analyzed by using an optical microscope with higher magnification to analyze the grooves. The grooves in the 150 nm–300 nm interval under the experimental conditions of (a) laser power of 0 W, cutting speed of 500 mm/min, (b) laser power of 10 W, cutting speed of 500 mm/min, (c) laser power of 25 W, cutting speed of 500 mm/min, and (d) laser power of 25 W, cutting speed of 100 mm/min were selected as the objects of analysis, as illustrated by the white areas in Figures 5 and 6. As shown in Figure 8a, in the absence of a laser, large grains undergo cleavage, forming pits and flaky flakes on the material surface. Twinned crystals undergo cleavage, producing discrete diagonal banding patterns on the surface. As shown in Figure 8b, under the action of a 10 W laser, the material softens at a high temperature, the hardness decreases, the fracture toughness increases and the crater formed by the fracture becomes smaller. The material undergoes cleavage along the direction perpendicular to the cutting direction, and the cleavage cracks are mostly craters and flaky spalling. As shown in Figure 8c, in 25 W laser-assisted cutting, the ductile domain area in the grooves increased, and the main form of damage to the material changed from pits to radial cleavage cracks. The radial cleavage cracks in the grooves became smaller under the experimental conditions of 25 W–100 mm/min, as shown in Figure 8d. In normal cutting without a laser, the material is removed as a mixture of brittle and ductile in the depth of cut in the interval 150 nm–300 nm. As the hardness of individual grains changes with different grain orientations, the form of removal varies greatly from grain to grain. Currently, the main form of damage to the material is large pits and flake spalling, with a high degree of brittle fracture. In in situ laser-assisted cutting, the material is still removed as a brittle–ductile mix, consistent with normal cutting without the addition of a laser. However, the main form of damage to the material changes. The stiffness of the material decreases at high temperatures, and the grains are more susceptible to transcrystalline fracture rather than intergranular fracture. The material undergoes cleavage mainly in the direction perpendicular to the cutting direction, producing radial cleavage cracks and block spalling in a few grains. As the laser power increased, the craters decreased, the length and width of the cleavage cracks decreased, and the material was removed more in a ductile mode. The reduction of microhardness and Young's modulus enhanced the ductile processing properties of HIP-ZnS. The high temperature induced by laser irradiation reduces the resistance of the material to local ductile deformation and weakens the grain boundaries, which helps to suppress the effects of crack extension and grain orientation variability on processing. Therefore, a large critical undeformed chip thickness can be applied to the HIP-ZnS turning process with laser assistance, which is conducive to improving the material-removal rate.

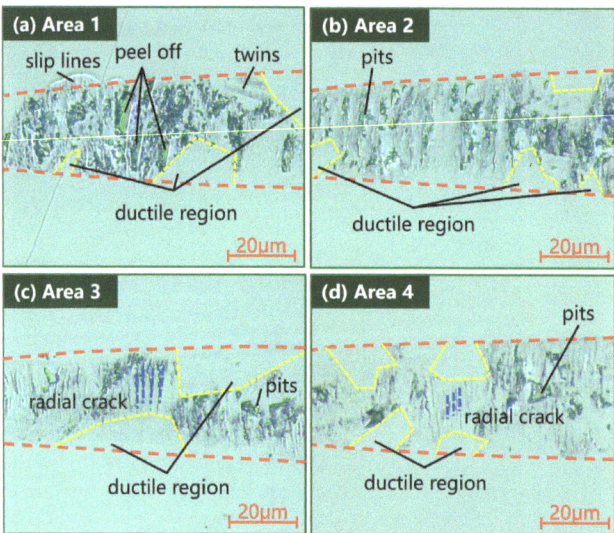

Figure 8. Results of the grooving experiment: (**a**) laser power of 0 W, cutting speed of 500 mm/min; (**b**) laser power of 10 W, cutting speed of 500 mm/min; (**c**) laser power of 25 W, cutting speed of 500 mm/min; (**d**) laser power of 25 W, cutting speed of 100 mm/min.

To demonstrate laser-assisted enhancement of material machinability, planning experiments of ZnS workpieces were carried out at 0 W and 20 W laser power, respectively.

The surface morphology after machining was observed by a white light interferometer in a field of view of 174 μm × 174 μm, as shown in Figure 9. Without the laser, the surface quality was poor, with a roughness Sa of 12.452 nm. The material was removed mainly by crack expansion in brittle mode, with a large number of craters distributed on the surface. In the laser-assisted case, the processed surface quality was better, with a roughness Sa of 2.885 nm. Intermediate cracks, transverse cracks, and pits were suppressed, but a small number of submicron craters and submicron stepped structures were still present on the surface. The reason for this phenomenon is that, in polycrystalline material processing, due to the existence of grain–grain orientation and size variability, ductile material removal of different thicknesses occurs on the crystal boundaries, resulting in the formation of grain boundary steps. The roughness of the tested surface originates from submicron pits and grain boundary steps. The experimental results show that the surface Sa obtained by in situ LADC is reduced by 55.6% compared to conventional single-point diamond turning for surface quality at 20 W laser power, verifying that the ductility of the multispectral ZnS crystals is improved during the in situ LADC process.

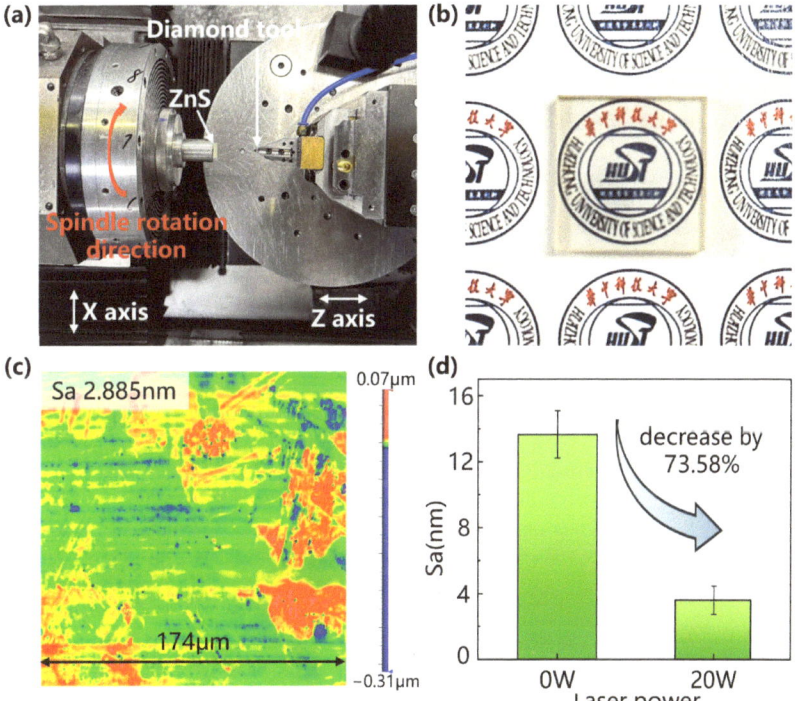

Figure 9. Planning experiments: (**a**) experimental setup; (**b**) processed sample; (**c**) surface quality after experiment; (**d**) surface quality comparison.

Figure 10 illustrates a microscopic view of the tool morphology after half an hour of laser-assisted cutting with a distance of 300 m. Figure 10a shows that, due to the low hardness of HIP-ZnS, no significant tool edge fragmentation was observed even for normal cutting without laser. In the picture of the tool after laser-assisted machining shown in Figure 10b, the cutting edge is very sharp, while the front and back cutting surfaces are very smooth. No visible wear bands are seen on the rear tool surface. In situ LADC is a stable cutting process that enables long-term HIP-ZnS machining.

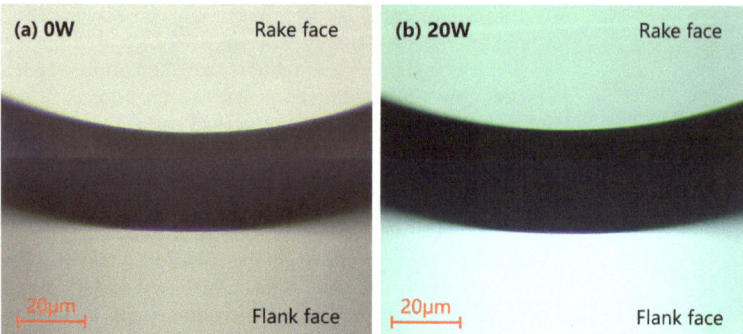

Figure 10. Wear of SCD tool (**a**) under OC and (**b**) under in-LAC.

4. Conclusions

In this paper, the machinability of ZnS micro-laser-assisted machining was verified. Compared to conventional diamond cutting, the surface quality of ZnS was significantly improved under laser-assisted machining. High-temperature nanoindentation experiments were carried out to investigate the evolution of the crystal properties of polycrystalline HIP-ZnS. The results show that high temperatures can enhance the ductile deformation of HIP-ZnS. The brittle–ductile transition process of the material was investigated by performing OC and in-LAC grooving experiments, and the removal mechanism of the material under in-LAC and OC was investigated using a characterization method. The effects of the thermal field, mechanical properties, and dynamic behavior of the brittle–ductile transition on the surface quality were investigated. Finally, uniform surface quality was experimented by using in situ laser-assisted machining methods. The main conclusions are summarized below:

(1) Thermal effects had a significant impact on the mechanical properties of HIP-ZnS. The depth of ductile deformation of HIP-ZnS increased from 395.53 nm at 25 °C to 624.89 nm at 400 °C, an increase of 57.99%. The microhardness and Young's modulus decreased with the increase in temperature by 47.69% and 64.25% at 400 °C, respectively. This shows the softening effect of a high temperature on the material and its ability to enhance the ductile deformation of the material.

(2) By OC and in-LAC grooving experiments, it was shown that compared with the conventional cutting, when the laser power was increased to 25 W, the large chipping initiation depth increased from 156.27 nm to 651.74 nm, which is an increase of 318.96%, and the defect density in the interval of 150–300 nm decreased from 66.35% to 54.67%.

(3) The main damage form of HIP-ZnS was changed from lamellar spalling and cratering to small-size cracking under the action of in-LAC. This change is conducive to the formation of surfaces with higher machining quality, indicating that the laser is able to suppress the removal pattern variability of different grain crystallographic directions to a certain extent and ensure the processing of ductile regions.

(4) The results of end face turning verified the superior machining quality of in-LAC over conventional single-point diamond turning. During surface shaping, intermediate cracks, transverse cracks, and pits are suppressed by the additional in situ laser, and the roughness of the measured surface originates from submicron pits and grain boundary steps. At a laser power of 20 W, the surface Sa obtained by in situ LADC was reduced by 73.58%.

In this paper, experiments on ZnS grooving were conducted at different laser powers as well as at different cutting speeds, mainly focusing on the influence of temperature effects on ZnS. The effect of laser assistance on the cutting process of ZnS was explored. To address the shortcomings of this study, the following future work will be proposed: (1) to explore the effect of laser-assisted processing at different wavelengths on the cutting process/removal mechanism of the material; (2) to carry out cutting experiments on ZnS produced by different methods of preparation with different grain sizes or grain orientations; to explore the effect of laser-assisted processing on the cutting process/removal mechanism of ZnS. This study involved ZnS cutting experiments with different grain sizes or grain orientations made by different preparation methods in order to explore the effect of laser-assisted machining and to expand the universality of laser-assisted machining technology to improve the cutting performance of ZnS.

Author Contributions: Conceptualization, L.Q. and H.Z.; methodology, H.L.; validation, J.X. and J.Z.; investigation, D.Z. and H.L.; resources, X.W.; data curation, S.M. and H.L.; writing—original draft preparation, H.L.; writing—review and editing, J.X.; visualization, H.L.; supervision, J.X.; project administration, X.W.; funding acquisition, J.X. All authors have read and agreed to the published version of the manuscript.

Funding: This research was funded by JCKY2023205B002.

Data Availability Statement: The original contributions presented in the study are included in the article, further inquiries can be directed to the corresponding author.

Conflicts of Interest: Author X.W., H.Z. and D.Z. was employed by the company Aviation Industry Corporation of China. The remaining authors declare that the research was conducted in the absence of any commercial or financial relationships that could be construed as a potential conflict of interest.

References

1. Zhang, X.; Huang, W.; Hu, X. Light scattering to investigate the surface quality of silicon substrates. In Proceedings of the Annual Conference of the Chinese-Society-for-Optical-Engineering on Applied Optics and Photonics, China (AOPC), Beijing, China, 5–7 May 2015.
2. Sun, H.; Tan, W.; Ruan, Y.; Bai, L.; Xu, J. Surface roughness classification using light scattering matrix and deep learning. *Sci. China-Technol. Sci.* **2024**, *67*, 520–535. [CrossRef]
3. Ramavath, P.; Biswas, P.; Johnson, R.; Reddy, G.J.; Laxminarayana, P. Hot Isostatic Pressing of ZnS Powder and CVD ZnS Ceramics: Comparative Evaluation of Physico-chemical, Microstructural and Transmission Properties. *Trans. Indian Ceram. Soc.* **2014**, *73*, 299–302. [CrossRef]
4. Ramavath, P.; Mahender, V.; Hareesh, U.S.; Johnson, R.; Kumari, S.; Prasad, N.E. Fracture behaviour of chemical vapour deposited and hot isostatically pressed zinc sulphide ceramics. *Mater. Sci. Eng. A-Struct. Mater. Prop. Microstruct. Process.* **2011**, *528*, 5030–5035. [CrossRef]
5. Geng, R.; Xie, Q.; Yang, X.; Zhang, R.; Zhang, W.; Qiu, H.; You, J.; Jiao, S. Experimental study of the feasibility of micro-laser-assisted machining in polycrystal ZnSe ceramics. In Proceedings of the Eighth Symposium on Novel Photoelectronic Detection Technology and Applications, Kunming, China, 9–11 November 2021; pp. 3350–3360.
6. Li, W.; Tong, Y.; Lian, W.; Liu, D.; Zhang, H. Experiment study on the cutting property of hot press Zinc Sulfide by single point diamond turning. In Proceedings of the 7th International Symposium on Advanced Optical Manufacturing and Testing Technologies (AOMATT)—Advanced Optical Manufacturing Technologies, Harbin, China, 26–29 April 2014.
7. Zhang, Y.; Sun, J.; Han, Y.; Xiao, Q.; Zhang, L. Study of stress-assisted nano-cutting mechanism of gallium arsenide. *Appl. Surf. Sci.* **2024**, *672*, 160813. [CrossRef]
8. Arif, M.; Zhang, X.; Rahman, M.; Kumar, S. A predictive model of the critical undeformed chip thickness for ductile-brittle transition in nano-machining of brittle materials. *Int. J. Mach. Tools Manuf.* **2013**, *64*, 114–122. [CrossRef]
9. Yang, Y.; Chen, Y.; Zhao, C. On a Novel Modulation Cutting Process for Potassium Dihydrogen Phosphate with an Increased Brittle-Ductile Transition Cutting Depth. *Machines* **2023**, *11*, 961. [CrossRef]
10. Yao, T.; Yang, X.; Kang, J.; Guo, Y.; Cheng, B.; Qin, Y.; Wang, Y.; Xie, Q.; Du, G. Ductile-brittle transition and ductile-regime removal mechanisms in micro-and nanoscale machining of ZnS crystals. *Infrared Phys. Technol.* **2024**, *138*, 105096. [CrossRef]
11. Navare, J.; Kang, D.; Zaytsev, D.; Bodlapati, C.; Ravindra, D.; Shahinian, H. Experimental investigation on the effect of crystal orientation of diamond tooling on micro laser assisted diamond turning of zinc sulfide. In Proceedings of the 48th SME North American Manufacturing Research Conference (NAMRC), Cincinnati, OH, USA, 22–26 June 2020; pp. 606–610.
12. Zong, W.J.; Cao, Z.M.; He, C.L.; Xue, C.X. Theoretical modelling and FE simulation on the oblique diamond turning of ZnS crystal. *Int. J. Mach. Tools Manuf.* **2016**, *100*, 55–71. [CrossRef]
13. Salzman, S.; Romanofsky, H.J.; Giannechini, L.J.; Jacobs, S.D.; Lambropoulos, J.C. Magnetorheological finishing of chemical-vapor deposited zinc sulfide via chemically and mechanically modified fluids. *Appl. Opt.* **2016**, *55*, 1481–1489. [CrossRef]
14. Guo, Y.; Yang, X.; Li, M.; Kang, J.; Zhang, W.; Xie, Q. Cutting Surface Quality and Cutting Conditions Optimization of Zinc Sulfide Crystal. *Chin. J. Rare Met.* **2023**, *47*, 315–321.
15. Chen, J.; Gong, Q.; Song, G.; Zhou, W.; Zhang, T.; An, Q.; Chen, M. Transition of material removal mechanism in cutting of unidirectional SiCf/SiC composites. *Int. J. Adv. Manuf. Technol.* **2024**, *133*, 391–408. [CrossRef]

16. Mohammadi, H.; Ravindra, D.; Kode, S.K.; Patten, J.A. Experimental work on micro laser-assisted diamond turning of silicon (111). *J. Manuf. Process.* **2015**, *19*, 125–128. [CrossRef]
17. Shahinian, H.; Di, K.; Navare, J.; Bodlapati, C.; Zaytsev, D.; Ravindra, D. Ultraprecision laser-assisted diamond machining of single crystal Ge. *Precis. Eng. J. Int. Soc. Precis. Eng. Nanotechnol.* **2020**, *65*, 149–155. [CrossRef]
18. Bodlapati, C.; Navare, J.; Zaytsev, D.; Kang, D.; Mohammadi, H.; Ravindra, D.; Arlt, M.; Shahinian, H. Effect of laser-assisted diamond turning (micro-LAM) on non-optimally oriented calcium fluoride crystal. In Proceedings of the 34th ASPE Annual Meeting, Pittsburgh, PA, USA, 28 October–1 November 2019.
19. Wang, L.; Wu, M.; Chen, H.; Hang, W.; Wang, X.; Han, Y.; Chen, H.; Chen, P.; Beri, T.H.; Luo, L. Damage evolution and plastic deformation mechanism of passivation layer during shear rheological polishing of polycrystalline tungsten. *J. Mater. Res. Technol.* **2024**, *28*, 1584–1596. [CrossRef]
20. Chen, H.; Wang, L.; Peng, F.; Xu, Q.; Xiong, Y.; Zhao, S.; Tokunaga, K.; Wu, Z.; Ma, Y.; Chen, P. Hydrogen retention and affecting factors in rolled tungsten: Thermal desorption spectra and molecular dynamics simulations. *Int. J. Hydrogen Energy* **2023**, *48*, 30522–30531. [CrossRef]
21. You, K.; Fang, F.; Yan, G.; Zhang, Y. Experimental Investigation on Laser Assisted Diamond Turning of Binderless Tungsten Carbide by In-Process Heating. *Micromachines* **2020**, *11*, 1104. [CrossRef]
22. Zhang, J.; Fu, Y.; Chen, X.; Shen, Z.; Zhang, J.; Xiao, J.; Xu, J. Investigation of the material removal process in in-situ laser-assisted diamond cutting of reaction-bonded silicon carbide. *J. Eur. Ceram. Soc.* **2023**, *43*, 2354–2365. [CrossRef]
23. Fan, M.; Zhou, X.; Song, J. Experimental investigation on cutting force and machining parameters optimization in in-situ laser-assisted machining of glass-ceramic. *Opt. Laser Technol.* **2024**, *169*, 110109. [CrossRef]
24. Song, H.; Dan, J.; Li, J.; Du, J.; Xiao, J.; Xu, J. Experimental study on the cutting force during laser-assisted machining of fused silica based on the Taguchi method and response surface methodology. *J. Manuf. Process.* **2019**, *38*, 9–20. [CrossRef]
25. He, Y.; Xiao, G.; Liu, Z.; Ni, J.; Liu, S. Subsurface damage in laser-assisted machining titanium alloys. *Int. J. Mech. Sci.* **2023**, *258*, 108576. [CrossRef]
26. Ma, Z.; Wang, Q.; Dong, J.; Wang, Z.; Yu, T. Experimental investigation and numerical analysis for machinability of alumina ceramic by laser-assisted grinding. *Precis. Eng.* **2021**, *72*, 798–806. [CrossRef]
27. Pu, Y.; Zhao, Y.; Zhang, H.; Zhao, G.; Meng, J.; Song, P. Study on the three-dimensional topography of the machined surface in laser-assisted machining of Si3N4 ceramics under different material removal modes. *Ceram. Int.* **2020**, *46*, 5695–5705. [CrossRef]
28. Kong, X.; Hu, G.; Hou, N.; Liu, N.; Wang, M. Numerical and experimental investigations on the laser assisted machining of the TC6 titanium alloy. *J. Manuf. Process.* **2023**, *96*, 68–79. [CrossRef]
29. Yang, S.; Zhang, L.; Wu, Z. Effect of anisotropy of potassium dihydrogen phosphate crystals on its deformation mechanisms subjected to nanoindentation. *ACS Appl. Mater. Interfaces* **2021**, *13*, 41351–41360. [CrossRef] [PubMed]
30. Yang, S.; Zhang, L.; Wu, Z. Subsurface damage minimization of KDP crystals. *Appl. Surf. Sci.* **2022**, *604*, 154592. [CrossRef]
31. Yang, S.; Zhang, L.; Wu, Z. An investigation on the nano-abrasion wear mechanisms of KDP crystals. *Wear* **2021**, *476*, 203692. [CrossRef]
32. Liu, Q.; Liao, Z.; Axinte, D. Temperature effect on the material removal mechanism of soft-brittle crystals at nano/micron scale. *Int. J. Mach. Tools Manuf.* **2020**, *159*, 103620. [CrossRef]

Disclaimer/Publisher's Note: The statements, opinions and data contained in all publications are solely those of the individual author(s) and contributor(s) and not of MDPI and/or the editor(s). MDPI and/or the editor(s) disclaim responsibility for any injury to people or property resulting from any ideas, methods, instructions or products referred to in the content.

Article

Laser Direct Writing of Setaria Virids-Inspired Hierarchical Surface with TiO$_2$ Coating for Anti-Sticking of Soft Tissue

Qingxu Zhang [1], Yanyan Yang [2], Shijie Huo [1], Shucheng Duan [1], Tianao Han [1], Guang Liu [1,*], Kaiteng Zhang [3], Dengke Chen [3], Guang Yang [1] and Huawei Chen [4,*]

[1] School of Mechanical Engineering, Hebei University of Science and Technology, Shijiazhuang 050018, China; 15631817192@163.com (Q.Z.); 17367837787@163.com (S.H.); 15174890829@163.com (S.D.); hf0402101211@163.com (T.H.); y_guang@126.com (G.Y.)
[2] 960 Hospital of the PLA, Tai'an 271000, China; yang1665319@163.com
[3] College of Transportation, Ludong University, Yantai 264025, China; zkteng90@163.com (K.Z.); 716316hay@163.com (D.C.)
[4] School of Mechanical Engineering and Automation, Beihang University, Beijing 100191, China
* Correspondence: liuguang0701@hebust.edu.cn (G.L.); chenhw75@buaa.edu.cn (H.C.)

Abstract: In minimally invasive surgery, the tendency for human tissue to adhere to the electrosurgical scalpel can complicate procedures and elevate the risk of medical accidents. Consequently, the development of an electrosurgical scalpel with an anti-sticking coating is critically important. Drawing inspiration from nature, we identified that the leaves of Setaria Virids exhibit exceptional non-stick properties. Utilizing this natural surface texture as a model, we designed and fabricated a specialized anti-sticking surface for electrosurgical scalpels. Employing nanosecond laser direct writing ablation technology, we created a micro-nano textured surface on the high-frequency electrosurgical scalpel that mimics the structure found on Setaria Virids leaves. Subsequently, a TiO$_2$ coating was deposited onto the ablated scalpel surface via magnetron sputtering, followed by plasma-induced hydrophobic modification and treatment with octadecyltrichlorosilane (OTS) to enhance the surface's affinity for silicone oil, thereby constructing a self-lubricating and anti-sticking surface. The spreading behavior of deionized water, absolute ethanol, and dimethyl silicone oil on this textured surface is investigated to confirm the effectiveness of the self-lubrication mechanism. Furthermore, the sticking force and quality are compared between the anti-sticking electrosurgical scalpel and a standard high-frequency electrosurgical scalpel to demonstrate the efficacy of the nanosecond laser-ablated micro-nano texture in preventing sticking. The findings indicate that the self-lubricating anti-sticking surface fabricated using this texture exhibits superior anti-sticking properties.

Keywords: setaria virids; laser direct writing; hierarchical micro texturing; TiO$_2$ coatings; anti-sticking; soft tissue

Citation: Zhang, Q.; Yang, Y.; Huo, S.; Duan, S.; Han, T.; Liu, G.; Zhang, K.; Chen, D.; Yang, G.; Chen, H. Laser Direct Writing of Setaria Virids-Inspired Hierarchical Surface with TiO$_2$ Coating for Anti-Sticking of Soft Tissue. *Micromachines* **2024**, *15*, 1155. https://doi.org/10.3390/mi15091155

Academic Editors: Nam-Trung Nguyen and Yao Liu

Received: 20 August 2024
Revised: 11 September 2024
Accepted: 13 September 2024
Published: 15 September 2024

Copyright: © 2024 by the authors. Licensee MDPI, Basel, Switzerland. This article is an open access article distributed under the terms and conditions of the Creative Commons Attribution (CC BY) license (https://creativecommons.org/licenses/by/4.0/).

1. Introduction

High-frequency electrosurgical scalpels have emerged as indispensable instruments in contemporary operating theaters, thanks to their superior performance. They are extensively utilized across various surgical disciplines, including general, thoracic and cardiovascular, neuro, urological, and plastic surgeries [1–5]. The deployment of high-frequency electrosurgical scalpels significantly reduces surgical duration and effectively decreases the incidence of post-operative complications [6–8], thereby enhancing overall surgical outcomes and expediting patient recovery. However, their use is not without challenges. The scalpels' operation relies on tissue cutting through heat generation by high-frequency current, leading to an extremely high temperature. Consequently, tissue sticking, which can result in enlarged wounds and bleeding, remains an issue, posing a threat to patient safety [9]. To mitigate this issue, developing a functionalized surface with exceptional anti-sticking properties on the electrosurgical scalpel is a key strategy.

Efforts to enhance the anti-sticking attributes of electrosurgical scalpels encompass physical and chemical approaches. Physical methods predominantly involve the use of nanosecond/picosecond lasers to fabricate micro-nano textures on the scalpel's surface [10–21]. For instance, Zhang Li and her team employed laser texturing technology to create bionic micro-nano composite structures on the surface of Co28Cr6Mo alloy. Furthermore, they successfully developed a functionalized surface with outstanding corrosion resistance and superlubricity characteristics by utilizing a modification process involving polydimethylsiloxane solution and silicone oil [22]. Following this, Zhou Haonan and colleagues ingeniously designed an anti-icing slippery liquid-infused porous surface (SLIPSs), drawing inspiration from the sleek inner surface of pitcher plants. Utilizing a dip-coating technique, they successfully created a hydrophobic silica layer that is grafted with long-chain organic molecules. Following the infusion of an oil phase, a three-tiered structure emerges, mirroring the internal architecture of pitcher plants, and this innovative surface demonstrates exceptional capabilities in resisting ice formation [23]. Chemical methods include the application of polymer coatings and sol-gel processes [24–28], such as PET [29], fluorinated polymers [30], silicones [31–34], and cyclosilazanes [35,36] with low surface energy. Research indicates that both approaches have limitations. Micro-nano textures produced by nanosecond/picosecond laser technology, while providing some level of anti-sticking, lack durability and are prone to wear during scalpel use, leading to the loss of anti-sticking capabilities. Additionally, traditional chemical methods for anti-sticking coatings may offer initial protection but are unreliable, particularly in high-temperature conditions where coatings may degrade and release toxic substances. This not only endangers patients but also risks the health of medical professionals. Therefore, there is an urgent need to investigate a novel method that can effectively address the anti-sticking challenges associated with electrosurgical scalpels [37,38].

Emulating nature, or learning from the natural world, is a pivotal strategy in scientific innovation. It involves a meticulous examination and study of the myriad phenomena and principles found in nature to harness inspiration and apply it to our design endeavors, with the goal of achieving more optimal results. Through careful observation of the leaves of Setaria Virids, we have identified a distinctive microgroove–columnar texture on their surfaces (Figure 1c,e). This texture enables liquid droplets to roll effortlessly across the leaf surface without lingering, substantially reducing the buildup of soil and stains [39]. This attribute not only maintains the cleanliness of the Setaria Virids leaves but also ensures their capacity for photosynthesis, thereby sustaining the energy required for growth.

In our quest to develop new methods for combating sticking in electrosurgical scalpels, we have drawn inspiration from the micro-nano textures of the natural world. We are exploring the application of these textures to the surface of electrosurgical scalpels using nanosecond laser technology. Subsequently, magnetron sputtering was employed to uniformly deposit TiO_2 onto the functionalized surface, forming a TiO_2 thin film, combined with the infusion of silicone oil to achieve anti-sticking and antimicrobial properties. TiO_2 thin films have emerged as a highly promising and valuable asset in the realm of disinfection and sterilization. This innovative material has captured extensive interest owing to its exceptional photocatalytic sterilization properties. When exposed to ultraviolet light, TiO_2 thin films are capable of generating a substantial number of electrons and holes, which trigger a cascade of photocatalytic reactions that effectively deactivate bacteria, viruses, and a range of other pathogenic micro-organisms. The distinctive photo-catalytic attributes of TiO_2 thin films pave the way for extensive applications in the food processing and disinfection industries. Especially within the healthcare sector, the utilization of TiO_2 thin films offers an efficient method for sterilizing medical instruments and hospital settings, markedly diminishing the incidence of hospital-acquired infections and ensuring a more secure environment for both patients and healthcare personnel. We have engineered a uniform array of microgroove textures, with five rows of microcolumns uniformly spaced at the base of each groove, designed to facilitate the directional and spontaneous diffusion of the lubricant along the microgrooves. This creates a lubricating film when the scalpel interacts with tissue, thereby reducing sticking [40,41]. Dimethyl silicone oil was chosen as the lubricant due to its ability to spread easily at high temperatures, its biologically inert nature, and its excellent hydrophobic properties [42–44]. To enhance the sticking of dimethyl silicone oil to the micro-nano textured surface, we employed a chemical technique, grafting an OTS (octadecyltrichlorosilane) self-assembled coating onto the self-lubricating anti-sticking surface. This forms a nanometers-thick siloxane bond, which not only improves the stability of the lubricant at high temperatures but also increases its longevity on the scalpel surface [45].

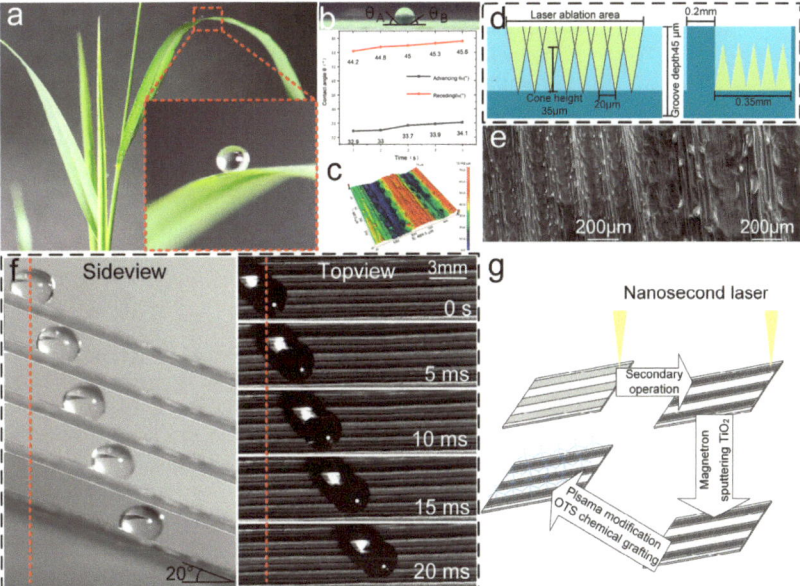

Figure 1. (a) The morphology of Setaria Virids and the sticking state of dewdrops on the surface of Setaria Virids leaves. (b) The front contact angle θ_A and the rear contact angle θ_B of a droplet on the surface of a horizontally placed Setaria Virids leaf. (c) White light interference morphology characterization of micro-nano textures processed by nanosecond laser. (d) Ideal illustration of micro-nano textures prepared using nanosecond laser. (e) Scanning Electron Microscopy (SEM) characterization of the microstructure on the surface of Setaria Virids leaves. (f) The flow of a droplet on the surface of a Setaria Virids leaf placed at an inclination of 20°. (g) A mechanism illustration showing the preparation of a functionalized surface with anti-sticking properties by creating a microgroove–micropillar composite texture using a nanosecond laser, followed by Plasma modification and a self-assembled molecular layer coating.

To validate the anti-sticking efficacy of the functionalized surface with micro-nano textures, we conducted tests on the unidirectional flow and reverse gravity spreading of liquids, and employed a push–pull force gauge to measure sticking force and sticking quantity. These experiments are aimed at thoroughly assessing the anti-sticking capabilities of the developed self-lubricating anti-sticking surfaces, providing a scientific foundation for further refinement and application. Through these extensive research efforts and testing, we aspire to deliver an efficient and dependable anti-sticking solution for electrosurgical scalpels, thereby enhancing the safety and efficiency of surgical procedures.

2. Materials and Methods

2.1. Materials

The experimental apparatus utilized in this study comprises a nanosecond laser-marking machine (Beijing Leijieming Laser Technology Development Company Limited. (Beijing, China)), a microscope, a Canon 90D camera (Canon Inc., Ota City, Japan), a high-speed camera (Hefei Zhongke Junda Vision Technology Co., Ltd. (Hefei, China)), a water bath heating platform, a Plasma ion cleaner (PlasmaTechnology GmbH, Herrenberg, Germany), a magnetron sputtering apparatus, a precision electronic balance (with a precision of 0.001 g), an in-house-built platform for measuring pushing and pulling forces, related equipment for high-frequency electrosurgical scalpels (Beijing Yingjiahua Technology Company Limited. (Beijing, China)), and a nitrogen gas generator. The materials include fresh, complete Setaria Virids leaves, 316 L stainless steel electrosurgical scalpel blades (with dimensions of 50 mm × 2.6 mm × 0.8 mm), and 316 L stainless steel sheets (with dimensions of 50 mm × 50 mm × 1 mm), The experimental substrates were provided by Beijing ENJOY Technologies Company Limited (Beijing, China). The chemicals involved are HCl (hydrochloric acid), H_2SO_4 (concentrated sulfuric acid), OTS (octadecyltrichlorosilane), dimethyl silicone oil with a viscosity of 10 cs, nitrogen gas, deionized water, anhydrous ethanol, fluorescein sodium, and Rhodamine B. All the chemical reagents used in the experiment were supplied by Beijing Chemical Works (Beijing, China). Fresh, homogeneous in vitro pig liver was selected as the soft tissue material for testing anti-sticking force and quantity.

The Setaria Virids leaves were cut into rectangular blocks measuring 0.7 cm × 0.5 cm and placed under a scanning electron microscope (SEM) for imaging at magnifications of 150× and 500× to capture the SEM images of the surface's microgroove–columnar texture (Figure 1e). The observation revealed that the surface of the Setaria Virids leaves exhibited a microgroove–columnar composite texture, with microcolumns evenly distributed on the convex platforms. It is speculated that this texture is designed to optimize the liquid flow properties of the leaf surface, such as reducing the sticking force of water droplets and promoting rainwater runoff, thus minimizing the accumulation of soil stains on the leaves and ensuring sufficient photosynthesis.

2.2. Fabrication of Electrodes with Hierachical Micro-Nano Array Structure

Utilizing a nanosecond laser direct-write ablation technique, a uniform microgroove-microcolumn composite texture was fabricated on the surface of an electrosurgical scalpel blade crafted from 316 L stainless steel, with microcolumns arrayed evenly at the base of each groove.

The process began by fully submerging the scalpel in anhydrous ethanol and subjecting it to an ultrasonic cleaning cycle at room temperature for three minutes to eliminate surface impurities. The scalpel was then sequentially rinsed with anhydrous ethanol and deionized water, followed by thorough drying under a stream of moving air. After cleaning, the scalpel was positioned at various angles (0°, 5°, 15°, 30°) relative to the horizontal plane on the laser processing table (Figure 2a).

The laser marking machine was initiated with the following settings: a laser power of 40%, a frequency of 40 KHz, a speed of 1000 mm/s, a line spacing of 0.02 mm, and a cycle count of 40. These parameters were used to create the primary microgroove texture. The sample, after its initial processing, was immersed in a solution composed of concentrated sulfuric acid, concentrated hydrochloric acid, and water in a ratio of 7:2:1, and then subjected to a water bath at 60 °C for 20 min. This step was crucial for the removal of surface irregularities and any molten residue resulting from the laser ablation.

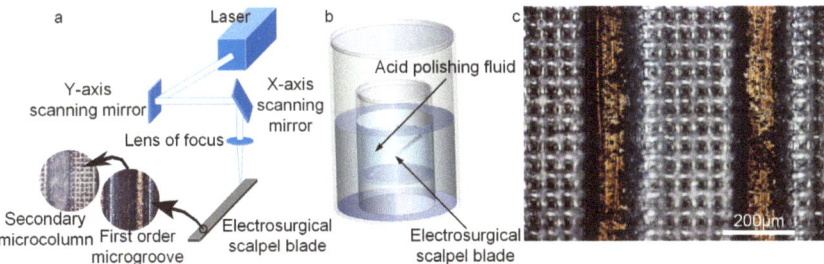

Figure 2. (**a**) Mechanism diagram of laser processing surface micro-nano texture. (**b**) Removal of sharp tips by water bath heating. (**c**) Microscopic observation of the processed surface micro-nano texture.

Following the etching process, the sample was returned to the same position on the laser processing table. A secondary array of uniformly distributed microcolumns was then ablated using refined laser parameters: a power of 72%, a frequency of 50 KHz, a speed of 2500 mm/s, a line spacing of 0.05 mm, and a cycle count of 12, resulting in the desired composite texture (Figure 2c).

Utilizing magnetron sputtering technology, a magnetron ion sputtering system was employed to conduct ion sputtering treatment on the electrosurgical scalpel, which features a functionalized surface. Before the sputtering process, the scalpel was meticulously cleaned in an ultrasonic cleaning machine to ensure the surface was free of contaminants that could compromise the adhesion of TiO_2 to the functionalized surface. This step was crucial for the formation of a uniform and continuous TiO_2 thin film. Experimental studies revealed that the optimal sputtering parameters were as follows: an oblique target sputtering configuration with a power setting of 90 w, a rotational speed of 20 rpm/min, and a coating duration of 50 min, resulting in a copper film thickness of roughly 50 nm for the most desirable outcomes. The TiO_2-coated sample was then introduced into a Plasma cleaner, where it underwent a 3-min modification process under an environment characterized by 60 W RF power and a vacuum pressure of 60 Pa (Figure 3b). This modification step was employed to augment the surface's content of hydroxyl groups (-OH), thereby enhancing the binding force of the micro-nano texture on the electrosurgical scalpel surface with OTS.

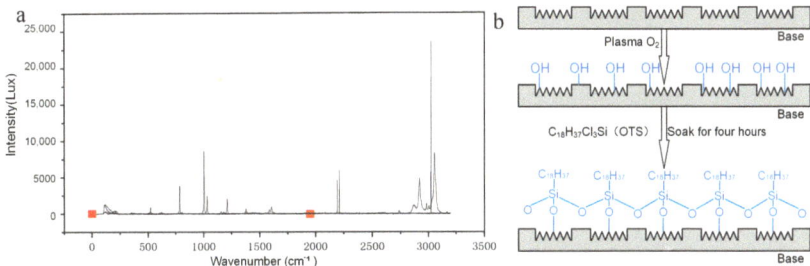

Figure 3. (**a**) Successful OTS chemical grafting was confirmed by Raman spectroscopy. (**b**) Chemical mechanism diagram of OTS self-assembly.

The Plasma-modified sample was then fully immersed in a toluene solution containing OTS at a concentration of 1 mmol/L for a duration of 4 h. Utilizing in situ Raman spectroscopy, this investigation elucidated the grafting mechanism of octadecyl trichlorosilane (OTS) molecules onto the functionalized substrate. The spectral data demonstrated a significant congruence in the vibrational peak intensities and wavenumbers between the OTS solution and the surface post-grafting, indicative of the formation of covalent silane bonds between OTS and the substrate interface. Furthermore, these observations corroborated the alterations in the surface morphology, which were consistent with the anticipated structural modifications. The spectroscopic analysis also confirmed the successful implementation of surface functionalities, such as the enhancement of hydrophobicity, and the reaction's completeness and specificity were ascertained, with the absence of residual OTS monomers or side products, thereby furnishing direct molecular-level corroboration of the surface modification protocol (Figure 3a). Subsequently, any residual OTS on the surface was rinsed away with anhydrous toluene, and the sample was fully dried in an environment of nitrogen gas [38]. Finally, 5 µL of dimethyl silicone oil with a viscosity of 10cs was evenly dispensed onto the sample surface using a pipette, ensuring that the oil completely filled the voids within the micro-nano texture to establish a functionalized surface endowed with ultra-lubricant and anti-sticking properties.

2.3. Morphological Characterization

In the experimental phase, nanosecond laser ablation technology was utilized to create intricate micro-nano textures on the surface of an electrosurgical scalpel, endowing it with enhanced functionality. The primary microgrooves, etched by the nanosecond laser, exhibited precise dimensions of 350 µm in width and 50 µm in depth, engineered to facilitate the flow of lubricant.

Nested within each primary microgroove, an array of five secondary microcolumns was meticulously positioned. These microcolumns showcased a consistent height range of 30 to 40 µm and a base diameter varying between 45 and 65 µm. The strategic arrangement of these microcolumns was intended to improve the retention and directed spread of lubricant, thereby enhancing the scalpel's performance. Moreover, the inter-groove ridges were designed to safeguard the integrity of the microcolumns during the scalpel's operation, preventing damage that could arise from extended use and potentially compromising the non-stick properties.

To meticulously assess the formed micro-nano textures, state-of-the-art characterization techniques were employed, including a white light interferometer and a scanning electron microscope (SEM). The white light interferometer revealed the textures' three-dimensional morphology, while the SEM provided detailed surface imagery at high resolution. By integrating the capabilities of both instruments, a thorough understanding of the microtextures' microscopic architecture and geometric attributes was achieved. These characterization methods ensured the fabricated textures met the stringent requirements for non-stick functionality on electrosurgical scalpel surfaces, and they provided critical data for future performance and refinement efforts (Figure 4).

2.4. Unidirectional Transport Properties and Antigravity Flow Properties

In the experimental setup, dimethyl silicone oil with a viscosity of 10cs was selected as the lubricant of choice and was infused into the intricate voids of the micro-nano textures. The objective of this procedure was to harness the specialized design of the micro-nano textures to facilitate the effective retention and distribution of the lubricant across the surface of the electrosurgical scalpel. The dimethyl silicone oil was chosen for its outstanding stability at elevated temperatures, biocompatibility, and hydrophobic nature, which make it an ideal lubricating agent.

To ensure that the lubricant remained continuously and effectively engaged within the micro-nano textures, a novel reverse gravity spreading mechanism was developed. This mechanism enables the lubricant to autonomously refill the areas within the textures that are depleted. By leveraging the capillary action and surface tension inherent in the micro-nano structures, the lubricant is encouraged to flow spontaneously between the microgrooves and microcolumns. This self-regulating process ensures the formation of a persistent and even lubricating film during tissue contact, thereby minimizing sticking and friction, and enhancing the performance of the scalpel (Figure 5).

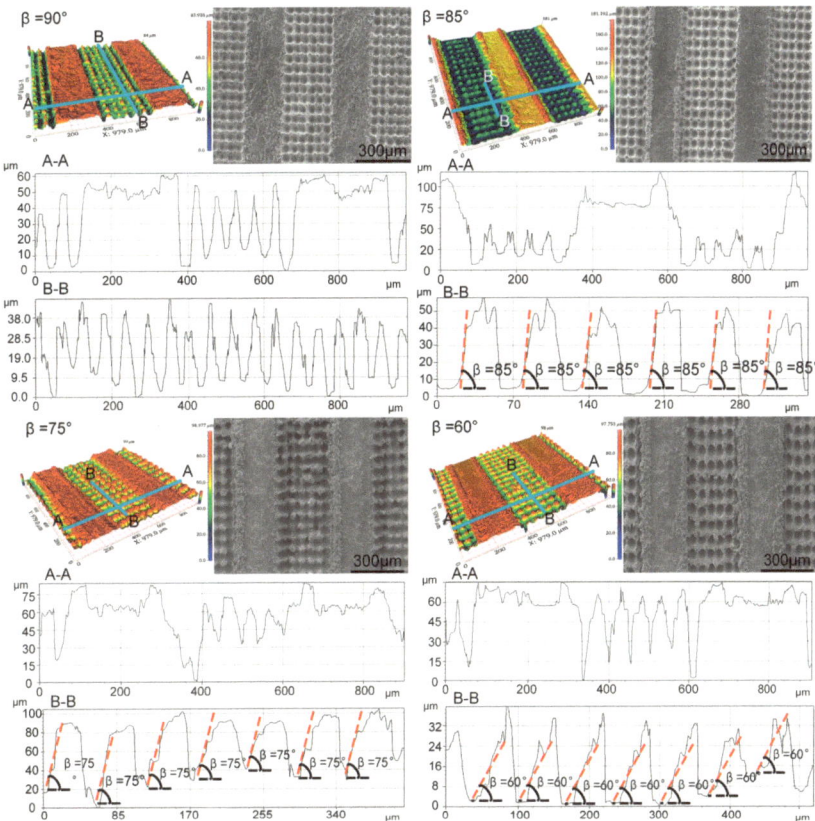

Figure 4. The white light interference morphology, longitudinal and transverse profile characterization, as well as SEM images of micro/nano texture were obtained at inclination angles β of 90°, 85°, 75°, and 60° between the oblique column and the horizontal plane. "A-A" and "B-B" denote the cross-sectional illustrations for the respective micro-nano textures, with "A-A" showcasing the horizontal profile and "B-B" the vertical profile.

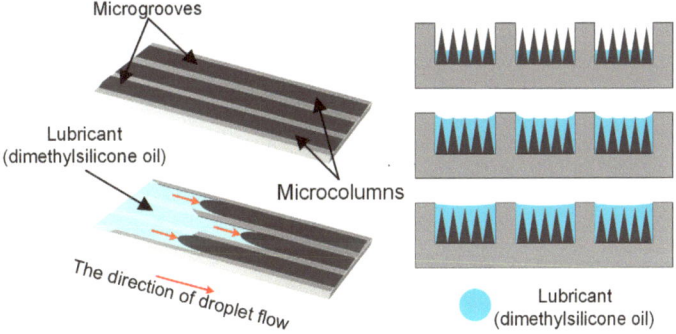

Figure 5. Self-lubricating anti-stick surface with added lubricant.

To investigate the reverse gravity spreading effect and unidirectional flow characteristics of functional surfaces enhanced with micro-nano textures under various liquids, we systematically examined these surfaces at a constant temperature of 25 °C. The experiment involved the use of deionized water, anhydrous ethanol, and dimethyl silicone oil as representative liquids. Drops of equal volume were introduced to the micro-nano textured surfaces to observe their reverse gravity spreading and unidirectional flow behaviors at different microcolumn angles. The goal was to elucidate how the geometric parameters of the micro-nano textures, the nature of the liquids, and the ambient conditions collaboratively affected the reverse gravity spreading and unidirectional flow dynamics. Deionized water and anhydrous ethanol were chosen for their widespread and typical use in biomedical contexts, while dimethyl silicone oil was considered due to its potential application as a lubricant.

The assessment of surface tension was based on the Young–Laplace pressure equation, in which Δp denotes the pressure differential across the liquid interface, R_1 and R_2 are the radii of curvature of the liquid surface in planes orthogonal to the horizontal, and σ represents the surface tension of the liquid [46].

$$\Delta p = \sigma(\frac{1}{R_1} + \frac{1}{R_2}) \tag{1}$$

Young's equation delineates the equilibrium condition at the interface between liquid, solid, and gaseous phases, where: γ represents the liquid's surface tension, θ denotes the contact angle, γ_{SG} characterizes the interfacial tension between the solid and the gas, and γ_{SL} signifies the interfacial tension between the solid and the liquid.

$$\gamma \cdot \cos\theta = \gamma_{SG} - \gamma_{SL} \tag{2}$$

On an ideal anti-sticking surface, the contact angle approaches 180 degrees, resulting in $\cos(\theta)$ approaching 0. This implies that γ_{SL} is nearly equal to γ_{SG}, indicating that the liquid almost does not wet the surface. The smaller the surface energy E, the poorer the hydrophilic properties of the surface. The formula for surface energy E is as follows:

$$E = \gamma \cdot A_{drop} \tag{3}$$

Assuming the contact radius of the droplet on the blade surface is r, the surface area of the droplet (A_{drop}) is calculated as follows:

$$A_{drop} = \pi \cdot r^2 \tag{4}$$

The calculated surface energy is below 10^{-6} J/m^2, which suggests that the hydrophobic properties of the functionalized surface treated with dimethyl silicone oil are commendable.

The observed reverse gravity spreading rates for the three liquids followed the sequence: $V_{\text{(anhydrous ethanol)}} > V_{\text{(deionized water)}} > V_{\text{(dimethyl silicone oil)}}$. As the angle of inclination of the microcolumns was further reduced, the reverse gravity spreading velocities of all three liquids significantly increased (Figure 6). We also noticed that with an increase in the angle β between the slanted microcolumns and the horizontal plane, the unidirectional flow of the liquids became more distinct, always flowing in the direction of the column's tilt. These findings not only provide critical experimental support for understanding how micro-nano textures regulate liquid flow but also offer essential guidance for refining the design of functional surfaces featuring such textures.

Our research revealed a strong correlation between the flow velocity of liquids and their surface tension σ. In particular, as the surface tension of a liquid rose, the corresponding radius of curvature diminished, resulting in a slower flow velocity. Since the surface tension order is $\sigma_{\text{(anhydrous ethanol)}} < \sigma_{\text{(deionized water)}} < \sigma_{\text{(dimethyl silicone oil)}}$, the flow velocities followed the pattern $V_{\text{(anhydrous ethanol)}} > V_{\text{(deionized water)}} > V_{\text{(dimethyl silicone oil)}}$. Moreover, variations in the angle β between the microcolumns and the horizontal plane had a pronounced impact on the liquids' spreading velocities. As the microcolumn inclination decreased, the spreading velocities of all three liquids rose. Notably, on a 15° inclined surface, whether flowing with or against the column's tilt, the order of reverse gravity spreading velocities was consistent with $V_{\text{(anhydrous ethanol)}} > V_{\text{(deionized water)}} > V_{\text{(dimethyl silicone oil)}}$; and as the microcolumn inclination decreased further, the reverse gravity spreading velocities of all three liquids accelerated markedly (Figure 6). With an increase in the angle β, the unidirectional flow became increasingly evident, with all flows in the direction of the column's tilt (Figure 7).

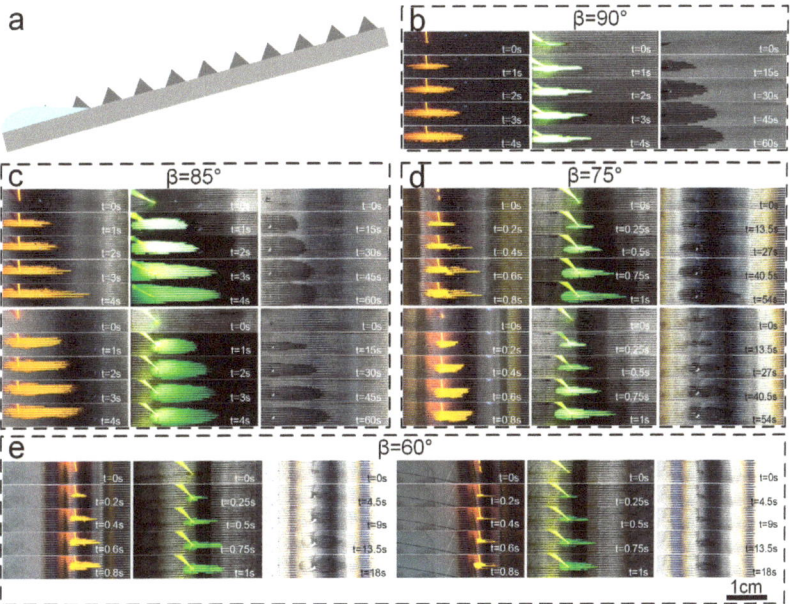

Figure 6. Antigravity spreading of deionized water (labeled with rhodamine B as orange yellow), anhydrous ethanol (labeled with fluorescein sodium as yellow green), and dimethylsilicone oil (colorless and transparent) on the texture of oblique columns with different inclination angles. (a) Antigravity spreading mechanism of liquid on micro/nano texture. (b) Antigravity spreading of liquid on a functional surface with a slant column angle of 90°. (c) Antigravity spreading of liquid on a functional surface with a slant column angle of 85°. (d) Antigravity spreading of liquid on a functional surface with a slant column angle of 75°. (e) Antigravity spreading of liquid on a functional surface with a slant column angle of 60°.

To delve into the intricacies of the reverse gravity spreading behavior of diverse liquid droplets on functionalized micro-nano textured surfaces with anti-sticking properties, a detailed schematic was developed using Visio software (2019) (Figure 6a). This illustration offers a clear visual representation of the droplet's journey across the textured surface. Upon contact with the microgrooves at the surface, the droplets initially diffuse along the grooves' bases. As they navigate through the array of microcolumns, the droplets are propelled against gravity by the increased Laplace pressure and the influence of additional capillary forces [46]. The droplets ascend along the direction of the microcolumn array. Moreover, the droplets experience a greater Laplace pressure in the direction of the microcolumn tilt compared to the opposite direction, resulting in faster spreading against gravity along the tilt of the microcolumns. This hypothesis is not only theoretically supported but is also borne out in practical applications (Figure 6). Experiments were conducted using Rhodamine B-labeled deionized water, fluorescein sodium-labeled anhydrous ethanol, and dimethyl silicone oil with a viscosity of 10cs to evaluate the reverse gravity spreading effects on the prepared functionalized surfaces. The experiment validated the calculated flow rates of different liquids (Figure 6b–e).

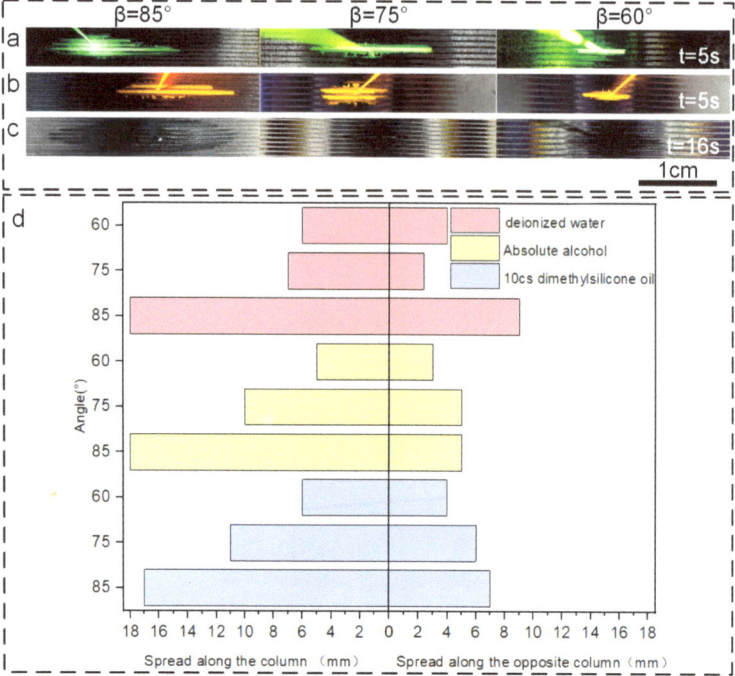

Figure 7. Unidirectional flow of the bulk on the texture of different slant column angles (90°, 85°, 75°, 60°). (**a**) Absolute ethanol (labeled with fluorescein sodium in yellow green). (**b**) Deionized water (labeled with rhodamine B in orange yellow). (**c**) 10cs dimethylsilicone oil (colorless and transparent). (**d**) Summary plot of one-way spreading data of the three liquids.

The examination of reverse gravity spreading has determined that an electrosurgical scalpel with a functionalized surface featuring micro-nano textures, produced via nanosecond laser technology, exhibits reverse gravity spreading properties when positioned at an angle α of 15° with respect to the horizontal plane. By comparing the reverse gravity spreading behavior of different liquids across micro-nano textures with varying inclined column angles β, it is evident that the smaller the β value, the more rapid the flow rate of the liquid. This corroborates the ability of the micro-nano textured functionalized surface, prepared using nanosecond laser technology, to facilitate the self-lubrication of the lubricant during surgical procedures when the electrosurgical scalpel is positioned at an angle less than 15° from the horizontal plane.

2.5. Cutting Experiment and Anti-Sticking Test

To assess the anti-sticking properties of the micro-nano textured functionalized surfaces, tissue cutting experiments were performed. Ensuring the precision and replicability of the results was paramount, hence stringent control over the experimental conditions was exercised to minimize any extraneous errors or influences. Fresh pork liver was initially trimmed into cubic samples measuring 2 cm on each side and then secured to an electrode plate positioned directly beneath the push–pull force gauge. The electrosurgical scalpel, equipped with the micro-nano textured functionalized surface, was mounted onto the measuring probe of the force gauge. To mimic actual surgical scenarios, the power supply for the high-frequency electrosurgical scalpel was adjusted to 50 W. The push–pull force gauge was utilized to determine the sticking force of the self-lubricating, anti-sticking surface (Figure 8).

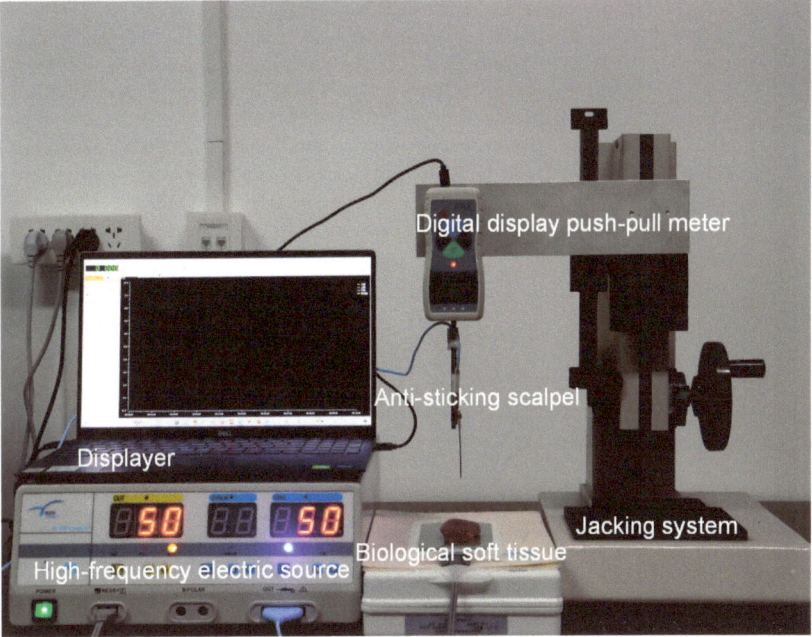

Figure 8. Sticking force measured by digital explicit push–pull dynamometer.

In the course of the experiment, the handle of the push–pull force gauge was operated to control the electrosurgical scalpel, which was activated, as it made contact with the pork liver surface at a rate of 1 mm/s. The scalpel was then withdrawn at the same speed. The force registered by the gauge during the withdrawal phase constituted the sticking force. This approach facilitated a quantitative evaluation of the anti-sticking efficacy of the micro-nano textured functionalized surfaces.

During the application of the anti-sticking electrosurgical scalpel, a segment of the biological soft tissue in contact with it underwent vaporization, concurrently accompanied by the discharge of blood. Quantitative measurements reveal that: the scalpel width (D) is 2 mm, the scalpel travel speed (v) is 5 mm/s, the blood density (ρ) is 1050 kg/m³, the blood kinematic viscosity (μ) is 3.5 mPa·s.

Reynolds number formula:

$$Re = \frac{\rho \cdot v \cdot D}{\mu} \tag{5}$$

Calculate the Reynolds number $Re < 2000$, is laminar flow, then the shear force calculation formula is:

$$\tau = \frac{F}{A} = \frac{6 \cdot \mu \cdot v}{D} \tag{6}$$

The formula for calculating the pressure P of the fluid on the scalpel surface is:

$$P = \frac{1}{2}\rho v^2 \tag{7}$$

By calculation, the shear force and pressure are below the safe value, which proves that the anti-stick electric scalpel with functional surface has good performance.

Through meticulous measurements of sticking force and sticking quantity, it was discovered that traditional high-frequency electrosurgical scalpels encounter substantial sticking issues during surgical procedures. However, this problem was significantly alleviated with the use of an electrosurgical scalpel featuring a micro-nano textured functionalized surface, which was fabricated using nanosecond laser technology. To further corroborate this improvement, a comparative study was conducted using both a standard high-frequency electrosurgical scalpel and anti-sticking scalpels with microcolumn angles of 90°, 85°, 75°, and 60°. These scalpels were employed to perform cyclic cutting experiments on fresh pork liver, with 1, 10, and 20 cycles, under identical environmental conditions. The sticking quantity, sticking force, and the size of thermal damage wounds left on the fresh pork liver were then carefully compared and observed (Figure 9).

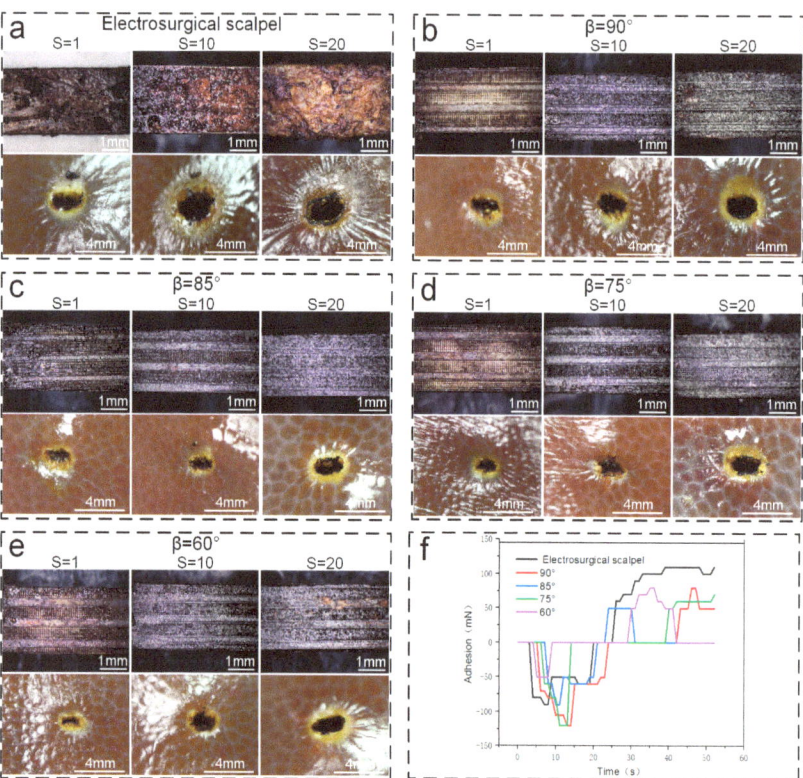

Figure 9. Sticking force and adhesion scale characteristics of high frequency electric scalpel and the inclination Angle β = 90°, 85°, 75°, and 60° between the upper oblique column of the electric scalpel and the horizontal plane. (**a**) When the operating power of the electric scalpel was 50 w, the amount of adhesion and thermal injury wound of fresh pig liver was cut with a high frequency electric scalpel (the number of electric cutting S = 1, 10, 20). (**b**) When the operating power of the electric scalpel was 50 w, the amount of adhesion and thermal injury wound of fresh pig liver was cut by the anti-stick electric scalpel with the angle of microcolumn β = 90° (the number of electric cutting S = 1, 10, 20). (**c**) When the electric scalpel working power was 50 w, the amount of adhesion and thermal injury wound of fresh pig liver was cut by the anti-stick electric scalpel with the angle of microcolumn β = 85° (the number of electric cutting S = 1, 10, 20). (**d**) When the operating power of electric scalpel was 50 w, the amount of adhesion and thermal injury wound of fresh pig liver was cut by the anti-stick electric scalpel with the angle β = 85°. The amount of adhesion and thermal injury wound of fresh pig liver was cut by the anti-stick electric scalpel with the angle of microcolumn β = 75° (the number of electric cutting S = 1, 10, 20). (**e**) When the operating power of the electric scalpel was 50 w, the amount of adhesion and thermal injury wound (electrotomy times S = 1, 10, 20) of fresh pig liver was cut with the anti-stick electrotome with micropillar angle β = 60°. (**f**) Comparison of adhesion forces between high frequency electrotome and different micropillar angle (90°, 85°, 75°, 60°).

The experimental findings reveal that the micro-nano textured functionalized surface fabricated using nanosecond laser technology demonstrates a marked advantage in reducing sticking force and sticking quantity, with this benefit being particularly pronounced after multiple cutting cycles. For example, the lifespan of an electrosurgical scalpel equipped with anti-sticking characteristics is extended by over three-fold compared to a conventional high-frequency electrosurgical scalpel, while the frequency of cleaning needed during its usage is significantly reduced. Moreover, the variation in the microcolumn angle also influences the anti-sticking performance, with a general trend indicating that the smaller the angle β, the more effective the anti-sticking.

Through adhesion force measurement experiments, we arrived at several pivotal findings: when performing a single electrosection on soft tissue, the functionalized surface of the anti-sticking electrosurgical scalpel with an inclined column angle of β at 85° demonstrated the most superior performance, achieving a minimum adhesion force of 50 mN. This represents a reduction of approximately 62.4%, compared to the high-frequency electrosurgical scalpel. In contrast, the surfaces with less effective anti-sticking properties had inclined column angles of β at 90° and 60°, yet still exhibited a decrease of around 45.6%, compared to the high-frequency variant. Upon the 20th electrosection of the biological tissue, the 85° inclined column angle continued to outperform, achieving a reduction over 90% versus the high-frequency electrosurgical scalpel. The 90° inclined column angle remained less effective, showing only a 52.4% reduction in adhesion force compared to the high-frequency scalpel. These data suggest an exponential growth trend in the differential adhesion force between the anti-sticking electrosurgical scalpel and the high-frequency electrosurgical scalpel. It is anticipated that after 40 cycles, the adhesion force of the anti-sticking electrosurgical scalpel will be diminished by 130%, providing robust evidence that the functionalized surface, prepared with nanosecond laser technology, possesses pronounced anti-sticking characteristics. Further examination of the thermal damage area at the incision site of fresh pork liver under different electrosurgical cutting cycles (S) revealed a positive correlation between the thermal damage area and the inclined column angle β. Specifically, the minimum thermal damage area is achieved when the inclined column angle β is 75°. Additionally, it was observed that the sticking amount increases as the inclined column angle β decreases. Compared to the traditional high-frequency electrosurgical scalpel, the sticking amount decreases by 316%, 425%, 503%, and 648%, respectively, for β values of 90°, 85°, 75°, and 60° (Figure 10).

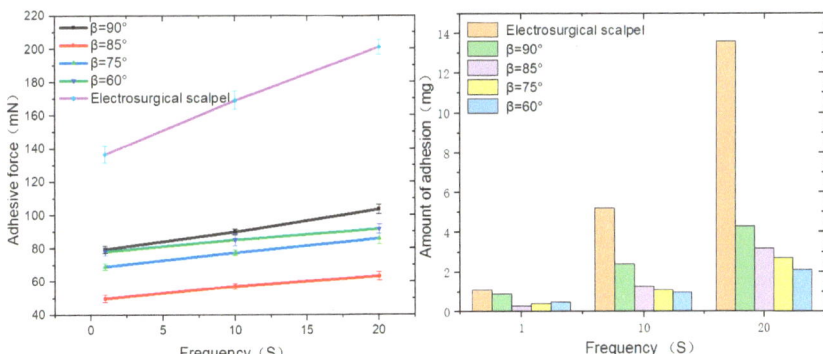

Figure 10. Comparison curves of adhesion amount versus adhesion force when the number of cycles are 1, 10, and 20.

These experimental results clearly demonstrate that the micro-nano textured functionalized surface prepared using nanosecond laser technology can significantly reduce sticking phenomena during surgical procedures with electrosurgical scalpels. It also diminishes the thermal damage caused by high-frequency currents. These findings confirm that the composite texture of microgrooves and microcolumns processed on the surface of electrosurgical scalpel blades using nanosecond laser technology, along with surface modification and the addition of lubricants, yields a functionalized surface with excellent anti-sticking properties and a small area of thermal damage to biological soft tissues.

Through cutting cycle tests, it was observed that the sticking force of the surface tissue significantly decreased, along with a substantial reduction in sticking quality. This result fully proves that the anti-sticking electrosurgical scalpel prepared using nanosecond laser technology possesses superior anti-sticking performance, maintaining stability over extended use and meeting the needs for sticking prevention in electrosurgical devices.

The excellent durability and efficient anti-sticking properties not only enhance surgical efficiency and safety but also reduce maintenance costs and surgical duration. This provides a more efficient and safe option for medical surgery.

3. Results and Discussion

By integrating the aforementioned research findings, we have leveraged nanosecond laser technology to design and fabricate a novel self-lubricating anti-sticking functionalized surface on the surface of a high-frequency electrosurgical scalpel. The achievements are as follows:

(1) This surface is composed of an array of microgrooves, each containing five rows of microcolumns, forming a unique micro-nano structure. On this micro-nano structure, we successfully grafted an OTS self-assembled coating, which plays a pivotal role in the effective diffusion, stable adhesion, and long-term retention of dimethyl silicone oil on the functionalized surface. The choice of dimethyl silicone oil as an efficient lubricant is due to its strong coating and stability, which are crucial for ensuring that the scalpel maintains good anti-sticking performance over extended use.

(2) The experiments also studied the unidirectional flow and reverse gravity spreading effects of three different liquids on the functionalized surface. The experimental results revealed that the functionalized surface not only possesses self-lubrication properties but can also achieve self-lubrication at a smaller angle with the horizontal plane. These experiments strongly confirm the feasibility of the lubricant in the prepared micro-nano texture for self-lubrication and self-supplementation.

(3) We measured the adhesion force and adhesion quantity of the anti-sticking electrosurgical scalpel compared to the high-frequency electrosurgical scalpel. The experiments showed that the adhesion force of the anti-sticking electrosurgical scalpel with the functionalized surface decreased up to 90% compared to the high-frequency electrosurgical scalpel, and the adhesion quantity reduced by 300–650%. These experimental results consistently demonstrate that the high-frequency electrosurgical scalpel with a self-lubricating anti-sticking surface prepared using nanosecond laser technology exhibits excellent anti-sticking performance.

These research findings not only provide new insights and methods for the design and fabrication of electrosurgical scalpels but also contribute significantly to enhancing the performance and safety of scalpels in practical applications. Looking forward, we anticipate that these discoveries will further advance the field of medical devices and bring more innovation and improvements to surgical procedures.

Author Contributions: Conceptualization, Q.Z., G.L. and H.C.; methodology, Q.Z.; software, S.H.; validation, G.L., Q.Z. and Y.Y.; formal analysis, S.D. and T.H.; investigation, Q.Z. and K.Z.; resources, S.H.; data curation, Y.Y. and D.C.; writing—original draft preparation, Q.Z.; writing—review and editing, G.L. and G.Y.; visualization, S.D.; supervision, G.L. and G.Y.; project administration, G.L. and H.C.; funding acquisition, G.L., D.C. and H.C. All authors have read and agreed to the published version of the manuscript.

Funding: This research was funded by the National Natural Science Foundation of China (Grant Nos. 52205306, 51935001, 52305311), Hebei Natural Science Foundation (Grant No. E2022208039), Science and Technology Project of Hebei Education Department (Grant No. BJK2023028), Shandong Natural Science Foundation (Grant No. ZR2023QE018).

Data Availability Statement: The original contributions presented in the study are included in the article, further inquiries can be directed to the corresponding author.

Acknowledgments: We extend our gratitude to Shang and HF AGILE DEVICE CO., LTD. for their assistance in filming the movies.

Conflicts of Interest: The authors declare no conflicts of interest.

References

1. Lu, T.; Lu, W. Application of burning treatment with high-frequency electric knife of clearing tumor in non-small cell lung cancer patients with adjacent organ invasion: A single-central experience. *Asian J. Surg.* **2021**, *44*, 1114–1115. [CrossRef] [PubMed]
2. Kawamura, A.; Akiba, Y.; Nagasawa, M.; Takashima, M.; Arai, Y.; Uoshima, K. Bone heating and implant removal using a high-frequency electrosurgical device. An in vivo experimental study. *Clin. Oral Implant. Res.* **2021**, *32*, 989–997. [CrossRef] [PubMed]
3. Göktay, F.; Erfan, G.; Çelik, N.S.; Öztürk, C.; Doruk, T.; Albayrak, H.; Yanık, M.E.; Albayrak, U. Early Cosmetic Results and Midterm Follow-up Findings of Rhinophyma Patients Treated With High-Frequency Electrosurgery and a Discussion on the Severity Assessment of the Disease. *J. Cutan. Med. Surg.* **2017**, *21*, 221–226. [CrossRef]
4. Bai, Y.; Yang, F.; Liu, C.; Li, F.; Wang, S.; Lin, R.; Ding, Z.; Meng, W.B.; Li, Z.S.; Linghu, E.Q.; et al. Expert consensus on the clinical application of high-frequency electrosurgery in digestive endoscopy. *J. Dig. Dis.* **2021**, *23*, 2–12. [CrossRef] [PubMed]
5. Antonio, C.R.; Trídico, L.A.; Marchi CM, G.; Antonio, J.R.; D'Ávila, S.C.G.P. High-frequency electrosurgery in ice-pick scars: Pre and post treatment comparative study. *Surg. Cosmet. Dermatol.* **2017**, *9*, 123–126. [CrossRef]
6. Zhang, K.; Liu, G.; Zhao, Z.; Zhang, S.; Yang, C.; Yang, J.; Zhang, L.; Chen, H. Liquid-Infused bionic microstructures on High-Frequency electrodes for enhanced spark effects and reduced tissue adhesion. *Chem. Eng. J.* **2024**, *485*, 149907. [CrossRef]
7. Li, K.; Xie, Y.; Yang, S.; Ritasalo, R.; Mariam, J.; Yu, M.; Bi, J.; Ding, H.; Lu, L. Synergetic Effects of Nanoscale ALD-HfO$_2$ Coatings and Bionic Microstructures for Antiadhesive Surgical Electrodes: Improved Cutting Performance, Antibacterial Property, and Biocompatibility. *ACS Appl. Mater. Interfaces* **2023**, *15*, 43550–43562. [CrossRef]
8. Chen, H.; Zhang, Y.; Zhang, L.; Ding, X.; Zhang, D. Applications of bioinspired approaches and challenges in medical devices. *Bio-Des. Manuf.* **2020**, *4*, 146–148. [CrossRef]
9. Liu, G.; Zhang, P.; Chen, H.; Han, Z.; Zhang, D. Bio-inspired Anti-adhesion Surfaces of Electrosurgical Scalpel. *J. Mech. Eng.* **2018**, *54*, 21–27. [CrossRef]
10. Martínez-Calderon, M.; Martín-Palma, R.J.; Rodríguez, A.; Gómez-Aranzadi, M.; García-Ruiz, J.P.; Olaizola, S.M.; Manso-Silván, M. Biomimetic hierarchical micro/nano texturing of TiAlV alloys by femtosecond laser processing for the control of cell adhesion and migration. *Phys. Rev. Mater.* **2020**, *4*, 056008. [CrossRef]
11. Wang, N.; Wang, Q.; Xu, S.; Zheng, X. Mechanical Stability of PDMS-Based Micro/Nanotextured Flexible Superhydrophobic Surfaces under External Loading. *ACS Appl. Mater. Interfaces* **2019**, *11*, 48583–48593. [CrossRef] [PubMed]
12. Ellinas, K.; Gogolides, E. Ultra-low friction, superhydrophobic, plasma micro-nanotextured fluorinated ethylene propylene (FEP) surfaces. *Micro Nano Eng.* **2022**, *14*, 100104. [CrossRef]
13. Sarkiris, P.; Ellinas, K.; Gkiolas, D.; Mathioulakis, D.; Gogolides, E. Motion of Drops with Different Viscosities on Micro-Nanotextured Surfaces of Varying Topography and Wetting Properties. *Adv. Funct. Mater.* **2019**, *29*, 1902905. [CrossRef]
14. Dimitrakellis, P.; Ellinas, K.; Kaprou, G.D.; Mastellos, D.C.; Tserepi, A.; Gogolides, E. Bactericidal Action of Smooth and Plasma Micro-Nanotextured Polymeric Surfaces with Varying Wettability, Enhanced by Incorporation of a Biocidal Agent. *Macromol. Mater. Eng.* **2021**, *306*, 2000694. [CrossRef]
15. Yang, Y.; Jia, E.; Xie, C.; Hu, M. Single-exposure femtosecond laser direct writing of Salvinia-inspired micropatterned functional surface. *Opt. Laser Technol.* **2024**, *177*, 111199. [CrossRef]
16. Zhang, W.; Shi, Z.; Chen, C.; Yang, X.; Yang, L.; Zeng, Z.; Zhang, B.; Liu, Q. Super-resolution GaAs nano-structures fabricated by laser direct writing. *Mater. Sci. Semicond. Process.* **2018**, *84*, 119–123. [CrossRef]
17. Liu, G.; Yang, J.; Zhang, K.; Wu, H.; Yan, H.; Yan, Y.; Zheng, Y.; Zhang, Q.; Chen, D.; Zhang, L.; et al. Recent progress on the development of bioinspired surfaces with high aspect ratio microarray structures: From fabrication to applications. *J. Control. Release* **2024**, *367*, 441–469. [CrossRef] [PubMed]
18. Zhang, Y.; Wu, D.; Zhang, Y.; Bian, Y.; Wang, C.; Li, J.; Chu, J.; Hu, Y. Femtosecond laser direct writing of functional stimulus-responsive structures and applications. *Int. J. Extrem. Manuf.* **2023**, *5*, 042012. [CrossRef]
19. Hong, X.; Zheng, Z.; Gao, Y. Construction of Robust Hierarchical Micro-nanostructure by Laser Irradiation and Hydrothermal Treatment on Titanium Alloy for Superhydrophobic and Slippery Surfaces. *J. Mater. Eng. Perform.* **2023**, *33*, 1885–1897. [CrossRef]
20. Rajan, R.A.; Ngo, C.-V.; Yang, J.; Liu, Y.; Rao, K.S.; Guo, C. Femtosecond and picosecond laser fabrication for long-term superhydrophilic metal surfaces. *Opt. Laser Technol.* **2021**, *143*, 107241. [CrossRef]
21. Gao, J.; Wu, Y.; Zhang, Z.; Zhao, D.; Zhu, H.; Xu, K.; Liu, Y. Achieving amorphous micro-nano superhydrophobic structures on quartz glass with a PTFE coating by laser back ablation. *Opt. Laser Technol.* **2022**, *149*, 107927. [CrossRef]

22. Zhang, L.; Tan, Z.; Wu, T.; Zhang, L.; Hao, B. Investigation on the Corrosion Resistance Properties of Cobalt-Chromium-Molybdenum Alloy Artificial Human Body Components with Robust Biomimetic Superhydrophobic and Slippery Surfaces Based on Laser Texturing. *Langmuir* **2023**, *39*, 14996–15013. [CrossRef] [PubMed]
23. Zhou, H.; Xu, Q.; Zhao, J.; Luo, H.; Huang, X.; Huang, J. Biomimetic super slippery surface with excellent and durable anti-icing property for immovable heritage conservation. *Prog. Org. Coat.* **2023**, *184*, 107818. [CrossRef]
24. Liu, Y.; Zhang, L.; Hu, J.; Cheng, B.; Yao, J.; Huang, Y.; Yang, H. Facile preparation of a robust, transparent superhydrophobic ZnO coating with self-cleaning, UV-blocking and bacterial anti-adhesion properties. *Surf. Coat. Technol.* **2024**, *477*, 130352. [CrossRef]
25. Jiang, Y.; Geng, H.; Peng, J.; Ni, X.; Pei, L.; Ye, P.; Lu, R.; Yuan, S.; Bai, Z.; Zhu, Y.; et al. A multifunctional superhydrophobic coating with efficient anti-adhesion and synergistic antibacterial properties. *Prog. Org. Coat.* **2024**, *186*, 108028. [CrossRef]
26. Zhang, B.; Xu, W.; Xia, D.H.; Fan, X.; Duan, J.; Lu, Y. Comparison Study of Self-Cleaning, Anti-Icing, and Durable Corrosion Resistance of Superhydrophobic and Lubricant-Infused Ultraslippery Surfaces. *Langmuir* **2021**, *37*, 11061–11071. [CrossRef]
27. Li, Y.; Xia, B.; Jiang, B. Thermal-induced durable superhydrophilicity of TiO_2 films with ultra-smooth surfaces. *J. Sol-Gel Sci. Technol.* **2018**, *87*, 50–58. [CrossRef]
28. Song, T.; Liu, Q.; Zhang, M.; Chen, R.; Takahashi, K.; Jing, X.; Liu, L.; Wang, J. Multiple sheet-layered super slippery surfaces based on anodic aluminium oxide and its anticorrosion property. *RSC Adv.* **2015**, *5*, 70080–70085. [CrossRef]
29. Zhang, X.; Wang, Y.; Cui, Z.; Zhang, X.; Wang, H.; Zhang, K.; Shi, W. Design, fabrication, and sensing applications of MWCNT/PET terahertz metasurface. *Opt. Laser Technol.* **2024**, *170*, 110204. [CrossRef]
30. Ma, W.; Takahara, A. Superamphiphobic coating based on halloysite clay nanotubes and a catechol bearing fluorinated polymer. *Abstr. Pap. Am. Chem. Soc.* **2014**, *248*, 1700907.
31. Baggio, C.; Turani, M.; Olivero, M.; Salvo, M.; Pugliese, D.; Sangermano, M. Selective distributed optical fiber sensing system based on silicone cladding optical fiber and Rayleigh backscattering reflectometry for the detection of hydrocarbon leakages. *Opt. Laser Technol.* **2023**, *161*, 109158. [CrossRef]
32. Ben, J.; Wu, P.; Wang, Y.; Liu, J.; Luo, Y. Preparation and Characterization of Modified ZrO_2/SiO_2/Silicone-Modified Acrylic Emulsion Superhydrophobic Coating. *Materials* **2023**, *16*, 7621. [CrossRef] [PubMed]
33. Wang, H.; Li, X.; Zhang, E.; Shi, J.; Xiong, X.; Kong, C.; Ren, J.; Li, C.; Wu, K. Strong Thermo-tolerant Silicone-Modified Waterborne Polyurethane/Polyimide Pressure-Sensitive Adhesive. *Langmuir* **2023**, *39*, 17611–17621. [CrossRef] [PubMed]
34. Yu, Q.; Zhang, Z.; Tan, P.; Zhou, J.; Ma, X.; Shao, Y.; Wei, S.; Gao, Z. Siloxane-Modified UV-Curable Castor-Oil-Based Waterborne Polyurethane Superhydrophobic Coatings. *Polymers* **2023**, *15*, 4588. [CrossRef]
35. Zhang, H.; Ou, J.; He, Y.; Hu, Y.; Wang, F.; Fang, X.; Li, W.; Amirfazli, A. Robust solid slippery coating for anti-icing and anti-sticking. *Colloids Surf. A Physicochem. Eng. Asp.* **2024**, *682*, 132853. [CrossRef]
36. Ma, J.; Zhang, C.; Zhang, P.; Song, J. One-step synthesis of functional slippery lubricated coating with substrate independence, anti-fouling property, fog collection, corrosion resistance, and icephobicity. *J. Colloid Interface Sci.* **2024**, *664*, 228–237. [CrossRef]
37. Liu, G.; Yang, J.; Zhang, K.; Yan, H.; Yan, Y.; Zheng, Y.; Zhang, L.; Zhao, Z.; Wang, L.; Yang, G.; et al. Nepenthes alata inspired anti-sticking surface via nanosecond laser fabrication. *Chem. Eng. J.* **2024**, *483*, 149192. [CrossRef]
38. Liu, G.; Zhang, P.; Liu, Y.; Zhang, D.; Chen, H. Self-Lubricating Slippery Surface with Wettability Gradients for Anti-Sticking of Electrosurgical Scalpel. *Micromachines* **2018**, *9*, 591. [CrossRef]
39. Xie, H.; Xu, W.; Wu, T. Bioinspired preparation of regular dual-level micropillars on polypropylene surfaces with robust hydrophobicity inspired by green bristlegrass leaves. *Polym. Adv. Technol.* **2019**, *31*, 492–500. [CrossRef]
40. Meng, R.; Deng, J.; Duan, R.; Liu, Y.; Zhang, G. Modifying tribological performances of AISI 316 stainless steel surfaces by laser surface texturing and various solid lubricants. *Opt. Laser Technol.* **2019**, *109*, 401–411. [CrossRef]
41. Wang, A.H.; Xia, J.; Yang, Z.X.; Xiong, D.H. A novel assembly of MoS_2-PTFE solid lubricants into wear-resistant micro-hole array template and corresponding tribological performance. *Opt. Laser Technol.* **2019**, *116*, 171–179. [CrossRef]
42. Rakheja, T.; Pathak, P.; Grewal, H.S. Facile fabrication of super-slippery and transparent glass with anti-fogging and self-cleaning characteristics. *Mater. Lett.* **2024**, *366*, 136506. [CrossRef]
43. Tan, M.; Wang, F.; Yang, J.; Zhong, Z.; Chen, G.; Chen, Z. Hydroxyl silicone oil grafting onto a rough thermoplastic polyurethane surface created durable super-hydrophobicity. *J. Biomater. Sci. Polym. Ed.* **2024**, *35*, 1359–1378. [CrossRef] [PubMed]
44. Xing, Y.; Li, X.; Xu, J.; Zhang, F. Organosilicon-modified double crosslinked waterborne polyurethane coatings. *Prog. Org. Coat.* **2024**, *190*, 108372. [CrossRef]
45. Dong, Z.; Qu, N.; Jiang, Q.; Zhang, T.; Han, Z.; Li, J.; Zhang, R.; Cheng, Z. Preparation of polystyrene and silane-modified nano-silica superhydrophobic and superoleophilic three-dimensional composite fiber membranes for efficient oil absorption and oil-water separation. *J. Environ. Chem. Eng.* **2024**, *12*, 112690. [CrossRef]
46. Myronyuk, O. Determination of critical surface tension of wetting of textured water-repellent surfaces. *Technol. Audit. Prod. Reserves* **2023**, *2*, 10–13. [CrossRef]

Disclaimer/Publisher's Note: The statements, opinions and data contained in all publications are solely those of the individual author(s) and contributor(s) and not of MDPI and/or the editor(s). MDPI and/or the editor(s) disclaim responsibility for any injury to people or property resulting from any ideas, methods, instructions or products referred to in the content.

Article

Contact Hole Shrinkage: Simulation Study of Resist Flow Process and Its Application to Block Copolymers

Sang-Kon Kim

The Faculty of Liberal Arts, Hongik University, Seoul 04066, Republic of Korea; sangkona@hongik.ac.kr

Abstract: For vertical interconnect access (VIA) in three-dimensional (3D) structure chips, including those with high bandwidth memory (HBM), shrinking contact holes (C/Hs) using the resist flow process (RFP) represents the most promising technology for low-k_1 (where CD = $k_1\lambda$/NA, CD is the critical dimension, λ is wavelength, and NA is the numerical aperture). This method offers a way to reduce dimensions without additional complex process steps and is independent of optical technologies. However, most empirical models are heuristic methods and use linear regression to predict the critical dimension of the reflowed structure but do not account for intermediate shapes. In this research, the resist flow process (RFP) was modeled using the evolution method, the finite-element method, machine learning, and deep learning under various reflow conditions to imitate experimental results. Deep learning and machine learning have proven to be useful for physical optimization problems without analytical solutions, particularly for regression and classification tasks. In this application, the self-assembly of cylinder-forming block copolymers (BCPs), confined in prepatterns of the resist reflow process (RFP) to produce small contact hole (C/H) dimensions, was described using the self-consistent field theory (SCFT). This research paves the way for the shrink modeling of the enhanced resist reflow process (RFP) for random contact holes (C/Hs) and the production of smaller contact holes.

Keywords: computational lithography; shrinkage process; resist reflow; resist flow process; surface evolver; finite element method; deep learning; machine learning; directed self-assembly; block copolymer; self-consistent field theory

1. Introduction

Due to the three-dimensional (3D) chip structure, the high cost of extreme ultraviolet (EUV) exposure tools, and the difficulty of multi-patterning, pattern shrinkage processes have generated significant interest in finding effective methods to reduce the feature sizes in microelectronics and data storage devices. These methods are divided into a top-down approach in photolithography [1,2] and a bottom-up approach in self-assembly [3,4]. For technology below the 10-nm scale, top-down exposure tools face obstacles, such as diffraction-limited resolution and the high cost of ownership [5–7]. Directed self-assembly, as a bottom-up approach, has problems such as insufficient process support and challenges in mass production [8,9]. However, thermal treatment is a process extension technique that uses current lithography equipment and resists [10]. In lithography, the thermal processes include softbake (SB), post-exposure bake (PEB), and thermal processing after development. These three types of thermal processes are essentially the same in terms of heat treatment but affect critical dimensions (CDs) differently. The purpose of softbake (SB) is to remove excess solvent after spin coating, relieve strain in the solid film, and provide better adhesion to the substrate [11]. Post-exposure bake (PEB) aims to reduce the standing wave effect, thus enhancing the linewidth control and resolution [12,13]. Shrink technologies after development include the resist flow process (RFP) [14], shrink assist film for enhanced resolution (SAFIER™) [15], and resolution enhancement lithography assisted by chemical shrink (RELACS™) [16,17]. Shrink assist film for enhanced resolution (SAFIER™) is a thermal flow with a passive overcoat, which is coated on resist patterns, baked for thermal flow, and then removed by rinsing with water. Shrink assist film for enhanced resolution (SAFIER™) provides mechanical support to contact walls to avoid pattern profile degradation during thermal reflow. Resolution enhancement lithography assisted by chemical shrink (RELACS™) is an active shrink overcoat, which is coated on a patterned wafer and heated (mixed bake) to crosslink and react with the resist film. Resolution enhancement lithography assisted by chemical shrink (RELACS™) exhibits less pitch dependence than the resist flow process (RFP) and shrink assist film for enhanced resolution (SAFIER™) [18]. Finally, the resist flow process (RFP), as a resolution-enhancement technique, is an effective method that does not require additional complex process steps [19,20]. The resist flow process (RFP) reduces the pattern size of a resist by thermally heating it above its glass transition temperature after development. The bonding of the synthesized resist is reduced at temperatures above glass transition, and its mobility is improved. In the case of patterning fine contact holes (C/Hs), which have lower resolution performance compared to line and space patterns due to insufficient depth of focus (DOF) and low aerial-image contrast, the three-dimensional (3D) structure of the synthesized resist is altered. As a result of the additional thermal energy, the contact hole (C/H) patterns shrink. However, predicting the results of the resist flow process (RFP) is challenging because the optical proximity effects (OPEs) of the resist flow process (RFP) become quite severe as critical dimensions (CDs) shrink to sub-10-nm patterns. Many parameters affect the results, such as the baking temperature, the baking time, the original characteristics of the resist, the volume of the resist surrounding contact holes (C/Hs), the initial size and shape of contact holes (C/Hs), and the contact hole (C/H) array. Most empirical models use heuristic methods, such as linear regression, to predict the critical dimensions (CDs) of the reflowed structure. However, these models do not account for intermediate shapes. As a result, no general equations or methods are available for predicting contact hole shrinking.

The objectives of this research were to develop a physically accurate resist model based on an understanding of the mechanistic behaviors that drive photoresist imaging, to achieve the best prediction of resist images across multiple processing conditions and to create a general simulation approach to reduce the optical proximity effects (OPEs) of the resist flow process (RFP). The resist flow process (RFP) is described using the surface evolution method, the finite-element method (FEM), machine learning, and deep learning under various reflow conditions to simulate experimental results. Machine learning and deep learning are particularly useful for physical optimization problems that lack analytical solutions. For applications of the resist flow process (RFP), the self-assembly of cylinder-forming block copolymers (BCPs) confined in prepatterns of the resist flow process (RFP) is described using self-consistent field theory (SCFT).

2. Simulation Methods

Figure 1 shows the resist flow process (RFP) in lithography. The resist flow process (RFP) depicted in Figure 1c occurs after the following processes: spin coating (Figure 1a), softbake, exposure, post-exposure bake, and development (Figure 1b). The resist flow process (RFP) reduces the pattern size of the resist by heating it above its glass transition temperature following the development process. At this elevated temperature, the resist's molecular bonds weaken, enhancing its mobility. This alters the resist's three-dimensional (3D) structure and results in the shrinkage of the contact hole (C/H) pattern due to the additional thermal energy.

2.1. Surface Evolver (SE) Method

The reflow process can be understood as an energy imbalance or surface-tension-driven creep process, where the initial pattern deforms toward a minimal energy shape under constant force due to polymer's wetting. Surface Evolver is a computer program that minimizes the energy of a surface subject to constraints [21–23]. The surface is represented as a simplicial complex. The energy can include surface tension, gravity, and other forms.

Constraints may be geometrical constraints on vertex positions or constraints on integrated quantities such as body volumes. Minimization is achieved by evolving the surface along the energy gradient. The condition for a minimum using the gradient descent method and the correcting motion R_v of vertex v are

$$\nabla f = \lambda \nabla g, \qquad (1)$$

$$\vec{R}_v = \sum_k c_k \vec{W}_{vk}, \; \sum_v \vec{R}_v \vec{W}_{vk} = -\delta_k, \; \sum_k c_k \sum_v \vec{W}_{vk} \vec{W}_{v\tilde{k}} = -\delta_{\tilde{k}} \qquad (2)$$

where f, g, λ, $\vec{W}_{vk}$, and δ_k are a function, a constraint function, a constant known as the Lagrange multiplier, the gradient of the constraint k as a function of the position of vertex v, and the excess value of the quantity k, respectively.

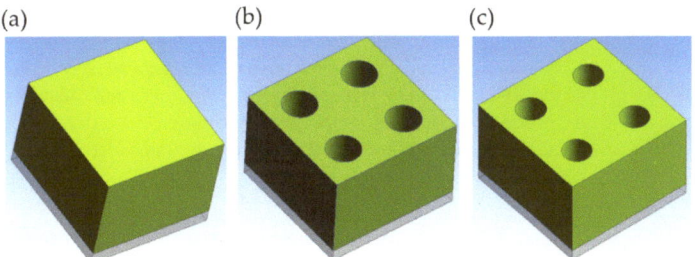

Figure 1. Schematic representation of the resist flow process (RFP): (**a**) spin coating; (**b**) contact hole (C/H) patterns after development process; and (**c**) contact hole (C/H) patterns after thermal reflow. The structure consists of a silicon substrate and resist.

2.2. Finite Element Method (FEM)

The finite element method (FEM) is a mathematical (numerical) technique for finding approximate solutions to partial differential equations. For a structure solver, the governing equation of structural analysis by Hooke's Law, the strain (ϵ)-displacement (u) relation, and the stress (σ)-strain relation are

$$E\frac{\partial^2 u}{\partial x^2} + Q = 0, \; \epsilon = \frac{\partial u}{\partial x}, \sigma = E\epsilon = E\frac{\partial u}{\partial x}, \qquad (3)$$

where E is the elastic modulus and Q is the axial force per unit length [24,25]. For a thermal conduction solver, which deals with heat transfer through molecular agitation within a material without any motion of the material, the governing equations of heat transfer analysis by Fourier's law, heat flux (q), thermal strain (ϵ_t), and the thermal stress (σ_t)-thermal strain relation are

$$\rho C_p \frac{\partial T}{\partial t} - \nabla \cdot (k \nabla T) = f, q = -k \nabla T = -k\frac{\partial T}{\partial n}, \; \epsilon_t = \alpha(\Delta T), \; \sigma_t = E \cdot \alpha(\Delta T), \qquad (4)$$

where T is temperature, k is the thermal conductivity, t is time, ρ is the material density, C_p is the specific heat, α is the coefficient of thermal expansion, E is the elastic modulus, and f is the heat generated inside the body [26,27]. The resist reflow above its glass transition temperature can be assumed to be an ideal fluid, which is incompressible and has a constant density, and the force exerted across a geometrical surface element within the fluid. For an incompressible fluid solver, the Navier–Stokes equation as the governing equation and the continuity equation are

$$\frac{\partial \vec{u}}{\partial t} + \left(\vec{u} \cdot \nabla\right)\vec{u} = -\frac{1}{\rho}\nabla p + \frac{1}{\rho}\nabla \cdot \vec{\tau} + f, \; \nabla \cdot \vec{u} = 0, \qquad (5)$$

where $\vec{u}$, t, $\vec{\tau}$, f, ρ, and p are the flow velocity, time, stress tensor, body force (force per unit mass), density, and pressure, respectively [28–30].

2.3. Orthogonal Fitting Functions

The resist flow process (RFP) bias (ΔCD), which represents the reduction of critical dimensions (CDs) before and after resist reflow, is influenced by the baking temperature (T_b), baking time (t_b), the original characteristics of the resist (K_r), the resist volume surrounding contact hole (C/H) (or the contact hole (C/H) array) (V_n), and the initial contact hole (C/H) size. Under a simple thermal assumption, the actual resist flow process (RFP) bias can be approximated as

$$\Delta CD = f(T_b, t_b) \cdot f(V_n) \cdot f(K_r), \qquad (6)$$

$$(\Delta CD)^2 \approx [\alpha_1 \cdot exp(\alpha_2/T)][\alpha_3 \cdot exp(\alpha_4 \cdot \Lambda)], \qquad (7)$$

where α_1, α_2, α_3, and α_4 are constant. The parameter V_n (resist volume surrounding contact hole (C/H)) can be replaced by a ratio, Λ, between the contact hole and pitch size. The bias function $f(T_b, t_b)$ of temperature and time is assumed to be an exponential function due to the relationship between diffusion length and viscosity.

2.4. Deep Learning and Machine Learning

For a multivariable linear regression, the linear model functions $f_{w,b}(x^{(i)})$ and the cost functions $J(w,b)$ with multiple variables are, respectively,

$$f_{w,b}(x^{(i)}) = w \cdot x^{(i)} + b, J(w,b) = \frac{1}{2m}\sum_{i=0}^{m-1}\left(f_{w,b}(x^{(i)}) - y^{(i)}\right)^2, \qquad (8)$$

where w and $x^{(i)}$ are vectors, $y^{(i)}$ is the target value, w_j and b are parameters, and m is the number of features [31]. For classification methods, a logistic regression model applies the sigmoid function to the linear regression model, and the cost functions $J(w,b)$ are, respectively,

$$f_{w,b}(x^{(i)}) = g(w \cdot x^{(i)} + b) = \frac{1}{1 + e^{-(w \cdot x^{(i)} + b)}}, \qquad (9)$$

$$J(w,b) = -\frac{1}{m}\sum_{i=1}^{m}\left[y^{(i)}log\left(f_{w,b}(x^{(i)})\right) + \left(1 - y^{(i)}\right)log\left(1 - f_{w,b}(x^{(i)})\right)\right] + \frac{\lambda}{2m}\sum_{j=1}^{n-1}w_j^2 \qquad (10)$$

where λ and w are a regularization constant and a parameter, respectively [32]. For multi-class classification, the softmax function a_j of neural networks converts a vector z into a probability distribution:

$$a_j = \frac{e^{z_j}}{\sum_{k=1}^{N}e^{z_k}}, \qquad (11)$$

where $z_j = w_j \cdot x + b_j$ ($j = 1, \cdots, N$) and N is number of features (or categories) in the output layer [33,34].

Decision trees (DTs) are a supervised learning approach used in data mining and machine learning [35]. For classification and regression, decision trees (DTs) are simple to understand and interpret by using visualized trees and require little data preparation. Decision trees (DTs) are represented as tree structures, where each internal node represents a feature, each branch represents a decision rule, and each leaf node represents a prediction. The algorithm works by recursively splitting the data into smaller and smaller subsets based on the feature values. At each node, the algorithm chooses the feature that best splits the data into groups with different target values. Gini impurity, as the criterion to measure the impurity or purity of a dataset, is calculated using the following formula:

$$Gini(t, D) = 1 - \sum_{l \in levels(t)} p(t = l)^2, \quad (12)$$

where p represents the proportion of data points belonging to class t in dataset D. Lower Gini impurity values indicate a purer dataset [36].

Support vector machines (SVMs) are a set of supervised learning methods used for classification and regression [37]. A support vector machine (SVM) constructs a hyperplane or set of hyperplanes in a high or infinite dimensional space, which can be used for classification and regression. For classification, the support vector classification (SVC) solves the following primal problem:

$$\min_{w,b,\zeta} \frac{1}{2} w^T w + C \sum_{i=1}^{n} \zeta_i, \, y_i\left(w^T \phi(x_i) + b\right) \geq 1 - \zeta_i, \, \zeta_i \geq 0, \, i = 1, \ldots, n \quad (13)$$

where $x_i \in \mathbb{R}^p$ is a training vector, $y \in \{1, -1\}^n$ is a vector, C is a regularization parameter, and ζ_i is the distance from their correct margin boundary [38].

2.5. Self-Consistent Field Theory (SCFT)

For block copolymer (BCP) self-assembly, self-consistent field theory (SCFT) is used to describe the self-assembly of diblock copolymers confined in prepatterns. Under the mean field approximation, the free energy (F) can be

$$\frac{F}{nK_BT} = -\ln\left[\frac{Q}{V}\right] - \frac{1}{V}\int d\vec{r}\left[\sum_k \omega_k(\vec{r})\phi_k(\vec{r}) + \sum \chi_{AB} N \phi_A(\vec{r})\phi_B(\vec{r}) + \xi(\vec{r})\left(\sum_k \phi_k(\vec{r}) - 1\right)\right], \quad (14)$$

where V is the dimensionless total volume of the system, ω_k is the mean field of types k (=A and B), χ_{AB} is the Flory–Huggins interaction parameter of the two types, $\xi(\vec{r})$ is the Lagrange multiplier enforcing the incompressibility constraint, N is the polymerization degree, and $Q\left(= (1/V)\int d\vec{r} q(\vec{r}, 1)\right)$ is a single-molecule partition function in mean fields [39–42]. Diffusion dynamics can model the time evolution of the system:

$$\frac{\partial}{\partial t}\phi_K(\vec{r}, t) = L\nabla^2 \mu_K(\vec{r}', t) + \eta_K, \quad (15)$$

where ϕ_K is the block concentration of type K (=A or B), L is the constant mobility, μ_K is the chemical potential, and η_K is the thermal noise [43–45]. By minimizing free energy for densities and mean fields,

$$\phi_A(\vec{r}) + \phi_B(\vec{r}) = 1, \, \omega_K(\vec{r}) = \chi_{AB} N \phi_K(\vec{r}) - H(\vec{r}) + \xi(\vec{r}), \quad (16)$$

where χ is the Flory–Huggins interaction parameter, N is the polymerization degree, H is the polymer–surface interaction, and ξ is the Lagrange multiplier enforcing the incompressibility constraint. The monomer density, due to the partial partition function, is

$$\phi_K(\vec{r}) = \frac{V}{Z}\int_a^b q_A(\vec{r}, s) q_B(\vec{r}, s) ds, \quad (17)$$

where V is the dimensionless total volume of the system, Z is the single-chain partition function in the independent particle approximation imposed by mean field theory, $q_K(\vec{r}, s)$ is the diffusion equation of the partition function due to the contour length (s) of a copolymer chain, and the integral intervals (a, b) are $(0, f)$ (or $(f, 1)$) at $K = A$ (or B)

3. Experiment

For a 193-nm argon fluoride (ArF) chemically amplified resist (CAR), a 6% transmittance attenuated phase shift mask (PSM), quaterpole off-axis illumination (OAI), argon fluoride (ArF) 193-nm illumination, and a 0.75 numerical aperture (NA) were used. The chemically amplified resist (CAR) thickness was 0.35 μm with a bottom antireflection coating (BARC) of 0.082 μm. For thermal reflow, an antireflective layer of 80-nm thick resist was coated over the silicon wafer prior to the resist process. The coated thickness was 0.37 μm. An ethylvinylether-based polymer was coated and prebaked at 100 °C for 60 s. The exposure system was an ASML-700 with a numerical aperture (NA) of 0.6, a partial coherency (σ) of 0.4, and an attenuated phase-shift mask. Exposed wafers were baked at 110 °C for 60 s on a hot plate and developed in 2.38 wt% tetramethyl ammonium hydroxide (TMAH) aqueous solution for 60 s. The resist reflow time was 90 s. Figure 2 shows the experimental results from Hynix Semiconductor Inc. for various temperature and duty ratios. The contact holes (C/Hs) become smaller as the duty ratio and temperature increased [46].

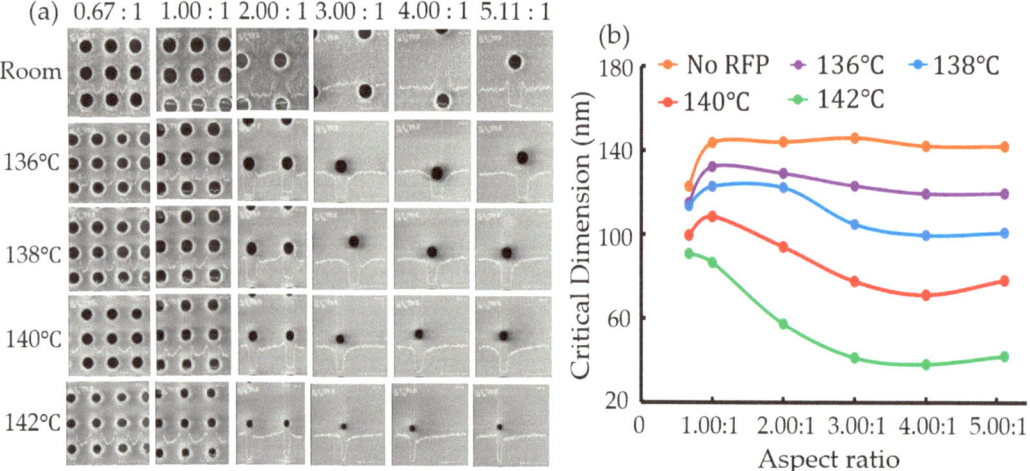

Figure 2. (**a**) Scanning electron microscope (SEM) images of contact holes (C/Hs) and (**b**) a graph of critical dimension (CD) with a function of temperature and duty ratio after the thermal resist process. The upper labels of the images in Figure 2a represent the aspect ratio, which is the ratio between the contact hole size and the pitch size.

4. Results and Discussion

4.1. Surface Evolver (SE)

Figure 3 shows the simulation outcomes generated using Surface Evolver (SE), a computer program by Kenneth Brake, using Equations (1) and (2). Surface Evolver (SE) treats three-dimensional (3D) structures as surfaces governed purely by energy optimization criteria. After the resist flow process (RFP), the surface energy and surface area of one contact hole and nine contact holes were reduced due to simulation runtimes, as shown in Figure 3a. Figure 3c,d display the reflowed images of one contact hole versus the simulation runtimes for the surface areas of 38.3685 and 32.9430, respectively. Figure 3f,g illustrate the reflowed images of nine contact holes for the surface areas of 155.0468 and 144.5697, respectively. However, the Surface Evolver (SE)-based model did not require explicit knowledge of material-specific parameters and predicted the dependence of shape evolution on the simulation runtime.

4.2. The Results of the Finite Element Method (FEM)

Figure 4 shows the simulation results generated through MATLAB to analyze the resist reflow through thermal cycling at hardbake. A mesh with elements no larger than 0.1 was generated, as illustrated in Figure 4a. The temperature distribution in Figure 4b was calculated from the steady-state solution of Equation (4). For both constant thermal conductivity and temperature-dependent thermal conductivity, Figure 4c illustrates the temperature–

time relationship in typical three-stage wafer proximity contact processing, including heating by the preheated hotplate, transfer by the wafer carrier from the hotplate to the chillplate, and cooling by the chillplate. The simulation conditions were mass density = 1, specific heat of the material = 1, constant thermal conductivity of the material = 1, and temperature-dependent thermal conductivity = 0.003 × temperature. Compared with the constant-conductivity case, the temperature on the right-hand edge is lower, as shown in Figure 4c. This is due to the lower conductivity in regions with lower temperatures.

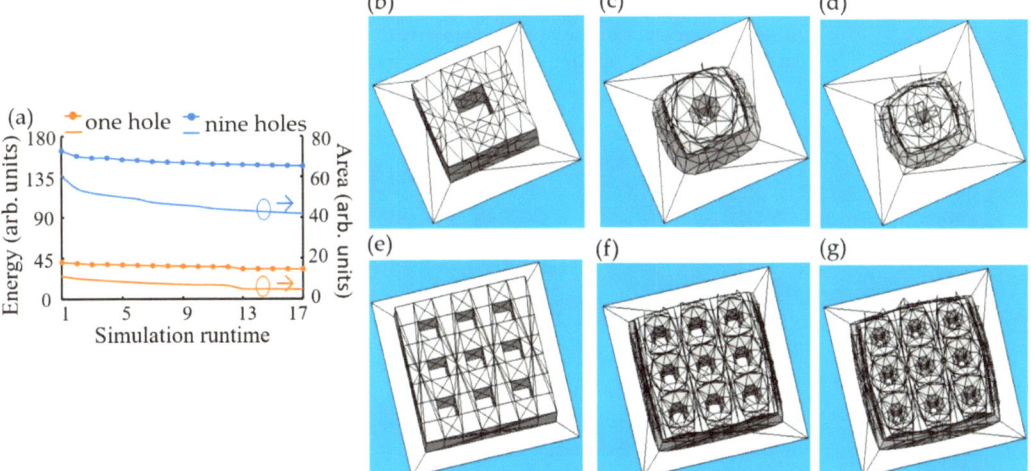

Figure 3. Simulation results of Surface Evolver: (**a**) surface energy and surface area for one contact hole pattern and nine contact holes in terms of simulation times, one contact hole and nine contact holes (**b**,**e**) before the resist flow process (RFP), at surface areas of (**c**) 38.3685, (**d**) 32.9430, (**f**) 155.0468, and (**g**) 144.5697 after the resist flow process (RFP), respectively.

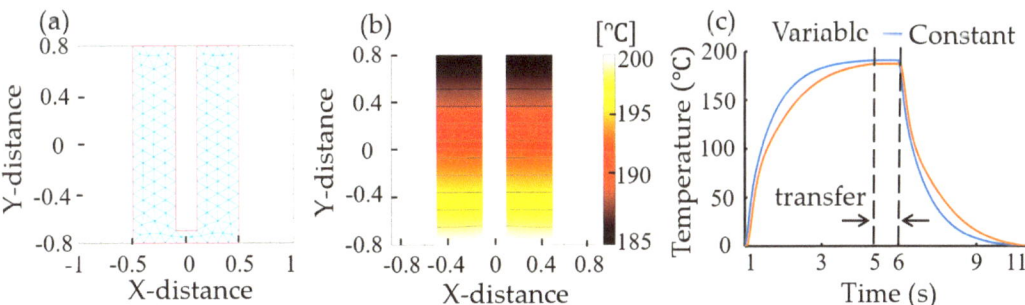

Figure 4. Heat transfer analysis generated through MATLAB: (**a**) a block with the finite element mash displayed, (**b**) a temperature contour with constant thermal conductivity, and (**c**) a plot of simulated bake cycle profiles with constant thermal conductivity (or constant) and temperature-dependent thermal conductivity (or variable), respectively. Transfer time is 1 s.

Figure 5 shows the contact hole (C/H) deformation obtained using the finite element method (FEM) in the static structural ANSYS program (Student Edition 2024) to analyze heating at the preheated hotplate for one contact hole (C/H) shape. Figure 5a presents a structure consisting of a 5 mm × 5 mm × 5.6 mm rectangular resist, a 5 mm × 5 mm × 0.8 mm rectangular silicon substrate, and an emptied cylinder with a height of 5 mm and a radius of 1.05 mm. The ANSYS model consisted of 83,878 nodes and 18,421 elements for a 200 µm mesh size, as shown in Figure 5b. Figure 5b,d show the three-dimensional (3D) plots of contact hole (C/H) thermal deformation at 0 s and 20 °C, 0.5 s and 74 °C, and 1 s and 168 °C. The maximum thermal deformation was 41.643 μm at 1 s and 168 °C, as shown in Figure 5c. The simulation conditions for a silicon substrate were density $\rho = 2330$ kg/m^3, Young's modulus $E = 150$ GPa, Poisson's ratio $v = 0.28$, and the coefficient of thermal expansion was $\alpha = 0.9$ ppm/K; the simulation conditions for the resist were density $\rho = 950$ kg/m^3, Young's modulus $E = 2$ GPa, Poisson's ratio $v = 0.33$, and the coefficient of thermal expansion was $\alpha = 60$ ppm/K. However, a limitation of ANSYS based on the finite element method (FEM) arises from meshing thicknesses greater than 10 μm and shear loading, which can cause an error in the finite element method (FEM) due to the impact on the stiffness matrix from variations in the length-to-thickness ratio of the beam element [47].

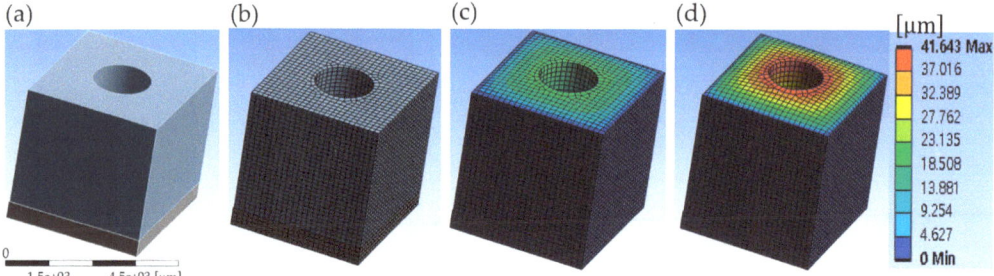

Figure 5. Static structural deformation of one contact hole (C/H) due to heat transfer, generated through ANSYS: (**a**) the structure of a 5 mm × 5 mm × 0.8 mm bottomplate, a 5 mm × 5 mm × 5.6 mm upper plate, and an emptied cylinder with a height of 5 mm and a radius of 1.05 mm, and plots of (**b**) mashing and deformation at (**c**) 0.5 s and (**d**) 1 s simulation times.

Figure 6 shows a phase change of a contact hole (C/H) sidewall due to a glass transition temperature. The melting boundaries, influenced by gravity, moved from the left side to the right side at 0, 1, 30, 50, 100, 150, and 200 simulation steps. The boundary behavior illustrates why the top contact hole (C/H) diameter was more variable than the bottom contact hole (C/H) diameter after the resist reflow. The simulation conditions were a simulation area of 8.89 mm × 6.35 mm; 3721 nodes and 3600 elements for a 0.546 mm mesh size after meshing; material properties (density of 750 kg/m^3, specific heat of 1800 J/(kg·K), thermal conductivity of 2 W/(m·K), viscosity of 0.00181 kg/(m·s), thermal expansion coefficient of 0.00012 K^{-1}, pure solvent melting heat of 174,000 J/kg, solidus temperature of 294 K, and liquidus temperature of 297 K); boundary conditions (left wall at 363 K and right wall at 294 K); solution controls under relaxation factors (pressure of 0.3, density of 1, body forces of 1, momentum of 0.8, turbulent kinetic energy of 0.8, specific dissipation rate of 0.8, turbulent viscosity of 1, liquid fraction update of 0.2, and energy of 0.7); and initial values (gauge pressure of 0 Pa, X-axis velocity of 0 m/s, Y-axis velocity of 0 m/s, turbulent kinetic energy of 1 m^2/s^2, specific dissipation rate of 1 s^{-1}, and temperature of 294 K).

4.3. Results of Orthogonal Fitting Functions and Deep Learning

Figure 7 shows the comparison between the experiment results and the simulation results of thermal reflow biases dependent on temperature and pitch sizes in a 193-nm chemically amplified resist (CAR). Depending on the aspect ratio, which is the ratio between the contact hole size and the pitch size, the simulated results from an orthogonal fitting function, linear regressions using a Python program (version 3.12.2) and Scikit-Learn, and a convolutional neural network (CNN) exhibited mismatches with the experimental results within different error ranges. Although the simulation parameters do not account for the chemical phenomenon of thermal reflow, the critical dimension bias after resist reflow can be predicted in a linear system using the fitting function of the experimental data. From

Equation (7), the simulated function of thermal bias for a 193-nm chemically amplified resist (CAR), using an orthogonal fitting functional method, was

$$(\Delta CD)^2_{Thermal\ reflow} = [\alpha_1 \times exp(\alpha_2/T)][\alpha_3 \times exp(\alpha_4 \times \Lambda)]$$
$$= [3.696 \times 10^{29} \times exp(-8347.64393/T)][1.88152 \times exp(-1.90217 \times \Lambda)] \quad (18)$$

where T is temperature and Λ is the ratio between the contact hole size and the pitch size, as shown in Figure 7a. Figure 7b,c show comparisons of linear regressions from a Python program and Scikit-Learn to the experimental results using square root scaling. For the Python program, after 200,000 iterations, the fitting function was $y = 0.69x_1 + 0.81x_2 - 107.739906$, in which y is $\log_{10}$(thermal reflow bias), x_1 is aspect ratio, and x_2 is temperature, as shown in Figure 7b. For Scikit-Learn, the fitting function was $y = 0.69x_1 + 0.81x_2 - 107.739888$, as shown in Figure 7c. For the convolutional neural network (CNN) regression, the network architecture comprised of four layers (Conv1D(finters = 64), Dense(64), Dense(16), and Dense(1)). The results of the convolutional neural network (CNN) regression were better than those of the fitting function and linear regressions, as shown in Figure 7d.

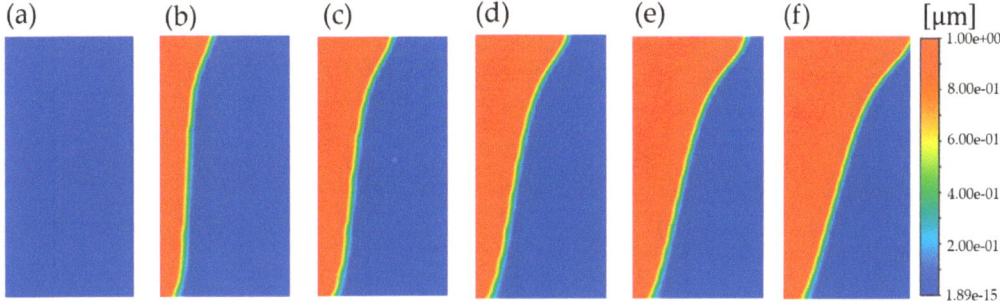

Figure 6. Phase change of a contact hole (C/H) sidewall due to gravity from solid to liquid during thermal reflow, generated using the ANSYS Fluent (Student Edition 2024): melting boundaries at (**a**) 0 steps, (**b**) 1 step, (**c**) 30 steps, (**d**) 50 steps, (**e**) 100 steps, and (**f**) 150 steps.

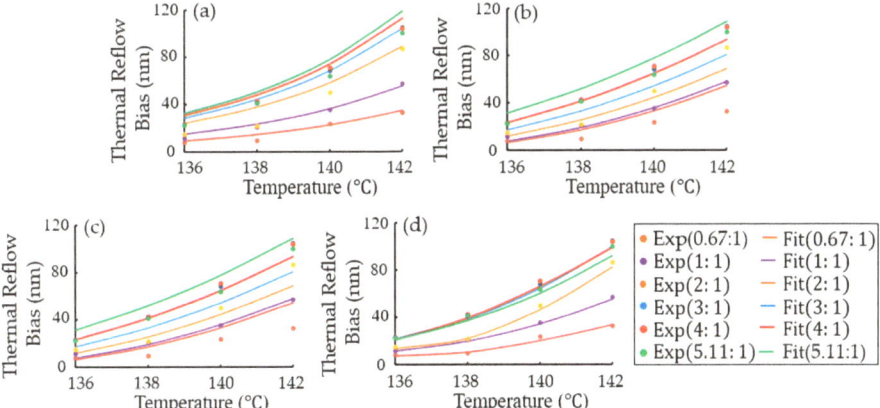

Figure 7. Experimental and simulation results for thermal reflow biases depending on temperatures and duty ratios in a 193-nm argon fluoride (ArF) chemically amplified resist (CAR): the results of (**a**) an orthogonal fitting function, linear regressions from (**b**) a Python program and (**c**) Scikit-Learn, and (**d**) a convolutional neural network (CNN). "Exp" and "Fit" represent the experimental results and simulation results, respectively.

Figure 8 shows the classifications of a logistic regression in deep learning using TensorFlow (version 2.17.0) and the decision surfaces of a decision tree using Scikit-Learn for binary and multiple classes. For binary classification, logistic regression with a sigmoid function in TensorFlow, after 100,000 epochs, resulted in a loss value of 0.4074 and a classified linear equation of $y = -(1/0.55) \times 1.22x - 79.41$ with one dense layer and three parameters, as shown in Figure 8a. For the multiple classification with logistic regression, after 250 epochs, the loss value was 1.0506, and the classified linear equations were $y = -(1/0.55) \cdot 1.22x - 79.41$ and $y = -(1/0.55) \times 1.21x - 81.4$ with two dense layers and 15 parameters, as shown in Figure 8b. For the decision surfaces of a decision tree using Scikit-Learn (version 1.3.2), the accuracies for the binary class with a max depth of two and the multiple class with a max depth of three were 0.875 and 0.7916, respectively, as shown in Figure 8b,d. Figure 8e shows a decision tree for multiple classes with a max depth of four.

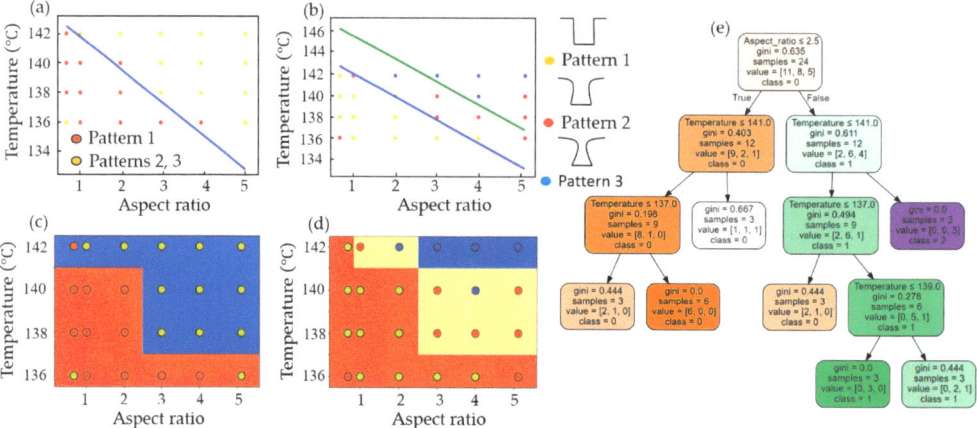

Figure 8. Classifications from a logistic regression in deep learning and a decision tree in machine learning: (**a**,**b**) are binary and multiple classifications from TensorFlow, respectively, (**c**,**d**) are the decision surfaces of a decision tree for the binary class with the max depth of two and the multiple class with a max depth of three, respectively, and (**e**) a decision tree for multiple classes with a max depth of four.

Figure 9 shows how to plot the decision surface for four support vector machine (SVM) classifiers with different kernels, which are a support vector classification (SVC) with a linear kernel, LinearSVC (linear kernel), a support vector classification (SVC) with a radial basis function (RBF) kernel, and a support vector classification (SVC) with a polynomial (degree 3) kernel. The linear models, such as support vector classification (SVC) (kernel = "linear") and LinearSVC(), yield insufficient decision boundaries for non-linear decision data, as shown in Figure 9a,b. The non-linear kernel models (Gaussian RBF and polynomial) provide more flexible non-linear decision boundaries, as shown in Figure 9c,d.

Figure 10 shows a comparison between the predicted pattern types from a convolutional neural network (CNN) and the experimental pattern types based on the side images of reflow resist. The network architecture comprised a dense layer with 25 units and a ReLU activation function, a second dense layer with 15 units and a ReLU activation function, and a third dense layer with 10 units and a linear activation function, totaling 1,638,975 parameters. A softmax function was used to calculate the probabilities. After 100 epochs, the loss value was 5.5164×10^{-5}, as shown in Figure 10. The simulated results agree well with the experimental results [48].

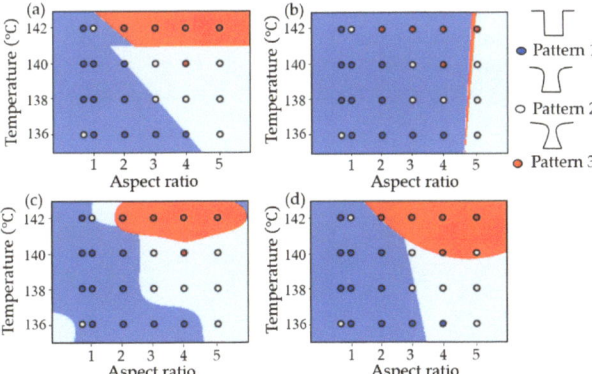

Figure 9. Multi-class classification using a support vector machine (SVM): (**a**) support vector classification (SVC) with a linear kernel, (**b**) LinearSVC (linear kernel), (**c**) support vector classification (SVC) with a radial basis function (RBF) kernel, and (**d**) support vector classification (SVC) with a polynomial (degree 3) kernel.

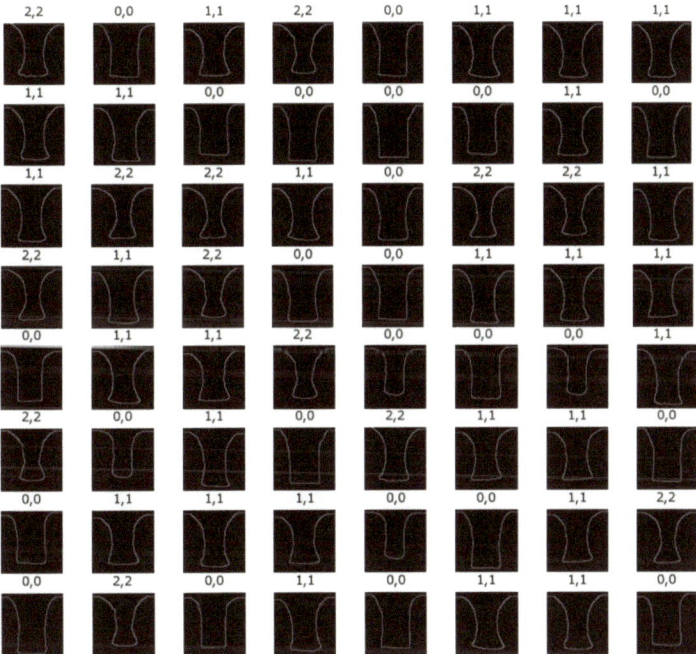

Figure 10. Classification of resist flow process (RFP) side images using a convolutional neural network (CNN). For the upper labels of the images, the first number represents the pattern type from experimental results, and the second number represents the predicted pattern type from a convolutional neural network (CNN).

4.4. Results of Self-Consistent Field Theory (SCFT)

Figure 11 illustrates the shrinking process of cylinder-forming block copolymers (BCPs). The PS-b-PMMA block copolymers (BCPs) are spin-coated into contact holes that have shrunk after thermal reflow. After thermal annealing and wet development in block copolymer (BCP) lithography processes [49], PMMA is removed from the contact hole.

The main benefits of this approach are smaller critical dimensions, improved contact hole uniformity, and reduced contact edge roughness [50]. Block copolymer (BCP) lithography can produce smaller contact hole (C/H) dimensions than other lithographic methods, as well as better contact hole uniformity and reduced contact edge roughness.

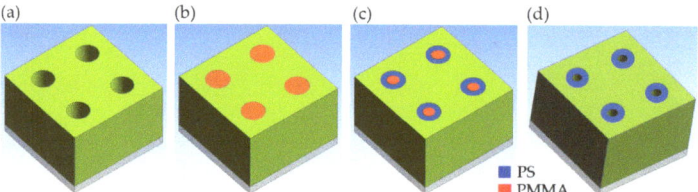

Figure 11. Schematic representation of the self-assembly of cylinder-forming block copolymers (BCPs) confined in cylindrical prepatterns: (**a**) contact hole (C/H) shrinkage patterns after thermal reflow (baking), (**b**) spin coating for PS-b-PMMA block copolymers (BCPs), (**c**) thermal annealing, and (**d**) wet development.

Figure 12 shows a comparison of the simulation results using rectangular guiding patterns $\left(\text{CD}_{\text{guiding}}\right)$ to the experimental results using cylindrical guiding patterns $\left(\text{CD}_{\text{guiding}}\right)$ from Reference 49 for the morphologies obtained after the directed self-assembly (DSA) of cylindrical PS-b-PMMA block copolymers (BCPs). The simulation results from self-consistent field theory (SCFT) were similar to the experimental results from Reference 49 for direct self-assembly (DSA) contact hole (C/H) shapes with the prepattern $\left(\text{CD}_{\text{guiding}}\right)$ and a BCP natural period (L_0). Simulation conditions included PMMA(7)PS(20) with $\chi N = 27$, PMMA(10)PS(27) with $\chi N = 35$, and PMMA(12) PS(34) with $\chi N = 46$. As one of simulation parameters, the numbers in brackets represent the lengths of the components. Specifically, the numbers of PMMA and PS correspond to their lengths as a proportion of the total number (natural period (L_0)) of the block copolymer (BCP)). These lengths are related to the molecular weights of the copolymers.

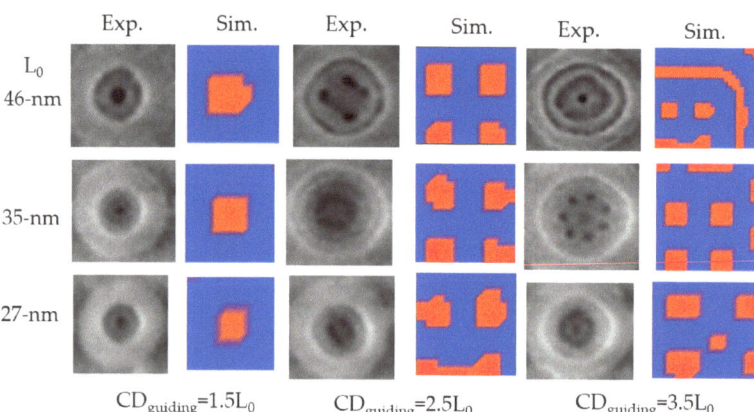

Figure 12. Comparison of the simulation results (Sim.) using rectangular guiding patterns ($\text{CD}_{\text{guiding}}$) to the experimental results (Exp.) using cylindrical guiding patterns ($\text{CD}_{\text{guiding}}$) from Reference [49] for the resulting morphologies obtained after the directed self-assembly (DSA) of cylindrical PS-b-PMMA block copolymers (BCPs). L_0 is a natural period of the block copolymer (BCP), which is directly correlated to the molecular weight of the block copolymer (BCP).

5. Conclusions

For contact hole shrinkage by resist reflow, the Surface Evolver method, which is a gradient descent technique, reduces the resist volume by decreasing the surface energy of the resist. A disadvantage of this method is the requirement to know the initial surface energy value. The finite element method (FEM) can be easily integrated with commercial tools such as ANSYS and SIEMENS. However, it necessitates the simultaneous implementation of a structural analysis, thermal analysis, and fluid mechanical analysis, for which the resist characteristics in each case must be known. Deep learning does not require knowledge of the resist characteristics. However, the error range should be below 5% for practical applications, necessitating more training and advanced algorithms. Although adding the block copolymer process increases the process complexity, using block copolymers after contact hole shrinkage by resist reflow can improve contact hole uniformity and reduced line edge roughness (LER). Thus, block copolymers can further reduce contact hole patterns. These models can aid in the shrinkage modeling of random contact holes and the fabrication of smaller contact holes.

Funding: This work was supported by the 2023 Hongik University Research Fund.

Data Availability Statement: The data, if required for reproduction, are available to obtain from the corresponding author.

Conflicts of Interest: The author declares no conflicts of interests.

References

1. Service, R.F. Optical lithography goes to extremes-and beyond. *Science* **2001**, *293*, 785–786. [CrossRef] [PubMed]
2. Randall, J.N.; Owen, J.H.G.; Lake, J.; Fuchs, E. Next generation of extreme-resolution electron beam lithography. *J. Vac. Sci. Technol. B* **2019**, *37*, 061605. [CrossRef]
3. Claveau, G.; Quemere, P.; Argoud, M.; Hazart, J.; Barros, P.P.; Sarrazin, A.; Posseme, N.; Tiron, R.; Chevalier, X.; Nicolet, C.; et al. Surface affinity role in graphoepitaxy of lamellar block copolymers. *J. Micro/Nanolithogr. MEMS MOEMS* **2016**, *15*, 031604. [CrossRef]
4. Cheng, J.Y.; Rettner, C.T.; Sanders, D.P.; Kim, H.-C.; Hinsberg, W.D. Dense self-assembly on sparse chemical patterns: Rectifying and multiplying lithographic patterns using block copolymers. *Adv. Mater.* **2008**, *20*, 3155–3158. [CrossRef]
5. Seisyan, R.P. Nanolithography in microelectronics: A review. *Tech. Phys.* **2011**, *56*, 1061–1073. [CrossRef]
6. Li, L.; Liu, X.; Pal, S.; Wang, S.; Ober, C.K.; Giannelis, E.P. Extreme ultraviolet resist materials for sub-7 nm patterning. *Chem. Soc. Rev.* **2017**, *46*, 4855–4866. [CrossRef] [PubMed]
7. Wu, B.; Kumar, A. Extreme ultraviolet lithography and three dimensional integrated circuit—A review. *Appl. Phys. Rev.* **2014**, *1*, 011104. [CrossRef]
8. Park, S.; Lee, D.H.; Xu, J.; Kim, B.; Hong, S.W.; Jeong, U.; Xu, T.; Russell, T.P. Macroscopic 10-terabit–per–square-inch arrays from block copolymers with lateral order. *Science* **2009**, *323*, 1030–1033. [CrossRef]
9. Pinto-Gomez, C.; Perez-Murano, F.; Bausells, J.; Villanueva, L.G.; Fernandez-Regulez, M. Directed self-assembly of block copolymers for the fabrication of functional devices. *Polymers* **2020**, *12*, 2432. [CrossRef]
10. Sekito, T.; Matsuura, Y.; Nagahara, T. Extension of 193 nm lithography by chemical shrink process. *J. Photopolym Sci. Technol.* **2016**, *29*, 761–764. [CrossRef]
11. Anhoj, T.A.; Jorgensen, A.M.; Zauner, D.A.; Hübner, J. The effect of soft bake temperature on the polymerization of SU-8 photoresist. *J. Micromech. Microeng.* **2006**, *16*, 1819–1824. [CrossRef]
12. Mülders, T.; Henke, W.; Elian, K.; Nölscher, C.; Sebald, M. New stochastic post-exposure bake simulation method. *J. Micro/Nanolithogr. MEMS MOEMS* **2005**, *4*, 043010.
13. Houle, F.A.; Hinsberg, W.D.; Sanchez, M.I. Kinetic model for positive tone resist dissolution and roughening. *Macromolecules* **2002**, *35*, 8591–8600. [CrossRef]
14. Kim, J.-S.; Jung, J.-C.; Lee, G.; Jung, M.-H.; Baik, K.-H. The extension of optical lithography to define contact holes required at advanced giga-bit-scale integration. *J. Photopolym. Sci. Technol.* **2000**, *13*, 471–476. [CrossRef]
15. Yang, X.; Gentile, H.; Eckert, A.; Brankovic, S.R. Electron-beam SAFIER™ process and its application for magnetic thin-film heads. *J. Vac. Sci. Technol. B* **2004**, *22*, 3339–3343. [CrossRef]
16. Terai, M.; Kumada, T.; Ishibashi, T.; Hanawa, T. Newly developed resolution enhancement lithography assisted by chemical shrink process and materials for next-generation devices. *Jpn. J. Appl. Phys.* **2006**, *45*, 5354–5358. [CrossRef]
17. Padmanaban, M.; Kudo, T.; Lin, G.; Hong, S.; Nishibe, T.; Takano, Y. Contact hole resist solutions for 45–90 nm node design rules. *J. Photopolym. Sci. Technol.* **2004**, *17*, 489–496. [CrossRef]

18. Chen, H.-L.; Ko, F.-H.; Li, L.-S.; Hsu, C.-K.; Chen, B.-C.; Chu, T.-C. Thermal flow and chemical shrink techniques for Sub-100 nm contact hole fabrication in electron beam lithography. *Jpn. J. Appl. Phys.* **2002**, *41*, 4163–4166. [CrossRef]
19. Thallikar, G.; Liao, H.; Cale, T.S.; Myers, F.R. Experimental and simulation studies of thermal flow of borophosphosilicate and phosphosilicate glasses. *J. Vac. Sci. Technol. B* **1995**, *13*, 1875–1878. [CrossRef]
20. Gong, S.; Shi, C.; Li, M. Flow performance and its effect on shape formation in PDMS assisted thermal reflow process. *Appl. Sci.* **2022**, *12*, 8282. [CrossRef]
21. Kirchner, R.; Schleunitz, A.; Schift, H. Energy-based thermal reflow simulation for 3D polymer shape prediction using Surface Evolver. *J. Micromech. Microeng.* **2014**, *24*, 055010. [CrossRef]
22. Brakke, K.A. Minimal surfaces, corners, and wires. *J. Geom. Anal.* **1992**, *2*, 11–36. [CrossRef]
23. Sidorov, F.; Rogozhin, A. Thermal reflow simulation for PMMA structures with nonuniform viscosity profile. *Polymers* **2023**, *15*, 3731. [CrossRef] [PubMed]
24. Alasfar, R.H.; Ahzi, S.; Barth, N.; Kochkodan, V.; Khraisheh, M.; Koç, M. A review on the modeling of the elastic modulus and yield stress of polymers and polymer nanocomposites: Effect of temperature, loading rate and porosity. *Polymers* **2022**, *14*, 360. [CrossRef]
25. Kim, S.-K. Transverse deflection for extreme ultraviolet pellicles. *Materials* **2023**, *16*, 3471. [CrossRef]
26. Ramu, A.T.; Bowers, J.E. A compact heat transfer model based on an enhanced Fourier law for analysis of frequency-domain thermoreflectance experiments. *Appl. Phys. Lett.* **2015**, *106*, 263102. [CrossRef]
27. Simoncelli, M.; Marzari, N.; Cepellotti, A. Generalization of Fourier's law into viscous heat equations. *Phys. Rev. X* **2020**, *10*, 011019. [CrossRef]
28. Foster, N.; Metaxas, D. Modeling water for computer animation. *Commun. ACM* **2000**, *43*, 60–67. [CrossRef]
29. Kasiman, E.H.; Kueh, A.B.H.; Yassin, A.Y.; Amin, N.S.; Amran, M.; Fediuk, R.; Kotov, E.V.; Murali, G. Mixed finite element formulation for Navier–Stokes equations for magnetic effects on biomagnetic fluid in a rectangular channel. *Materials* **2022**, *15*, 2865. [CrossRef]
30. Wang, S. Extensions to the Navier–Stokes equations. *Phys. Fluids* **2022**, *34*, 053106. [CrossRef]
31. Krzywinski, M.; Altman, N. Multiple linear regression. *Nat. Methods* **2015**, *12*, 1103–1104. [CrossRef] [PubMed]
32. Dreiseitl, S.; Ohno-Machado, L. Logistic regression and artificial neural network classification models: A methodology review. *J. Biomed. Inform.* **2002**, *35*, 352–359. [CrossRef]
33. Zhu, Q.; He, Z.; Zhang, T.; Cui, W. Improving classification performance of softmax loss function based on scalable batch-normalization. *Appl. Sci.* **2020**, *10*, 2950. [CrossRef]
34. Cardarilli, G.C.; Di Nunzio, L.; Fazzolari, R.; Giardino, D.; Nannarelli, A.; Re, M.; Spanò, S. A pseudo-softmax function for hardware-based high speed image classification. *Sci. Rep.* **2021**, *11*, 15307. [CrossRef] [PubMed]
35. Bansal, M.; Goyal, A.; Choudhary, A. A comparative analysis of k-nearest neighbor, genetic, support vector machine, decision tree, and long short term memory algorithms in machine learning. *Decis. Anal. J.* **2022**, *3*, 100071. [CrossRef]
36. Kushwah, J.S.; Kumar, A.; Patel, S.; Soni, R.; Gawande, A.; Gupta, S. Comparative study of regressor and classifier with decision tree using modern tools. *Mater. Today Proc.* **2022**, *56*, 3571–3576. [CrossRef]
37. Salcedo-Sanz, S.; Rojo-Álvarez, J.L.; Martínez-Ramón, M.; Camps-Valls, G. Support vector machines in engineering: An overview. *WIREs Data Min. Knowl. Discov.* **2014**, *4*, 234–267. [CrossRef]
38. Brereton, R.G.; Lloyd, G.R. Support vector machines for classification and regression. *Analyst* **2010**, *135*, 230–267. [CrossRef]
39. Drolet, F.; Fredrickson, G.H. Combinatorial screening of complex block copolymer assembly with self-consistent field theory. *Phys. Rev. Lett.* **1999**, *83*, 4317–4320. [CrossRef]
40. Ye, X.; Edwards, B.J.; Khomami, B. Elucidating the formation of block copolymer nanostructures on patterned surfaces: A self-consistent field theory study. *Macromolecules* **2010**, *43*, 9594–9597. [CrossRef]
41. Man, X.; Andelman, D.; Orland, H. Block copolymer at nano-patterned surfaces. *Macromolecules* **2010**, *43*, 7261–7268. [CrossRef]
42. Man, X.; Andelman, D.; Orland, H.; Thebault, P.; Liu, P.-H.; Guenoun, P.; Daillant, J.; Landis, S. Organization of block copolymers using nanoImprint lithography: Comparison of theory and experiments. *Macromolecules* **2011**, *44*, 2206–2211. [CrossRef]
43. Drolet, F.; Fredrickson, G.H. Optimizing chain bridging in complex block copolymers. *Macromolecules* **2001**, *34*, 5317–5324. [CrossRef]
44. Xie, O.; Olsen, B.D. A self-consistent field theory formalism for sequence-defined polymers. *Macromolecules* **2022**, *55*, 6516–6524. [CrossRef]
45. Ly, D.Q.; Honda, T.; Kawakatsu, T.; Zvelindovsky, A.V. Kinetic pathway of gyroid-to-cylinder transition in diblock copolymer melt under an electric field. *Macromolecules* **2007**, *40*, 2928–2935. [CrossRef]
46. Kim, J.-S.; Jung, J.C.; Kong, K.-K.; Lee, G.; Lee, S.-K.; Hwang, Y.-S.; Shin, K.-S. Contact hole patterning performance of ArF resist for 0.10 μm technology node. In Proceedings of the SPIE Advances in Resist Technology and Processing XIX, San Jose, CA, USA, 3–8 March 2002; Volume 4690, pp. 577–585.
47. Kim, S.-K. Theoretical study of extreme ultraviolet pellicles with nanometer thicknesses. *Solid State Electron.* **2024**, *216*, 108924. [CrossRef]
48. Chia, C.; Martis, J.; Jeffrey, S.S.; Howe, R.T. Neural network-based model of photoresist reflow. *J. Vac. Sci. Technol. B* **2019**, *37*, 061604. [CrossRef]

49. Gharbi, A.; Tiron, R.; Argoud, M.; Chevalier, X.; Belledent, J.; Pradelles, J.; Barros, P.P.; Navarro, C.; Nicolet, C.; Hadziioannu, G.; et al. Contact holes patterning by directed self-assembly of block copolymers: What would be the bossung plot? In Proceedings of the SPIE Alternative Lithographic Technologies VI, San Jose, CA, USA, 23–27 February 2014; Volume 9049, p. 90491N.
50. Yu, B.; Jin, Q.; Ding, D.; Li, B.; Shi, A.-C. Confinement-induced morphologies of cylinder-forming asymmetric diblock copolymers. *Macromolecules* **2008**, *41*, 4042–4054. [CrossRef]

Disclaimer/Publisher's Note: The statements, opinions and data contained in all publications are solely those of the individual author(s) and contributor(s) and not of MDPI and/or the editor(s). MDPI and/or the editor(s) disclaim responsibility for any injury to people or property resulting from any ideas, methods, instructions or products referred to in the content.

Communication

Off-Stoichiometry Thiol-Ene (OSTE) Micro Mushroom Forest: A Superhydrophobic Substrate

Haonan Li [1], Muyang Zhang [1], Yeqian Liu [2], Shangneng Yu [2], Xionghui Li [2], Zejingqiu Chen [3], Zitao Feng [2], Jie Zhou [1], Qinghao He [1], Xinyi Chen [2], Huiru Zhang [4], Jiaen Zhang [5], Xingwei Zhang [5,*] and Weijin Guo [2,*]

1 Department of Electrical Engineering, Shantou University, Shantou 515063, China
2 Department of Biomedical Engineering, Shantou University, Shantou 515063, China
3 Department of Biology, Shantou University, Shantou 515063, China
4 Guangdong University Research Findings Commercialization Center, Foshan 528000, China
5 Department of Mechanical Engineering, Shantou University, Shantou 515063, China
* Correspondence: zhangxw@stu.edu.cn (X.Z.); guoweijin@stu.edu.cn (W.G.)

Abstract: Superhydrophobic surfaces have been used in various fields of engineering due to their resistance to corrosion and fouling and their ability to control fluid movement. Traditionally, superhydrophobic surfaces are fabricated via chemical methods of changing the surface energy or mechanical methods of controlling the surface topology. Many of the conventional mechanical methods use a top-to-bottom scheme to control the surface topology. Here, we develop a novel fabrication method of superhydrophobic substrates using a bottom-to-top scheme via polymer OSTE, which is a prototyping polymer material developed for the fabrication of microchips due to its superior photocuring ability, mechanical properties, and surface modification ability. We fabricate a superhydrophobic substrate by OSTE–OSTE micro mushroom forest via a two-step lithography process. At first, we fabricate an OSTE pillar forest as the mushroom stems; then, we fabricate the mushroom heads via backside lithography with diffused UV light. Such topology and surface properties of OSTE render these structures superhydrophobic, with water droplets reaching a contact angle of 152.9 ± 0.2°, a sliding angle of 4.1°, and a contact angle hysteresis of less than 0.5°. These characteristics indicate the promising potential of this substrate for superhydrophobic applications.

Keywords: OSTE; superhydrophobic; backside lithography; micro mushroom; surface topology; diffused UV light

Citation: Li, H.; Zhang, M.; Liu, Y.; Yu, S.; Li, X.; Chen, Z.; Feng, Z.; Zhou, J.; He, Q.; Chen, X.; et al. Off-Stoichiometry Thiol-Ene (OSTE) Micro Mushroom Forest: A Superhydrophobic Substrate. *Micromachines* **2024**, *15*, 1088. https://doi.org/10.3390/mi15091088

Academic Editors: Yao Liu and Jinjie Zhou

Received: 8 August 2024
Revised: 25 August 2024
Accepted: 27 August 2024
Published: 28 August 2024

Copyright: © 2024 by the authors. Licensee MDPI, Basel, Switzerland. This article is an open access article distributed under the terms and conditions of the Creative Commons Attribution (CC BY) license (https://creativecommons.org/licenses/by/4.0/).

1. Introduction

Usually, when the contact angle of a water droplet on a surface is greater than 150 degrees in nature, we can call such surface a superhydrophobic surface [1]. There are many superhydrophobic surfaces in nature, such as the lotus leaf and leg of Aquarium paludum Fabricius. Superhydrophobic substrates have been applied in various fields of engineering, such as self-cleaning, anti-corrosion, oil or water separation, anti-icing, antifogging, and antifouling [2]. People have been working on various methods of fabricating superhydrophobic substrates [3,4]. In general, there are two ways to fabricate superhydrophobic substrates. One is via chemical methods for surface energy modification, of which the purpose is to decrease the surface energy of the substrate, such as superhydrophobic spray. Yu et al. developed a hydrophobic spray based on silica particles [5]. Xie et al. provided a superamphiphobic coating surface with bionic microstucture via FPU/PMMA mixture [6]. Asadollahi et al. used an atmospheric pressure plasma jet to develop an organosilicon-based superhydrophobic coating [7]. The fabrication of superhydrophobic coatings through electrodeposition on metals that are readily oxidizable is also a typical strategy [8]. Another method for the fabrication of superhydrophobic substrates is through manipulation of the surface topology using mechanical methods. For researchers in the field of microsystems, there is an interest in changing the surface topology to construct a superhydrophobic surface. In contrast to surface energy modification, the manipulation of surface topography involves the design of composite microstructures on the substrate, thereby altering the surface properties of the substrate, which can improve the stability and durability of the substrate at the same time. Etching has often been used to fabricate superhydrophobic surfaces by metal, such as aluminium and titanium [9–11]. The fabrication of superhydrophobic substrates by designing silica templates and utilizing soft lithography has also been demonstrated to be an effective solution [12]. In addition, Wang et al. created an integrated topological superhydrophobic structure on aluminium alloy via femtosecond laser ablation [13].

Among various surface topologies, there is a typical kind for superhydrophobic substrates: the mushroom-like microstructure. This structure exhibits re-entrant characteristics and generally shows favorable superhydrophobic properties. Researchers have used different methods for fabrication of these mushroom-like microstructures. Liu et al. fabricated a mushroom-like superhydrophobic microstructure via the etching of a silicon substrate [14]. Cumont et al. prepared nickel micro mushroom structures using UV-assisted nanoimprint lithography and tested their antimicrobial properties [15]. Nevertheless, the fabrication procedures of both silicon etching and nanoimprinting are complex. With the fast development of 3D printing technology, many researchers have employed 3D printing for the fabrication of mushroom-like superhydrophobic microstructures [16,17]. However, the precision and working principle of the 3D printer impose limitations on the surface properties of microstructures, including surface roughness, which in turn affects the surface properties of the substrate. In short, the fabrication of superhydrophobic substrates with micro mushroom structures is a highly active research area, but the existing techniques still have some limitations. The majority of these techniques, including etching and laser burnishing, employ a top-to-bottom fabrication process, which can increase the overall time required for fabrication. Additionally, the need for specialized instruments, the presence of cumbersome procedures, and the generation of non-negligible surface roughness, all cause inconveniences in the fabrication process and prevent further applications of these techniques.

In this work, we propose a novel bottom-to-top strategy for the fabrication of a superhydrophobic substrate using polymer OSTE–OSTE micro mushroom forest. OSTE is formed by thiol monomers and ene monomers with an off-stoichiometry mix ratio and cured via click chemistry under UV irradiation. It is a transparent photocurable polymer that exhibits favorable mechanical properties and has reactive surface groups upon curing [18]. OSTE is compatible with a multitude of traditional microfabrication techniques, including casting, reaction injection molding, photolithography, and micromachining. OSTE has been used in the fabrication of a wide variety of microfluidic chips [19–23]. Moreover, OSTE photolithography is highly adaptable and can be employed to fabricate intricate microstructures [24]. Recently, a fabrication method utilizing OSTE backside lithography with a light diffuser has been developed for the manufacturing of convex microstructures [25]. We plan to fabricate OSTE micro mushroom forest using photolithography of OSTE.

We fabricate OSTE micro mushroom forest via a two-step lithography. The initial step is to construct an OSTE pillar forest on a flat substrate. Subsequently, a light diffuser is employed to disperse the parallel UV light. OSTE pillars are utilized as optical waveguides to reflect the diffused UV light. Finally, OSTE is cured on top of the pillars to form the mushroom heads via backside lithography [25,26]. Contact angle and sliding angle tests are used to characterize the surface properties of the OSTE micro mushroom forest.

2. Materials and Methods

Chrome glass mask is from Jixian Opto-electronic (Shenzhen, China). Isopropyl alcohol (IPA) and gelatin are from Macklin (Shanghai, China). Propylene glycol monomethyl ether acetate (PGMEA) is from Aladdin (Shanghai, China). Black dye is from Fleur Couleur (Zhejiang, China). Diffusing glass is from Edmund Optics (75 mm Diameter Opal Diffusing Glass, Barrington, NJ, USA). UV lithography machine is from the Institute of Optics and

Electronics, Chinese Academy of Sciences (URE-2000/35AL, Chengdu, China). Polymer OSTE is prepared according to a previous work [18].

The fabrication of the OSTE micro mushroom forest can be divided into two main parts: OSTE pillar forest and OSTE mushroom heads. Figure 1 shows a 3D schematic diagram of the entire fabrication process. The first step is the fabrication of OSTE pillar forest, and Figure 2a shows a cross-section view of the fabrication process of OSTE pillar forest. CAD software (AutoCAD 2022, Autodesk, San Rafael, CA, USA) is employed to design the chrome glass mask, and the mask design is shown in Figure 2b. Subsequently, the mask is placed in contact with uncured OSTE, and a plastic spacer of specific thickness is added between the substrate and the mask to define the height of the pillars. After UV exposure, the chromium mask is removed and developed using PGMEA, followed by cleaning with IPA to obtain the pillar forest. Then, the pillar forest is subjected to an oxygen plasma treatment.

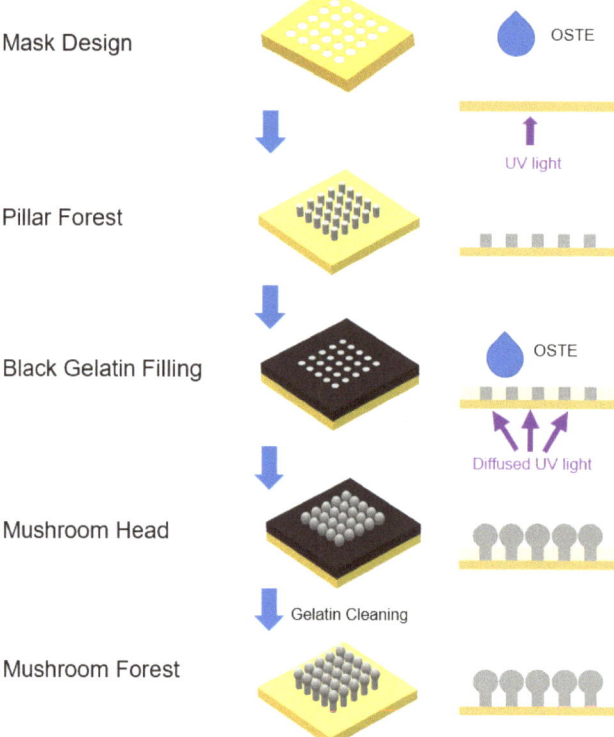

Figure 1. Schematic diagram of the fabrication of OSTE micro mushroom forest. At first, we fabricate the OSTE pillar forest on a flat substrate. Then, we use black gelatin solution to fill the space between pillars under capillary action and solidify gelatin. After that, diffused UV light is reflected through pillars and cures the OSTE contacting the pillar top, thereafter forming the OSTE micro mushroom forest. This diagram is not to scale.

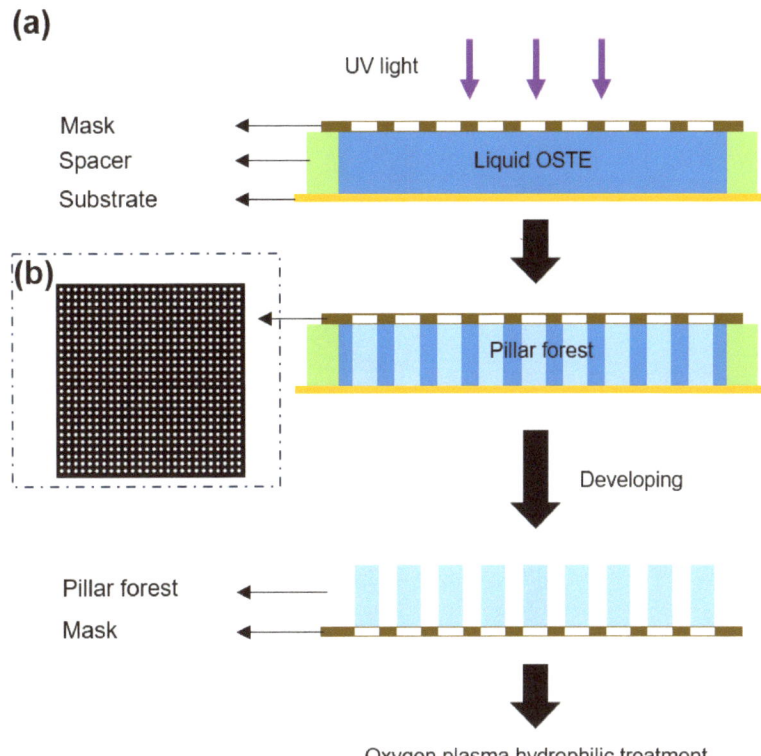

Figure 2. (**a**) The fabrication procedures of OSTE pillar forest in cross-section view. The pillars are in an orthogonal arrangement. (**b**) The design of the chrome glass mask.

The second step is the fabrication of mushroom heads, and a cross-section view is illustrated in Figure 3a. Gelatin solution (10% in DI water) with 1% black dye is prepared and dropped onto the pillar forest. The black gelatin solution is then drawn into the interstices of the pillars via capillary action. Subsequently, the gelatin is allowed to solidify at 4 °C. After that, the OSTE pillar forest with solidified gelatin is put in contact with uncured OSTE and then covered with a light diffuser. Subsequently, the pillars are subjected to UV light curing, development, and post-curing. As the light diffuser transforms the parallel UV light into diffused UV light, the diffused UV light is reflected in the pillars, thereby curing the OSTE on top of the pillars and forming spherical microstructures. The working principle of the UV light diffuser and the fabrication of the mushroom heads are shown in Figure 3b. Ultimately, the residual gelatin is removed by washing of water of 37 °C, resulting in the final OSTE micro mushroom forest.

The experimental images are captured by a stereomicroscope (Leica M205C, Wetzlar, Germany). Contact angle measurements are conducted via a contact angle measurement machine (Sindin SDC-200S, Dongguan, China).

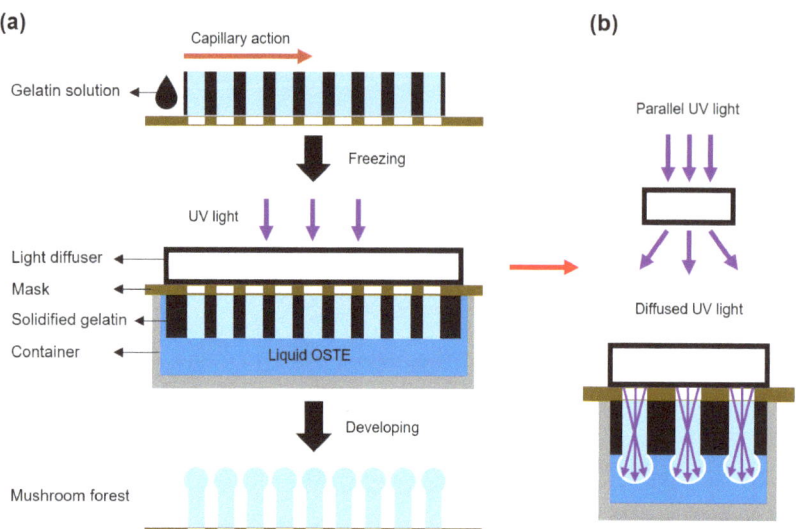

Figure 3. (**a**) The cross-section view of the fabrication process of the OSTE micro mushroom forest. (**b**) The working principle of the light diffuser and the fabrication of the mushroom head: diffused UV light is reflected in the pillars and emitted from the pillar top, curing a spherical microstructure.

3. Results and Discussion

We successfully fabricate OSTE micro mushroom forest of different dimensions with pillar diameters ranging from 100 μm to 400 μm. As shown in Figure 4, the diameters of the pillars are 300 μm (Figure 4a,b) and 400 μm (Figure 4c,d). It is clearly shown that there is a spherical head on the straight pillars.

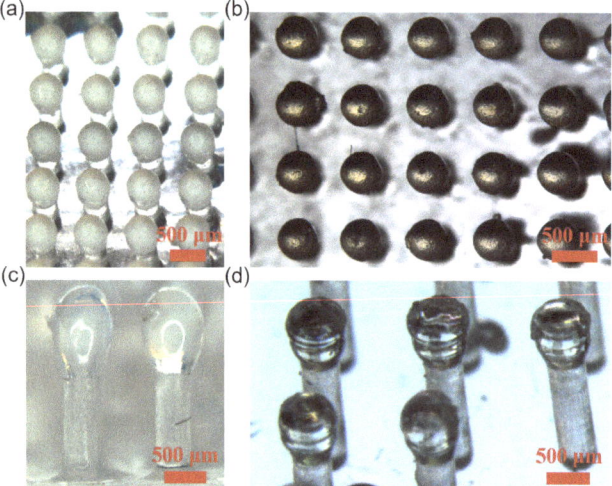

Figure 4. (**a**,**b**) OSTE micro mushroom forest with a pillar diameter of 300 μm and a distance of 400 μm. (**c**) OSTE micro mushroom forest with a pillar diameter of 400 μm and a distance of 600 μm. (**d**) OSTE micro mushroom forest with a pillar diameter of 400 μm and a distance of 1000 μm. The difference in brightness of the images is due to different imaging angle, illumination intensity, and exposure time.

The contact angle of water droplets on OSTE micro mushroom forest exceeds 150°, with a contact angle hysteresis of less than 0.5° and a sliding angle of less than 5°, as shown in Figures 5 and 6, indicating the superhydrophobicity of the substrate. Compared to a flat OSTE substrate, the enhancement of the micro mushroom forest in terms of hydrophobicity is significant, as shown in Figure 5. In addition, we conduct an adhesion test of water droplets on this substrate, and Figure 7 shows the low adhesion of OSTE micro mushroom forest to water droplets (also shown in Video S1). Furthermore, a series of stability tests have been conducted on OSTE micro mushroom forest, as illustrated in Figure 8. OSTE micro mushroom structure is shown to remain superhydrophobic with increasing storage time. These tests specifically check the alteration of the superhydrophobicity of OSTE micro mushroom forest over time after fabrication. In addition, the surface properties of polymer OSTE is quite stable in air environment, which could facilitate the storage of this substrate.

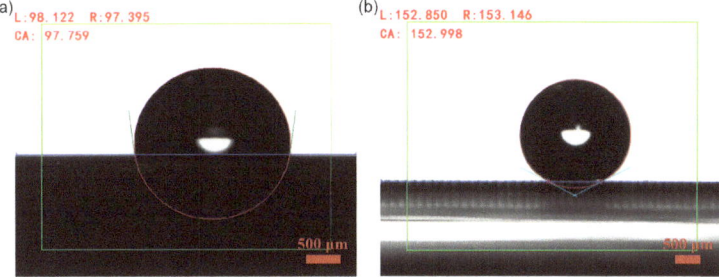

Figure 5. (**a**) The contact angle of a water droplet on a flat OSTE substrate. (**b**) The contact angle of a water droplet on the OSTE micro mushroom forest (with a pillar diameter of 100 μm and a distance of 100 μm), which is 152.9 ± 0.2°. The volume of the water droplet in both images is 5.0 μL.

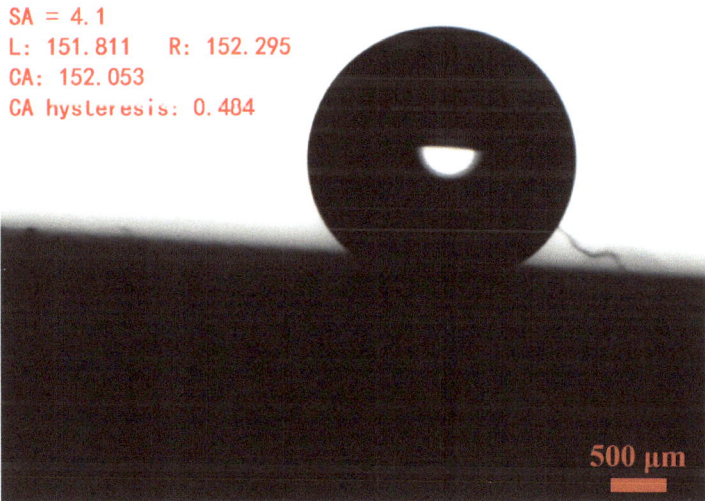

Figure 6. The sliding angle of a water droplet with a volume of 5.0 μL on OSTE micro mushroom forest (with a pillar diameter of 100 μm and a distance of 100 μm) is evaluated, with a sliding angle of 4.1° and a contact angle hysteresis of 0.484°.

This fabrication method has a high design flexibility in terms of the dimensions of the microstrucrures, including the diameter and height of the pillars. This two-step lithography method allows for the straightforward fabrication of a large-area OSTE micro mushroom forest in a short period of time. This fabrication method is compatible with a mask size of 5 inch × 5 inch, and the whole fabrication process takes less than 30 min. In theory, the area of OSTE micro mushroom forest that can be fabricated via this method is determined by the size of UV light source. Therefore, it is possible to fabricate a larger OSTE micro mushroom forest if the size of UV light source is big enough. In addition, the area of OSTE micro mushroom forest has an insignificant influence on the speed of fabrication. Moreover, since OSTE has adjustable mechanical properties, including the E-modulus [18], the robustness of OSTE micro mushroom forest can also be tuned for different application scenarios. We believe that this method has a significant potential in the field of microsystems, such as in the fabrication of substrates for liquid motion control and antibacterium [27].

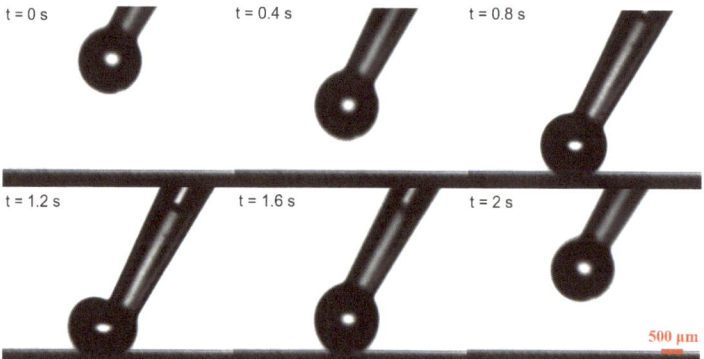

Figure 7. A series of images that show the low adhesion of water droplets on OSTE micro mushroom forest (with a pillar diameter of 100 μm and a distance of 100 μm). The volume of the water droplet is 5.0 μL.

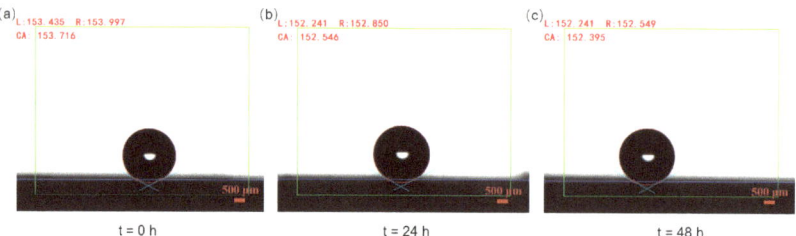

Figure 8. (a–c) The images illustrate the water droplet contact angle tests conducted on OSTE micro mushroom forest (with a pillar diameter of 100 μm and a distance of 50 μm) over a period of time after fabrication. The volume of the water droplet is 5.0 μL.

4. Conclusions and Outlook

A novel bottom-to-top strategy is developed for the fabrication of a superhydrophobic substrate by polymer OSTE–OSTE micro mushroom forest. This method is more efficient and straightforward than traditional microfabrication methods, and it allows for the fabrication of large-area superhydrophobic substrates in a relatively short period of time. Furthermore, the method is flexible since the dimensions of OSTE micro mushroom forest can be easily modified by adjusting the mask design. OSTE micro mushroom forest exhibits excellent superhydrophobic properties and low adhesion properties, with a contact angle of 152.9 ± 0.2°, a sliding angle of 4.1°, and a contact angle hysteresis of less than 0.5°. As to the future outlook, we believe that by scaling down the mask design and improving the fabrication process, we will eventually be able to fabricate micro mushroom structures with pillar diameters of tens of micrometers for further enhancement of superhydrophobic properties. It is also possible to fabricate tilted micro mushroom structures through this method, which can be used for directed droplet bouncing. Furthermore, we envision that this method should be compatible with certain other photocurable polymers. This approach has a significant potential for the research of superhydrophobic substrates in the near future.

Supplementary Materials: The following are available online at https://www.mdpi.com/article/10.3390/mi15091088/s1, Video S1: Adhesion test.

Author Contributions: H.L.: Investigation, Methodology, Data curation, Visualization, Writing—original draft; M.Z.: Investigation, Methodology, Data curation, Visualization; Y.L.: Investigation, Methodology; S.Y.: Investigation, Methodology; X.L.: Investigation, Methodology; Z.C.: Investigation, Methodology; Z.F.: Investigation, Methodology; J.Z. (Jie Zhou): Investigation, Methodology; Q.H.: Investigation, Methodology; X.C.: Investigation, Methodology; H.Z.: Investigation, Methodology, Supervision; J.Z. (Jiaen Zhang): Investigation, Methodology; X.Z.: Investigation, Methodology, Supervision; W.G.: Conceptualization, Funding acquisition, Investigation, Methodology,Writing, Supervision. All authors have read and agreed to the published version of the manuscript.

Funding: This research was funded by Shantou University (STU Scientific Research Foundation for Talents: NTF20034), the Bureau of Education of Foshan City (Innovative Research Project), the Department of Education of Guangdong Province (Scientific Research Project: 2022KQNCX019), the Guangdong Basic and Applied Basic Research Foundation (2022A1515110855), and the National Natural Science Foundation of China (62341403).

Data Availability Statement: The original contributions presented in the study are included in the article, further inquiries are available per request addressed to guoweijin@stu.edu.cn.

Conflicts of Interest: Weijin Guo, Haonan Li, Muyang Zhang, Shangneng Yu, and Zejingqiu Chen have filed a patent application related to this work (Application No. 202311478248.4) in China. The remaining authors declare that the research was conducted in the absence of any commercial or financial relationships that could be construed as a potential conflict of interest.

References

1. Li, X.M.; Reinhoudt, D.; Crego-Calama, M. What do we need for a superhydrophobic surface? A review on the recent progress in the preparation of superhydrophobic surfaces. *Chem. Soc. Rev.* **2007**, *36*, 1350–1368. [CrossRef] [PubMed]
2. Das, S.; Kumar, S.; Samal, S.K.; Mohanty, S.; Nayak, S.K. A review on superhydrophobic polymer nanocoatings: Recent development and applications. *Ind. Eng. Chem. Res.* **2018**, *57*, 2727–2745. [CrossRef]
3. Guo, Z.; Liu, W.; Su, B.L. Superhydrophobic surfaces: From natural to biomimetic to functional. *J. Colloid Interface Sci.* **2011**, *353*, 335–355. [CrossRef]
4. Johnson, R.E., Jr.; Dettre, R.H. Contact angle hysteresis. III. Study of an idealized heterogeneous surface. *J. Phys. Chem.* **1964**, *68*, 1744–1750. [CrossRef]
5. Yu, X.; Liu, X.; Shi, X.; Zhang, Z.; Wang, H.; Feng, L. SiO_2 nanoparticle-based superhydrophobic spray and multi-functional surfaces by a facile and scalable method. *Ceram. Int.* **2019**, *45*, 15741–15744. [CrossRef]
6. Xie, Q.; Xu, J.; Feng, L.; Jiang, L.; Tang, W.; Luo, X.; Han, C.C. Facile creation of a super-amphiphobic coating surface with bionic microstructure. *Adv. Mater.* **2004**, *16*, 302–305. [CrossRef]
7. Asadollahi, S.; Profili, J.; Farzaneh, M.; Stafford, L. Development of organosilicon-based superhydrophobic coatings through atmospheric pressure plasma polymerization of HMDSO in nitrogen plasma. *Materials* **2019**, *12*, 219. [CrossRef]
8. Yin, X.; Mu, P.; Wang, Q.; Li, J. Superhydrophobic ZIF-8-based dual-layer coating for enhanced corrosion protection of Mg alloy. *ACS Appl. Mater. Interfaces* **2020**, *12*, 35453–35463. [CrossRef]
9. Yang, H.; Dong, Y.; Li, X.; He, W.; Liu, Y.; Mu, X.; Zhao, Y.; Wang, X.; Yang, F.; Fu, W.; et al. Enhancing tribological performance of AA3003 aluminum alloy via adjusting surface wettability: Synergistic effects of chemical etching and modification. *Colloids Surf. A Physicochem. Eng. Asp.* **2024**, 134330. [CrossRef]
10. Lu, Y.; Song, J.; Liu, X.; Xu, W.; Xing, Y.; Wei, Z. Preparation of superoleophobic and superhydrophobic titanium surfaces via an environmentally friendly electrochemical etching method. *ACS Sustain. Chem. Eng.* **2013**, *1*, 102–109. [CrossRef]
11. Liao, R.; Zuo, Z.; Guo, C.; Yuan, Y.; Zhuang, A. Fabrication of superhydrophobic surface on aluminum by continuous chemical etching and its anti-icing property. *Appl. Surf. Sci.* **2014**, *317*, 701–709. [CrossRef]
12. Wu, Y.; Zeng, J.; Si, Y.; Chen, M.; Wu, L. Large-area preparation of robust and transparent superomniphobic polymer films. *ACS Nano* **2018**, *12*, 10338–10346. [CrossRef] [PubMed]

13. Wang, W.; Chen, Y.; Sun, X.; Zhang, Y.; Chen, T.; Mei, X. Demonstration of an enhanced "interconnect topology"-based superhydrophobic surface on 2024 aluminum alloy by femtosecond laser ablation and temperature-controlled aging treatment. *J. Phys. Chem. C* **2021**, *125*, 24196–24210. [CrossRef]
14. Liu, T.; Kim, C.J. Turning a surface superrepellent even to completely wetting liquids. *Science* **2014**, *346*, 1096–1100. [CrossRef] [PubMed]
15. Cumont, A.; Zhang, R.; Corscadden, L.; Pan, J.; Zheng, Y.; Ye, H. Characterisation and antibacterial investigation of a novel coating consisting of mushroom microstructures and HFCVD graphite. *Mater. Des.* **2020**, *189*, 108498. [CrossRef]
16. Yin, Q.; Guo, Q.; Wang, Z.; Chen, Y.; Duan, H.; Cheng, P. 3D-printed bioinspired Cassie–baxter wettability for controllable microdroplet manipulation. *ACS Appl. Mater. Interfaces* **2020**, *13*, 1979–1987. [CrossRef]
17. Liu, Y.; Zhang, H.; Wang, P.; He, Z.; Dong, G. 3D-printed bionic superhydrophobic surface with petal-like microstructures for droplet manipulation, oil-water separation, and drag reduction. *Mater. Des.* **2022**, *219*, 110765. [CrossRef]
18. Carlborg, C.F.; Haraldsson, T.; Öberg, K.; Malkoch, M.; Van Der Wijngaart, W. Beyond PDMS: Off-stoichiometry thiol–ene (OSTE) based soft lithography for rapid prototyping of microfluidic devices. *Lab Chip* **2011**, *11*, 3136–3147. [CrossRef]
19. Sandström, N.; Shafagh, R.Z.; Vastesson, A.; Carlborg, C.; van der Wijngaart, W.; Haraldsson, T. Reaction injection molding and direct covalent bonding of OSTE+ polymer microfluidic devices. *J. Micromech. Microeng.* **2015**, *25*, 075002. [CrossRef]
20. Guo, W.; Gustafsson, L.; Jansson, R.; Hedhammar, M.; van der Wijngaart, W. Formation of a thin-walled spider silk tube on a micromachined scaffold. In Proceedings of the 2018 IEEE Micro Electro Mechanical Systems (MEMS), Belfast, UK, 21–25 January 2018; pp. 83–85.
21. Zandi Shafagh, R.; Vastesson, A.; Guo, W.; Van Der Wijngaart, W.; Haraldsson, T. E-beam nanostructuring and direct click biofunctionalization of thiol–ene resist. *ACS Nano* **2018**, *12*, 9940–9946. [CrossRef]
22. Guo, W.; Hansson, J.; Gustafsson, L.; van der Wijngaart, W. "Bend-and-Bond" Polymer Microfluidic Origami. In Proceedings of the 2021 IEEE 34th International Conference on Micro Electro Mechanical Systems (MEMS), Virtual, 25–29 January 2021; pp. 222–225.
23. Xiao, Z.; Sun, L.; Yang, Y.; Feng, Z.; Dai, S.; Yang, H.; Zhang, X.; Sheu, C.L.; Guo, W. High-performance passive plasma separation on OSTE pillar forest. *Biosensors* **2021**, *11*, 355. [CrossRef] [PubMed]
24. Guo, W.; Hansson, J.; van der Wijngaart, W. Synthetic paper separates plasma from whole blood with low protein loss. *Anal. Chem.* **2020**, *92*, 6194–6199. [CrossRef] [PubMed]
25. Zhang, M.; Li, H.; Xiao, Z.; Feng, Z.; Yu, S.; Chen, Z.; Zhang, H.; Guo, W. Fabrication of Concave Microwells and Microchannels by Off-stoichiometry Thiol-ene (OSTE) Backside Lithography. In Proceedings of the 2023 IEEE 18th International Conference on Nano/Micro Engineered and Molecular Systems (NEMS), Jeju Island, Republic of Korea, 14–17 May 2023; pp. 104–107.
26. Lee, J.H.; Choi, W.S.; Lee, K.H.; Yoon, J.B. A simple and effective fabrication method for various 3D microstructures: Backside 3D diffuser lithography. *J. Micromech. Microeng.* **2008**, *18*, 125015. [CrossRef]
27. Yang, Y.; Zhang, Y.; Li, G.; Zhang, M.; Wang, X.; Song, Y.; Liu, S.; Cai, Y.; Wu, D.; Chu, J.; et al. Directional rebound of compound droplets on asymmetric self-grown tilted mushroom-like micropillars for anti-bacterial and anti-icing applications. *Chem. Eng. J.* **2023**, *472*, 144949. [CrossRef]

Disclaimer/Publisher's Note: The statements, opinions and data contained in all publications are solely those of the individual author(s) and contributor(s) and not of MDPI and/or the editor(s). MDPI and/or the editor(s) disclaim responsibility for any injury to people or property resulting from any ideas, methods, instructions or products referred to in the content.

Article

The Synthesis and Assembly Mechanism of Micro/Nano-Sized Polystyrene Spheres and Their Application in Subwavelength Structures

Yeeu-Chang Lee [1,*], Hsu-Kang Wu [1], Yu-Zhong Peng [1] and Wei-Chun Chen [2]

1 Department of Mechanical Engineering, Chung Yuan Christian University, Chung Li 32023, Taiwan; pudding1214@hotmail.com (H.-K.W.); al950808@gmail.com (Y.-Z.P.)
2 National Applied Research Laboratories, Taiwan Instrument Research Institute (TIRI), Hsinchu 30076, Taiwan; weichun@narlabs.org.tw
* Correspondence: yclee@cycu.edu.tw

Abstract: The following study involved the utilization of dispersion polymerization to synthesize micron/nano-sized polystyrene (PS) spheres, which were then deposited onto a silicon substrate using the floating assembly method to form a long-range monolayer. Subsequently, dry etching techniques were utilized to create subwavelength structures. The adjustment of the stabilizer polyvinylpyrrolidone (PVP), together with changes in the monomer concentration, yielded PS spheres ranging from 500 nm to 5.6 µm in diameter. These PS spheres were suspended in a mixture of alcohol and deionized water before being arranged using the floating assembly method. The resulting tightly packed particle arrangement is attributed to van der Waals forces, Coulomb electrostatic forces between the PS spheres, and surface tension effects. The interplay of these forces was analyzed to comprehend the resulting structure. Dry etching, utilizing the PS spheres as masks, enabled the exploration of the effects of etching parameters on the resultant structures. Unlike traditional dry etching methods controlling RF power and etching gases, in the present study, we focused on adjusting the oxygen flow rate to achieve cylindrical, conical, and parabolic etched structures.

Keywords: colloidal lithography; polystyrene (PS) spheres; floating assembly method; subwavelength structures

Citation: Lee, Y.-C.; Wu, H.-K.; Peng, Y.-Z.; Chen, W.-C. The Synthesis and Assembly Mechanism of Micro/Nano-Sized Polystyrene Spheres and Their Application in Subwavelength Structures. *Micromachines* 2024, 15, 841. https://doi.org/10.3390/mi15070841

Academic Editors: Yao Liu and Jinjie Zhou

Received: 30 May 2024
Revised: 27 June 2024
Accepted: 27 June 2024
Published: 28 June 2024

Copyright: © 2024 by the authors. Licensee MDPI, Basel, Switzerland. This article is an open access article distributed under the terms and conditions of the Creative Commons Attribution (CC BY) license (https://creativecommons.org/licenses/by/4.0/).

1. Introduction

Colloidal lithography is a highly promising technique with numerous advantages. It allows for the fabrication of structures ranging from tens of micrometers to several nanometers in size, enabling the creation of periodic micro- and nanostructures. While there are many methods available for the generation of micro- and nano-patterns, such as optical lithography, electron beam lithography, focused ion beam lithography, and nanoimprint lithography, colloidal lithography offers relative simplicity in terms of both the process involved and the equipment required. Moreover, it has already found fascinating applications in various fields. A combination of etching processes can be utilized in biomimetic anti-reflective moth-eye structures, thereby reducing the loss of incident light, and consequently increasing the efficiency of photonic devices. For example, Chan et al. [1] and Chou et al. [2], respectively, fabricated moth-eye structures on the surfaces of LEDs and solar cells to enhance light emission and power generation efficiency. Colloidal particles possess the characteristic of a large ordered surface area. The use of surface-enhanced Raman scattering (SERS) leads to an increase in the scattering cross-section of molecules adsorbed on or near metal nanostructures. This increase in cross-section results in the amplified intensity of the measured Raman signal. Due to the size range for colloidal particles, which can vary from nanometers to tens of micrometers, and their ability to achieve long-range ordered arrangements through templating techniques, even creating micro/nanocomposite 3D structures, they are utilized as low-cost templates for SERS [3–6]. Colloidal particles can also serve a large number of different carriers, and can find applications in biomedical technology. In their study, L. Liang et al. [7] considered polystyrene (PS) spheres for their specific surface area, good mechanical properties, and high oxygen permeability. They employed Pt(II) octaethylporphyrin (PtOEP) as the oxygen sensing agent, and achieved novel proportional oxygen-sensitive microspheres using a polymerizable methacrylate-derived rhodamine derivative (Rhod-MA). These microspheres were utilized for air pressure measurement, the in situ monitoring of cellular oxygen respiration, and the analysis of dissolved oxygen concentrations in several liquids, including beverages.

To accommodate the various application sizes of colloidal particles, different synthesis methods have been proposed to prepare micro- and nanoparticles of different diameters [8]. Among them, dispersion polymerization is the most commonly employed method. The oil droplet nucleation theory of dispersion polymerization shares many similarities with the homogeneous nucleation theory of emulsion polymerization. However, dispersion polymerization predominantly involves the use of organic solvents, which exhibit a better affinity with the polymer than water. Consequently, the oligomers precipitate out of the solvent with longer chain lengths compared to those precipitating in emulsion polymerization. This results in the microspheres obtained via dispersion polymerization being larger in size than those produced using emulsion polymerization.

To enable the synthesized particles to form stacked arrangements to facilitate subsequent processes, the arrangement of micro/nanospheres has become a crucial technique. J. Chen et al. [9] deposited droplets of spheres onto a substrate, utilizing centrifugal force generated by spin-coating to remove excess droplets, and continued spinning to evaporate the solvent from the central part, ultimately achieving sphere alignment. By adjusting the rotational speed, the spin-coating method can be used to arrange double-layer to multilayer arrangements; however, presenting a densely packed monolayer hexagonal lattice over a large area is more challenging. S. Jeong et al. [10] wound stainless steel wire around a stainless steel rod and then deposited pre-prepared spherical droplets onto the surface of a substrate. The researchers used stainless steel fixtures to roll the substrate, squeezing the microspheres into alignment through the gaps between the stainless steel wires. Microspheres arranged in this manner are prone to forming grain boundaries, stacking, and voids, due to the difficulty in controlling and managing the spacing between the stainless steel wires. E. Sirotkin et al. [11] altered the gas–liquid electrostatic environment to achieve self-assembly alignment. Firstly, hydrophilic microspheres were modified with carboxylic acid to impart a negative charge onto the surface of the microspheres. Sodium ions (Na^+) were then introduced into the water, utilizing the repulsive force between the modified water and microspheres' positive and negative charges, enabling the microspheres to remain at the gas–liquid interface. To achieve large-area and single-crystal alignment of microspheres, it is crucial to control the concentration of sodium ions in the water. The findings presented in the literature [11] indicate that increasing the repulsive force between water and microspheres can aid in the avoidance of grain boundary formation.

From the above examples derived from the literature, it is evident that the scope of applications for micro/nanospheres is extensive, as a complete process can be formed through their synthesis, self-assembly alignment, and subsequent application. Dispersed polymerization methods can be used to produce particles of various sizes. In the study presented herein, we explored the impact of synthesis parameters on particle size through the use of synthetic methods. Subsequently, we utilized the floating assembly method to arrange the synthesized nanoparticles into a densely packed monolayer hexagonal lattice. Conventional photolithography technology involves the use of a photoresist on a substrate to define an etching mask layer. However, due to optical limits during exposure, smaller linewidth patterns may suffer from diffraction, resulting in incomplete pattern definition and the resolution being affected. This, in turn, leads to changes in linewidth dimensions during subsequent development processes. In contrast, colloidal lithography is not constrained by optical limitations, allowing for precise pattern size definition. In the present study, we employed colloidal particles as etching masks and, in conjunction with different plasma etching and etching processes, were able to etch out subwavelength cone, as well as cylindrical and parabolic structures.

2. Experiment

2.1. Synthesis of Micro/Nano PS Spheres

Due to the inherent tendency of styrene monomers to gradually polymerize at room temperature, commercial products are typically supplemented with inhibitors (such as hydroquinone or tertiary-butylcatechol) as stabilizers to delay their polymerization, allowing for storage. To prevent the inhibitors from negatively impacting the experimental results, it is necessary to purify the styrene monomer before conducting experiments. The experiment performed in the present study involved the use of vacuum distillation, utilizing reduced pressure to lower the boiling point and achieve fractionation based on the differing boiling points of components. The boiling points of styrene and hydroquinone are 145 °C and 287 °C, respectively. Under a vacuum environment of 75 Torr, the boiling points of styrene and hydroquinone decrease to 72 °C and 190 °C, respectively. Therefore, the distillation system was operated under a vacuum of 75 Torr and heated to 75 °C to obtain purified styrene through distillation.

The purified styrene monomer was then used to prepare PS spheres via dispersion polymerization. Unlike methods requiring the prior addition of surfactants, dispersion polymerization necessitates achieving a homogeneous phase in the solution before initiating polymerization. The preparation procedure is outlined as follows:

(1) Preparation of the homogeneous solution: Purified styrene monomer, initiator (2,2′-Azobis(2-methylpropionitrile), AIBN), and stabilizer polyvinylpyrrolidone (PVP, K-16) are added to a four-neck flask containing 99.5% ethanol. The mixture is then subjected to ultrasonic agitation to ensure complete dissolution, resulting in a homogeneous solution containing tiny oil droplets, serving as nucleation sites for subsequent processes.

(2) Isothermal polymerization: A silicone oil bath is placed on a heating plate and maintained at 95 °C. The aforementioned four-neck flask is then placed into the oil bath, and the polymerization reaction proceeds for 12 h. Upon completion of polymerization, the solution is allowed to cool naturally for approximately two hours to room temperature.

(3) Particle centrifugation and precipitation: Centrifugal separation is conducted at 9000 rpm for 45 min, resulting in the precipitation of solids. The precipitate is then washed with a mixture of methanol and water to remove any unreacted monomers, initiators, and solvents. Finally, the precipitate is placed in a vacuum oven to allow for drying.

Changing the dosage of the stabilizer, monomer, and initiator will affect the final particle size and distribution. Table 1 presents the process parameters used in this experiment.

Table 1. The weight of the solvent (ethanol), monomer, initiator, and stabilizer used in dispersion polymerization.

Exp.	Ethanol (g)	Styrene (g)	AIBN (g)	PVP (g)
(a)	47.5	6	0.18	40
(b)	47.5	6	0.18	27
(c)	47.5	9	0.18	4
(d)	67.2	30	0.8	2

2.2. Arrangement of Close-Packed PS Spheres

In the following study, the floating assembly method was employed to arrange PS spheres, and automated laboratory equipment for particle alignment was developed. The deposition process is outlined below.

(1) Preparation of the PS Sphere Suspension

The aim of preparing the sphere suspension is to ensure the uniform dispersion of spheres in the liquid. Dry micro/nanoparticle powder is mixed with ethanol and deionized water. The mixture is then sonicated for 1 h using an ultrasonic oscillator to ensure the uniform dispersion of micro/nanospheres in the liquid.

(2) PS Sphere Arrangement

   A. Place the silicon wafer on the platform of the lifting module and pour deionized water into the water tank.
   B. Draw the prepared sphere suspension into a syringe and place the syringe on the syringe holder of the injection pump module.
   C. Set the injection and baffle movement speed, baffle and platform movement distance, and baffle retraction distance.
   D. Continuously inject the suspension at the liquid surface. Slowly move the baffle forward until the injection is complete and then move the baffle backward to tightly arrange the PS beads on the liquid surface.
   E. Move the lift module upward to detach the substrate from the liquid surface, completing the PS sphere arrangement.

The schematic diagram of the PS spheres arrangement process of D and E is shown in Figure 1. For further details, please refer to ref. [12].

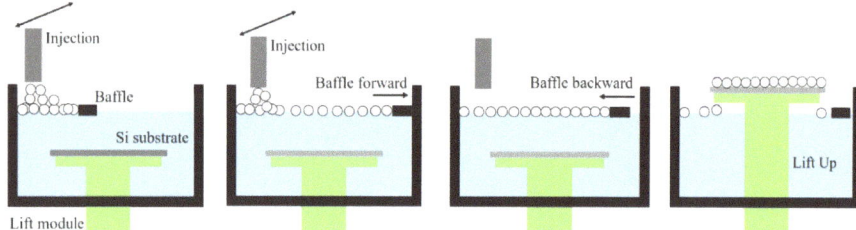

Figure 1. Arrangement of PS spheres using the floating assembly method.

2.3. Plasma Etching for Subwavelength Structure Fabrication

PS sphere deposition on a silicon substrate as etching masks was employed, coupled with a high-density plasma (HDP, Unaxis/Nextral 860L) etching system, to fabricate subwavelength structures. Figure 2 illustrates the proposed fabrication process.

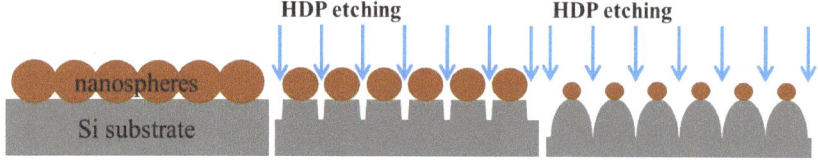

Figure 2. Schematic illustration of the HDP etching processes.

(1) In the dry etching system, the RF power affects the etching rate and surface roughness. In this experiment, 50 W and 100 W were used as the etching powers, respectively.
(2) The gases introduced into the system simultaneously are SF_6, Ar, and O_2. SF_6 is the primary etching gas for the silicon material in this process, with a fixed flow rate of

50 sccm. Ar serves to maintain plasma stability, with a fixed flow rate of 25 sccm. The O_2 flow rate plays a crucial role in the etching mechanism, as adding a small amount of O_2 can assist dissociation, increase free radical generation, and enhance the etching rate. However, excessive O_2 can reduce the etching rate due to dilution effects. Therefore, in the present study, O_2 flow rates of 15, 20, 35, 40, and 50 sccm were introduced to create various subwavelength structures.
(3) Clean with acetone, isopropanol, and deionized water sequentially for 3 min each, using an ultrasonic cleaner to remove any residual PS spheres.

3. Results and Discussion

3.1. Parameters Used in Dispersion Polymerization

Figure 3 shows PS spheres of different sizes produced using the parameters listed in Table 1. The results of the experiment demonstrate that PVP can stably control the particle size of the spheres. Naturally, as the size of the spheres increases, more monomer is required, and more initiator is needed to generate additional free radicals.

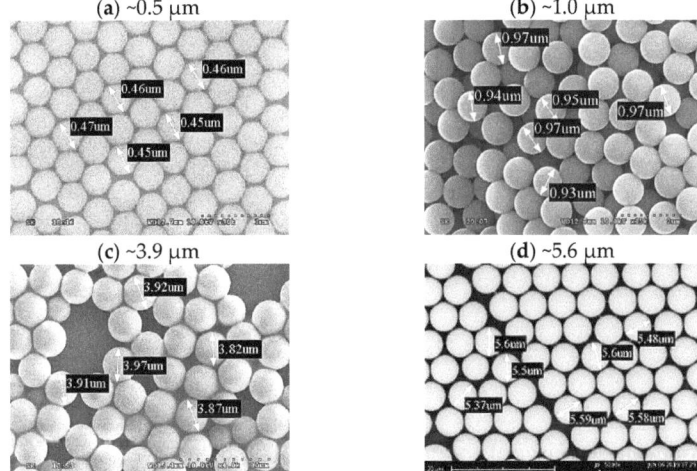

Figure 3. SEM images of PS spheres of different sizes prepared according to the perspective parameters listed in Table 1. For SEM inspection, the spheres are assembled on Si substrate at a temperature of 22 °C from a deionized water–methanol mixture, without attending long-range order.

Monomers, initiators, and stabilizers dissolve uniformly in a solvent to form a homogeneous solution. During polymerization, initiators decompose into primary radicals, which then polymerize with the monomers in the solution. When the polymer chain length exceeds a critical length, oligomeric radicals precipitate from the solvent to form nuclei. These nuclei formed by precipitation are unstable within the system, leading multiple nuclei to aggregate into stable growing microspheres. Stabilizers adsorb onto the surface of the growing microspheres, providing stability. The growing microspheres absorb monomers and radicals from the continuous phase, and undergo polymerization inside the microsphere. At this point, polymerization shifts from the continuous phase to inside the microsphere, continuing until completion [13,14]. Monomer concentration has a significant impact on the growth phase. A sufficient monomer concentration is required for the microspheres to continue growing. The double bonds of the styrene monomer react with the free radicals generated by the initiator, sustaining chain reaction polymerization to form polymers. If the monomer concentration is low, the concentration of the formed chain oligomers will also be low, making it difficult for the polystyrene spheres to absorb and grow, leading to a decrease in average particle size. However, if the monomer concentration is too high, the number of nuclei will be limited, and the microspheres' absorption rate will not keep pace with the monomer reaction rate. The excess monomers will react to form new nuclei, a phenomenon known as secondary nucleation. This will in turn result in an uneven particle size distribution and excess small spheres in the solution. Therefore, increasing the monomer concentration within an appropriate range can enhance the sphere diameter. The initiator concentration also influences the particle size distribution. When the initiator concentration is too low, the number of primary free radicals decomposed in the system decreases, reducing the rate at which polymer chains precipitate. This process significantly prolongs the nucleation period, resulting in a broader final particle size distribution. Conversely, when the initiator concentration is too high, the growing microspheres are larger, and the rate of formation of oligomer free radicals and inactive polymer chains (i.e., polymer chains that have reached the critical chain length) increases. Consequently, the efficiency of the growing microspheres in capturing these chains decreases. Since the capture efficiency is lower than the formation rate, secondary nucleation may occur within the system, leading to a broader particle size distribution.

In the present study, the stabilizer concentration was primarily controlled to alter the sphere diameter. The concentration of the stabilizer affects the reaction rate of the microspheres and the critical chain length at precipitation, leading to variations in size during the nucleation stage. Since the nucleation and growth stages occur in the continuous phase, the particles are in an unstable state and easily bond with other molecules. Stabilizer molecules, having both hydrophilic and hydrophobic groups, can react with free radicals to form copolymers that adsorb onto the surface of the microspheres. This process forms a protective layer that stabilizes the polymer particles, preventing further reactions or agglomeration. An appropriate amount of stabilizer helps in shaping the spheres and maintaining uniformity. However, adding too much stabilizer can lead to rapid adsorption, hindering the collision and reaction of monomers with the spheres, resulting in a smaller overall particle size and many small spheres that stabilize before reaching full growth. Table 1 shows that by keeping the contents of ethanol, styrene, and AIBN constant, and only reducing the weight of the stabilizer PVP from 40 g to 27 g, the size of the PS spheres increases from 0.5 μm to 1.0 μm. To further obtain PS spheres with a larger diameter, the PVP content was reduced to only 4 g. However, it was also necessary to increase the styrene content to 9 g to provide sufficient monomer, resulting in PS spheres with a diameter close to 4 μm. The use of a lower amount of stabilizer extends the growth phase of the spheres, allowing for continuous monomer adsorption and gradual stabilization. Nonetheless, this prolonged stabilization period can produce many new small spheres, and an insufficient amount of stabilizer can lead to an unstable polymerization system. Some microspheres may struggle to adsorb stabilizer molecules, significantly affecting uniformity. In the last experiment in Table 1, the PVP content was reduced to just 2 g, while the contents of ethanol, styrene, and AIBN were increased. Although the size of the PS spheres increased, the uniformity of the sphere diameters decreased.

3.2. Arrangement of PS Spheres

Using the floating assembly method, hydrophobic micro- and nanospheres were injected onto the surface of deionized water and then deposited. The surface tension of deionized water affects the flotation of the spheres on the water surface. Additionally, the composition of the suspension also influences whether the injected spheres stack or disperse on the water surface.

3.2.1. Effect of Surface Tension on the Arrangement of PS Spheres

For polystyrene microspheres, there exists a hydrophobic characteristic between them and water. The vertical component of the surface tension force (ST) balances with gravity (W) and buoyancy (B), while the horizontal component balances in the horizontal direction, as shown in Figure 4a. When hydrophobic microspheres approach one another, the convex water surface between the two microspheres tends to level out due to surface tension. This causes the surface tension to become horizontal, disrupting the balance of the horizontal components. The resultant horizontal force on microsphere A points to the right, and the resultant horizontal force on microsphere B points to the left, ultimately causing microspheres A and B to move closer to each other due to the direction of the resultant forces, as shown in Figure 4b.

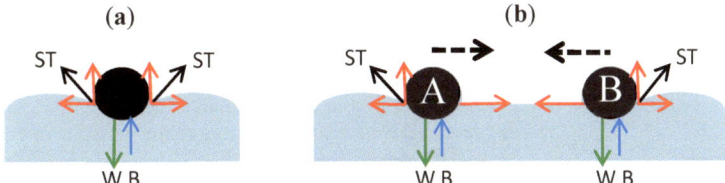

Figure 4. (**a**) Schematic diagram of one particle with vertical and horizontal components of surface supporting force in balance; (**b**) schematic diagram of particles A and B approaching each other due to imbalance in horizontal components of surface supporting force.

Factors affecting surface tension include pressure, liquid purity, and temperature [15]. Surface tension decreases with increasing temperature. As the temperature of the liquid rises, the momentum of the liquid molecules increases, reducing the mutual attractive forces between them. This process then leads to a decrease in the attractive forces pulling the liquid molecules within the surface layer toward the interior of the liquid, ultimately resulting in a reduction in surface tension. We achieved PS sphere arrangements using deionized water at two temperatures, 20 °C and 35 °C. At the lower temperature, the higher surface tension allowed the spheres to effectively float on the water surface, resulting in a closed arrangement. At the higher temperature, some spheres partially sunk into the water, while others floated on the surface but in a stacked form. This then led to the deposition process failing to achieve a monolayer hexagonal lattice arrangement. The schematic diagram and SEM images are shown in Figure 5.

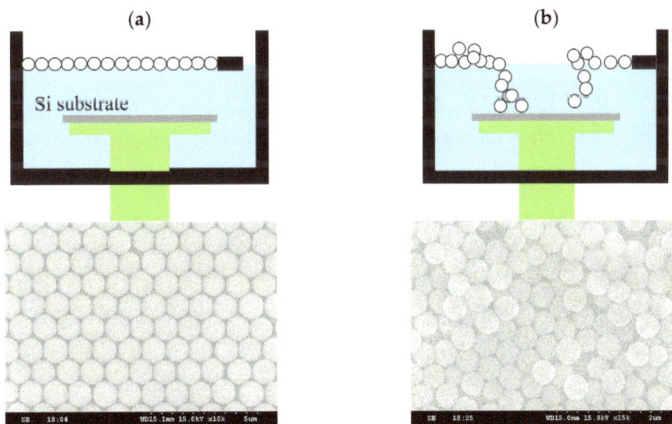

Figure 5. Results of PS sphere arrangement in deionized water at different temperatures, as follows: (**a**) 20 °C and (**b**) 35 °C.

3.2.2. Modification of Repulsive and Attractive Forces

The hydrophobic PS spheres synthesized in the experiment, described herein as colloidal particles, do not carry surface charges. Nevertheless, the sphere suspension consists of PS spheres, deionized water, and either methanol or ethanol. In our experiment, we fixed the ratio of PS spheres to deionized water and adjusted the amounts of methanol and ethanol. Figure 6 shows the results of spreading the spheres with a ratio of methanol

(ethanol)–PS spheres–deionized water = 1:2:2. From the results, it can be observed that using methanol causes the injected microspheres to be more dispersed on the liquid surface, resulting in the deposited PS spheres being unable to densely arrange, and appearing irregularly dispersed, as shown in Figure 6a. Conversely, using ethanol causes the injected microspheres to aggregate and stack more closely on the liquid surface, as shown in Figure 6b. The surface tensions of methanol/air and ethanol/air at 20 °C are 22.50 mN/m and 22.39 mN/m, respectively, showing a negligible difference. However, Figure 6 illustrates significant differences in suspensions of methanol and ethanol mixed with PS powder and deionized water at the same proportions. This result may be explained by methanol having higher polarity than ethanol, resulting in stronger Coulomb repulsive forces generated by the surface charge of the particles in the methanol.

Figure 6. Arrangement of PS spheres on Si from a suspension in deionized water and (**a**) methanol, (**b**) ethanol, obtained at 22 °C.

As the distance between two particles decreases, they experience van der Waals forces, leading to mutual attraction and the formation of a closely packed arrangement. This attraction is inversely proportional to the distance between particles, and is known as the induction dipole–induced dipole interaction within van der Waals forces, also referred to as London dispersion forces. Therefore, colloidal particles may be mutually attracted, leading to a closely packed arrangement. During the arrangement process of PS particles, appropriate baffles backward, applying external force to bring the microspheres closer together to facilitate self-assembly arrangement. The agitation from external forces causes water inside the container to ripple. Using the water's motion, the distance between the PS spheres around the void areas is reduced, allowing hydrophobic spheres to approach each other and be attracted, resulting in arrangement formation.

Due to the tendency of methanol solution to induce the dispersion of PS spheres, ethanol solution was adopted in the experiment described herein, with adjustments made to the ethanol ratio. Figure 7 shows the arrangement results for ethanol–PS spheres–deionized water ratios of 1:2:2, 2:2:2, and 3:2:2. As the ethanol concentration increases, the dispersion between PS spheres improves. When the ratio is 2:2:2, a hexagonal, closely packed arrangement is achieved.

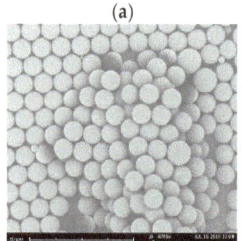 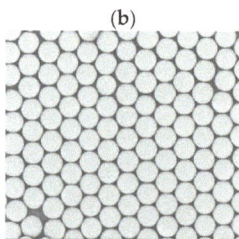 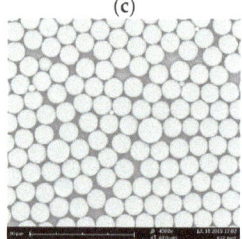

Figure 7. Arrangement of PS spheres at different ratios of ethanol to PS spheres to deionized water. (**a**) 1:2:2; (**b**) 2:2:2; (**c**) 3:2:2.

3.3. Plasma Etching for Subwavelength Structure

HDP etching, characterized by its high dissociative ion density, offers faster etching rates, better etching selectivity, and superior sidewall profile control [16]. The aim of the present study was to fabricate subwavelength structures while simultaneously considering both vertical and lateral etching capabilities. Using 0.5 µm PS spheres as etching masks, etching capability was enhanced by increasing the RF power, and the morphology of subwavelength structures was adjusted by varying the O_2 flow rate within the vacuum chamber.

RF power levels of 50 W and 100 W were employed for etching durations of 2 min and 5 min, respectively. The experimental results are shown in Table 2. Under the conditions of RF power of 50 W for 2 min, etching occurs through the gaps between PS spheres, resulting in flat-topped conical structures with edges. Similar cone-shaped structural profiles were obtained for both parameter sets, as follows: RF power of 100 W for 2 min of etching and RF power of 50 W for 5 min of etching. The etching rate with an RF power of 100 W was faster compared to that achieved with an RF power of 50 W, and stronger oxygen plasma etching capabilities were also exhibited to remove PS spheres. Prolonging the etching time may lead to excessive etching, resulting in the structures not retaining the desired geometry and height.

Table 2. SEM images of etching with different RF powers and etching times, with the same gas flow rates of SF6: 50 sccm, Ar: 25 sccm, O_2: 5 sccm, and chamber pressure: 25 mTorr.

E/T (min)	RF Power (W)	
	50	100
2		
5		

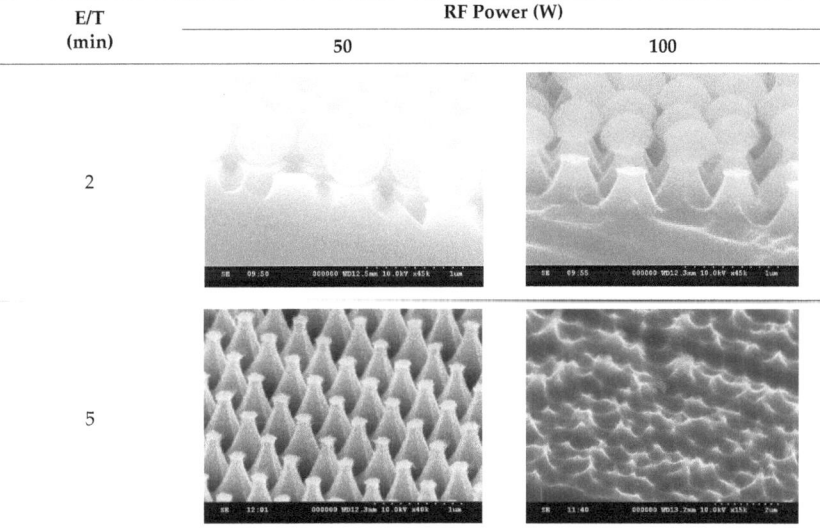

In the HDP system, the gases introduced were SF_6, Ar, and O_2. SF_6 is the primary gas used for etching silicon material in this process. Ar maintains plasma stability, while O_2 plays a crucial role in the etching mechanism. In a plasma environment, SF_6 generates fluorine radicals to etch silicon and form SiF_4. Subsequently, SiF_4 reacts with oxygen radicals to form SiOF, creating a passivation of the silicon surface, and thereby causing anisotropic etching [17,18]. To observe the above trend, in our experiment, an RF power of 50 W was selected to compare different O_2 flow rates and observe their effects on etching rate and structural morphology, respectively. Different O_2 flow rates of 5, 15, and 25 sccm were introduced, and the etching process was conducted for 5 min. Figure 8 shows the SEM images obtained after adjusting the O_2 flow rate individually. Observing Figure 8a,b, the vertical etching rates are similar. However, due to the lower oxygen content in Figure 8a, the lateral etching capability of Si is smaller compared to Figure 8b, resulting in a faster lateral etching rate. Furthermore, a smaller O_2 gas flow results in isotropic etching of Si. When the oxygen flow rate increases again in Figure 8c, the lateral etching capability becomes worse, and the vertical etching capability is also affected. With RF power = 50 W, the oxygen cannot be sufficiently dissociated when the oxygen flow increases from 15 sccm to 25 sccm, so the PS sphere itself remains more intact. Under the same etching conditions, the flat-top area of the structure in Figure 8c is larger.

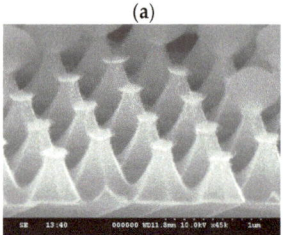 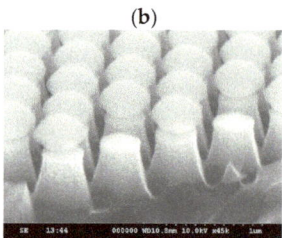 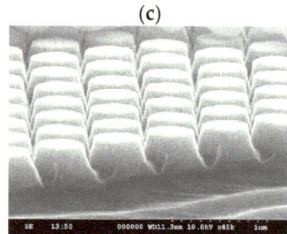

Figure 8. SEM images of etching for 5 min with fixed RF power at 50 W and chamber pressure at 25 mTorr; the gas flow rates of SF_6 and Ar were kept constant at 50 sccm and 25 sccm, with different O_2 flow rates of (**a**) 5 sccm, (**b**) 15 sccm, and (**c**) 25 sccm.

With the same etching duration and an O_2 flow rate of 5 sccm, it was inferred that there was a higher etching rate on the substrate, indicating stronger lateral etching capability. Analysis of Figure 9 reveals that, due to the gaps between the PS spheres (highlighted in yellow), plasma etching can freely penetrate between the PS spheres, resulting in the formation of conical structures with six-sided edges. As the O_2 flow rate increases, the etching gas concentration decreases, leading to a reduction in the silicon etching rate. Consequently, the lateral etching capability decreases, resulting in the structural profile approaching a columnar shape. Additionally, the results of previous studies [17,18] have indicated that oxygen plasma can effectively reduce the diameter of PS nanospheres. This process increases the gaps between adjacent PS spheres, allowing for uniform penetration of the plasma etching. As a result, the sidewalls of the structure do not develop edges.

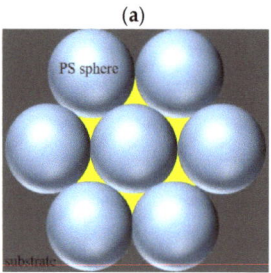 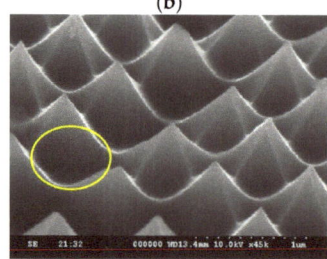

Figure 9. (**a**) Schematic diagram of the arrangement gaps between PS spheres and the (**b**) SEM image of the etching structure. The yellow area on the left shows the gaps between the tightly packed PS spheres, which are the regions prone to etching. The etching produces the indentations circled in yellow on the right.

Based on the experiments described above, it can be observed that increasing the RF power enhances silicon etching capability; in comparison, increasing the O_2 flow rate reduces silicon etching capability, but enables further etching of the PS spheres, resulting in structures without edges. In light of the above, we adjusted the RF power back to the value of 100 W and introduced O_2 flow rates of 15, 20, 35, 40, and 50 sccm for 5 min of etching, while keeping the other gas parameters unchanged. The etching results are shown in Figure 10. At an RF power of 100 W, the ability of the etching gas to dissociate into radicals is excellent; however, O_2 at 15 sccm is insufficient to form a complete passivation, resulting in strong etching of the silicon, forming a conical structure with edges, as shown in Figure 10a. When the O_2 flow rate is increased to 25 sccm, the SiOF passivation, as previously mentioned, again reduces the sidewall etching ability of the silicon, resulting in a steeper profile, as shown in Figure 10b. As the O_2 flow rate continues to increase, it will significantly erode the PS spheres due to the oxygen plasma, causing the protective ability of the etching mask to disappear. Consequently, the height of the etched parabolic structures is significantly reduced, as shown in Figure 10d,e.

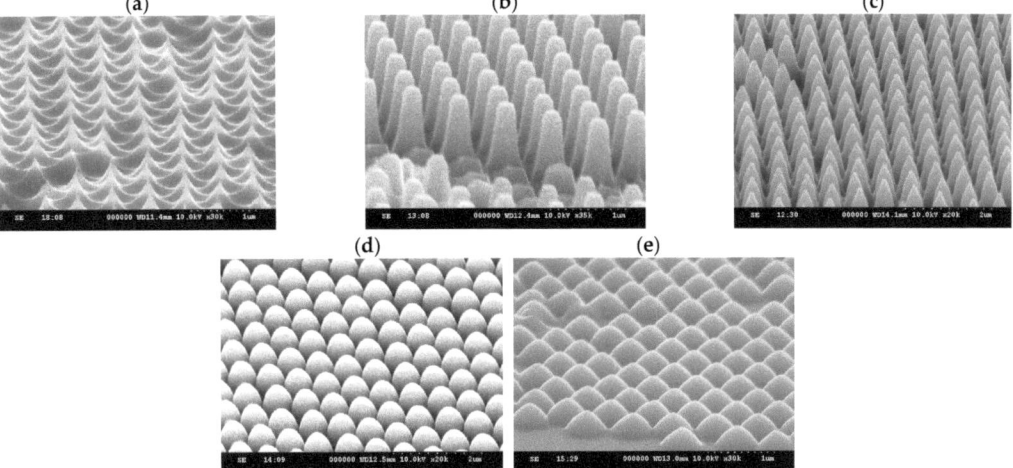

Figure 10. SEM images of etching for 5 min with fixed RF power at 100 W and chamber pressure at 25 mTorr; the gas flow rates of SF_6 and Ar were kept constant at 50 sccm and 25 sccm, with different O_2 flow rates of (**a**) 15, (**b**) 25, (**c**) 35, (**d**) 40, and (**e**) 50 sccm.

4. Conclusions

The following study primarily involved investigating the synthesis of polystyrene micro/nanospheres, their close packing arrangement, and the fabrication of subwavelength cone structures through plasma etching. The conclusions derived are as follows:

(1) Preparation of submicron- to micron-sized polystyrene microspheres through dispersion polymerization: The appropriate amount of stabilizer can aid in the formation of PS spheres and maintain a certain level of uniformity. Adding an excessive amount of stabilizer can lead to overly rapid adsorption, making it difficult for monomers to react with the microspheres upon collision, resulting in an overall decrease in particle size.

(2) Using the floating assembly method to arrange PS spheres: Surface tension effectively enables the microspheres to form a monolayer arrangement on the liquid surface, while van der Waals forces and Coulomb repulsion maintain a static equilibrium, allowing the particles to maintain a fixed distance of arrangement. Disturbing the floating system with external forces can effectively promote the tight arrangement of PS spheres.

(3) The fabrication of subwavelength cone structures can be achieved using HDP etching, where the control of oxygen flow significantly impacts the morphology of the subwavelength structures.

Author Contributions: Conceptualization, Y.-C.L.; methodology, H.-K.W. and Y.-Z.P.; validation, Y.-C.L. and W.-C.C.; resources, Y.-C.L. and W.-C.C.; writing—original draft preparation, Y.-C.L.; writing—review and editing, W.-C.C.; project administration, Y.-C.L. All authors have read and agreed to the published version of the manuscript.

Funding: The authors acknowledge the funding support provided by the National Science and Technology Council of Taiwan under grant no. MOST 111-2221-E-033-038-MY2.

Data Availability Statement: Data are contained within the article.

Conflicts of Interest: The authors declare no conflict of interest.

References

1. Chan, L.W.; Morse, D.E.; Gordon, M.J. Moth eye-inspired anti-reflective surfaces for improved IR optical systems & visible LEDs fabricated with colloidal lithography and etching. *Bioinspir. Biomim.* **2018**, *13*, 041001.
2. Chou, Y.Y.; Chang, C.C.; Lee, Y.C. Fabrication of sub-wavelength antireflective structure to enhance the efficiency of InGaAs solar cells. *Int. J. Autom. Smart Technol.* **2014**, *4*, 174–178.
3. Petruš, O.; Oriňak, A.; Oriňaková, R.; Králová, Z.O.; Múdra, E.; Kupková, M.; Kovaľ, K. Colloidal lithography with electrochemical nickel deposition as a unique method for improved silver decorated nanocavities in SERS applications. *Appl. Surf. Sci.* **2017**, *423*, 322–330. [CrossRef]
4. Mikac, L.; Ivanda, M.; Gotić, M.; Janicki, V.; Zorc, H.; Janči, T.; Vidaček, S. Surface-enhanced Raman spectroscopy substrate based on Ag-coated self-assembled polystyrene spheres. *J. Mol. Struct.* **2017**, *1146*, 530–535. [CrossRef]
5. Zhu, C.; Liu, D.; Yan, M.; Xu, G.; Zhai, H.; Luo, J.; Wang, G.; Jiang, D.; Yuan, Y. Three-dimensional surface-enhanced Raman scattering substrates constructed by integrating template-assisted electrodeposition and post-growth of silver nanoparticles. *J. Colloid Interface Sci.* **2022**, *608*, 2111–2119. [CrossRef] [PubMed]
6. Bi, J.; Wang, R. Electrodeposited superhydrophobic hierarchical structures as sensitive surface enhanced Raman scattering substrates. *Mater. Lett.* **2020**, *271*, 127738. [CrossRef]
7. Liang, L.; Li, G.; Mei, Z.; Shi, J.; Mao, Y.; Pan, T.; Liao, C.; Zhang, J.; Tian, Y. Preparation and application of ratiometric polystyrene-based microspheres as oxygen sensors. *Anal. Chim. Acta* **2018**, *1030*, 194–201. [CrossRef] [PubMed]
8. Zhanwang, X. Review on preparation of polystyrene microspheres. *Appl. Comput. Eng.* **2023**, *3*, 45–49. [CrossRef]
9. Chen, J.; Dong, P.; Di, D.; Wang, C.; Wang, H.; Wang, J.; Wu, X. Controllable fabrication of 2D colloidal-crystal films with polystyrene nanospheres of various diameters by spin-coating. *Appl. Surf. Sci.* **2013**, *270*, 6–15. [CrossRef]
10. Jeong, S.; Hu, L.; Lee, H.R.; Garnett, E.; Choi, J.W.; Cui, Y. Fast and scalable printing of large area monolayer nanoparticles for nanotexturing applications. *Nano Lett.* **2010**, *10*, 2989–2994. [CrossRef] [PubMed]
11. Sirotkin, E.; Apweiler, J.D.; Ogrin, F.Y. Macroscopic ordering of polystyrene carboxylate-modified nanospheres self-assembled at the water-air interface. *Langmuir* **2010**, *26*, 10677–10683. [CrossRef] [PubMed]
12. Huang, J.W.; Kao, Y.M.; Chiu, P.W.; Wu, T.H.; Lee, Y.C. Light diffusion film fabricated using colloidal lithography and embossing. *J. Nanopart. Res.* **2021**, *23*, 21. [CrossRef]
13. Fitch, R.M.; Tsai, C.H. Homogeneous Nucleation of Polymer Colloids, IV: The Role of Soluble Oligomeric Radicals. In *Polymer Colloids*; Fitch, R.M., Ed.; Springer: Boston, MA, USA, 1971.
14. Fitch, R.M.; Prenosil, M.B.; Sprick, K.J. The mechanism of particle formation in polymer hydrosols. I. Kinetics of Aqueous Polymerization of Methyl Methacrylate. *J. Polym. Sci.* **1969**, *27*, 95–118.
15. Barnes, G.; Gentle, I. *Interfacial Science: An Introduction*, 2nd ed.; Oxford University Press: Oxford, UK, 2011.
16. Lieberman, M.A.; Lichtenberg, A.J. *Principles of Plasma Discharges and Materials Processing*; John Wiley & Sons: Hoboken, NJ, USA, 2005.
17. Jansen, H.; de Boer, M.; Legtenberg, R.; Elwenspoek, M. The black silicon method: A universal method for determining the parameter setting of a fluorine-based reactive ion etcher in deep silicon trench etching with profile control. *J. Micromech. Microeng.* **1995**, *5*, 115–120. [CrossRef]
18. Nguyen, V.T.H.; Shkondin, E.; Jensen, F.; Hübner, J.; Leussink, P.; Jansen, H. Ultrahigh aspect ratio etching of silicon in SF_6-O_2 plasma: The clear-oxidize-remove-etch (CORE) sequence and chromium mask. *J. Vac. Technol. A* **2020**, *38*, 053002. [CrossRef]

Disclaimer/Publisher's Note: The statements, opinions and data contained in all publications are solely those of the individual author(s) and contributor(s) and not of MDPI and/or the editor(s). MDPI and/or the editor(s) disclaim responsibility for any injury to people or property resulting from any ideas, methods, instructions or products referred to in the content.

Review

A Review of Femtosecond Laser Processing of Silicon Carbide

Quanjing Wang, Ru Zhang *, Qingkui Chen and Ran Duan

School of Mechanical and Electronic Engineering, Shandong Jianzhu University, Jinan 250101, China; wqj338@163.com (Q.W.); cqk@sdjzu.edu.cn (Q.C.); duanran20@sdjzu.edu.cn (R.D.)
* Correspondence: zhangru21@sdjzu.edu.cn

Abstract: Silicon carbide (SiC) is a promising semiconductor material as well as a challenging material to machine, owing to its unique characteristics including high hardness, superior thermal conductivity, and chemical inertness. The ultrafast nature of femtosecond lasers enables precise and controlled material removal and modification, making them ideal for SiC processing. In this review, we aim to provide an overview of the process properties, progress, and applications by discussing the various methodologies involved in femtosecond laser processing of SiC. These methodologies encompass direct processing, composite processing, modification of the processing environment, beam shaping, etc. In addition, we have explored the myriad applications that arise from applying femtosecond laser processing to SiC. Furthermore, we highlight recent advancements, challenges, and future prospects in the field. This review provides as an important direction for exploring the progress of femtosecond laser micro/nano processing, in order to discuss the diversity of processes used for manufacturing SiC devices.

Keywords: femtosecond laser; silicon carbide; processing; application

Citation: Wang, Q.; Zhang, R.; Chen, Q.; Duan, R. A Review of Femtosecond Laser Processing of Silicon Carbide. *Micromachines* **2024**, *15*, 639. https://doi.org/10.3390/mi15050639

Academic Editor: Junyeob Yeo

Received: 21 April 2024
Revised: 4 May 2024
Accepted: 7 May 2024
Published: 10 May 2024

Copyright: © 2024 by the authors. Licensee MDPI, Basel, Switzerland. This article is an open access article distributed under the terms and conditions of the Creative Commons Attribution (CC BY) license (https:// creativecommons.org/licenses/by/ 4.0/).

1. Introduction

SiC has gained significant attention as a promising material for various applications due to its exceptional properties, including high hardness, excellent thermal conductivity, and chemical inertness [1]. Compared to silicon, SiC has many advantages, including higher operating temperature, lower conduction loss, higher bandgap, and higher critical breakdown field strength. SiC is a widely utilized advanced material and a representative of third-generation wide bandgap semiconductor materials. SiC is essential for exploring its potential in diverse fields such as electronics, photonics, aerospace, and microelectromechanical systems (MEMS) [2–4].

SiC exists in the forms of single crystal, polycrystalline, and amorphous [5]. At present, over 250 types of allotrope crystals of single crystal SiC have been discovered [6]. Figure 1 shows the semiconductor properties of several common single crystals SiC and other semiconductor materials [7–9]. The band gap of SiC is higher than that of silicon and Gallium arsenide, slightly lower than that of Gallium nitride, and the band gap of 4H-SiC is higher than that of 3C-SiC and 6H-SiC. Due to the characteristics of wide bandgap semiconductors, SiC exhibits high stability. The critical breakdown field of SiC is about 10 times that of silicon, the saturation electric drive rate of SiC is about 2 times that of silicon, the thermal conductivity of SiC is 3–4 times that of silicon and Gallium nitride, and the melting point of SiC is 1.5–1.8 times that of silicon and gallium nitride.

Other emerging materials, such as 2D semiconductor materials, have been investigated by many scholars. Antimonene, an emerging 2D semiconductor material, features a customizable bandgap, high carrier mobility, low thermal conductivity, and outstanding optical properties. These qualities make its interaction with photons highly promising in nanophotonic applications such as photodetectors, solar cells, photocatalysis, cancer treatment, surface plasmon resonance sensors, and nonlinear photon devices [10]. Graphdiyne displays environmentally friendly properties, increased chemical stability, a large specific surface area, a narrow bandgap, and high carrier mobility [11]. Compared to many other carbon materials, it has greater potential in nanophotonic applications. Its photoelectric performance can be enhanced through element doping [12] and hybridization with other nanostructures [13]. Due to its exceptional photovoltaic performance, graphdiyne is utilized in solar cells [14]. Selenium nanostructures, as narrow bandgap semiconductors, have been synthesized with controllable size, shape, and structure. However, under harsh conditions, their stability is poor, and their electronic/optoelectronic performance is suboptimal. Therefore, doping or epitaxial growth techniques are commonly employed to introduce heteroatoms and enhance the performance of selenium nanomaterials [15]. Black phosphorus is a typical layered two-dimensional material with high carrier mobility, an in-plane anisotropic structure, and an adjustable direct bandgap. However, its sensitivity to the environment necessitates improvements in its physical and chemical properties through doping methods [16,17]. Bismuth possesses physical and chemical properties including a tunable bandgap, excellent photoresponse, strong diamagnetism, and high photothermal conversion efficiency. The synthesis of bismuth-based hybrid materials can be employed to create ultra-small and ultra-sensitive photodetectors, facilitate the photocatalytic degradation of organic pollutants, and advance biophotonics applications [18,19]. Two-dimensional tellurium exhibits environmental stability, a bandgap dependent on thickness, piezoelectric properties, high carrier mobility, and light responsiveness. Consequently, it finds application in photodetectors, energy harvesting devices, piezoelectric devices, and other areas [20]. To address issues such as surface redox reactions, inadequate specific capacity, and unstable output potential in water-based zinc-ion batteries, a novel zinc tellurium battery cathode material utilizing a conversion-type ion storage mechanism can be employed [21].

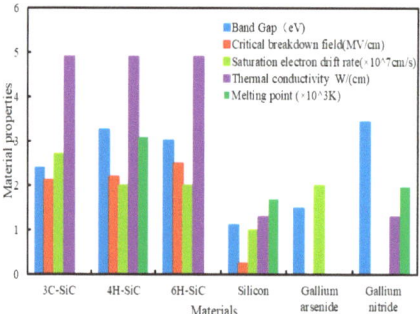

Figure 1. Comparison of material properties.

SiC is processed using various methods, with diamond wire saw cutting being a popular choice. However, SiC tends to experience crack damage during processing. Diamond wire saws may encounter challenges such as abrasive detachment, wire wear, surface blockage, and graphitization when utilized in SiC processing [22–24]. Ultra-precision polishing technology serves as the final step in the SiC machining process, employed to guarantee the surface roughness, flatness, and material removal rate of the SiC substrate [25,26]. Mechanical polishing is utilized to enhance the processing efficiency of polishing and improve the surface roughness of the substrate. Chemical mechanical polishing is employed to achieve ultra-smooth, defect-free, damage-free, and flat machined surfaces [27]. However, the ultra-precision polishing process of SiC single crystal substrates is complex and multifaceted, with the quality of polishing linked to the multi-dimensional motion trajectory of the polishing head. Consequently, further research is required to explore the polishing mechanism and process parameters. Yang et al. [28] studied ultrasonic-vibration-assisted electrochemical mechanical polishing to boost the effectiveness of conventional methods. This technique increased the contact force between abrasives and the SiC surface. Nevertheless, this method tends to raise the surface roughness of SiC. Wet etching technology comprises conventional wet etching and electrochemical etching. Conventional wet chemical etching of SiC typically involves high temperatures and is usually conducted in an alkaline solution of 488 K phosphoric acid or 373 K or above $K_3Fe(CN)_6$. Electrochemical etching is intricate, exhibiting low etching efficiency and poor uniformity [29]. Dry etching methods such as reactive ion etching are common in semiconductor processing [30]. However, the strong chemical stability of SiC requires a mask during dry etching, leading to low etching efficiency and a complicated process. Non-traditional, energy-assisted mechanical processing methods like vibration, laser, electrical, magnetic fields, chemical processes, advanced coolant systems, and others are employed for challenging materials such as hard and brittle substances [31–33]. These techniques improve machinability, enhance material removal rates, and boost surface quality. Laser processing can achieve efficient and high-precision machining of microstructures [34,35]. Zhao et al. proposed a novel method called laser-induced, oxidation-assisted milling to address surface damage issues commonly encountered in milling processes, such as pits, cracks, scratches, matrix coatings, and burrs [36,37]. In this method, an oxide layer is generated in the ablation zone through laser-induced oxidation, which is then easily removed by the milling tool. This process is repeated several times until the material is fully machined. Compared to conventional milling methods, no tool wear is observed in the milled oxide layer and the cutting forces are extremely low. Additionally, lower surface roughness can be achieved. However, laser processing can easily introduce thermal and physical damage.

Femtosecond laser technology has emerged as a powerful tool for the precise fabrication and modification of materials [38]. To achieve a uniform distribution of laser spot energy density, enable smaller aperture processing, and mitigate issues like edge collapse during processing, an attenuation module can be incorporated into the femtosecond laser precision micro-processing system [39,40]. By expanding and collimating the beam, placing attenuation modules along the propagation path, and adjusting the polarization state, a circularly polarized femtosecond laser beam with low power can be generated. The debris adsorption device in the femtosecond laser precision micromachining system is utilized to capture debris produced during the machining process. The ultrafast pulses generated by femtosecond lasers facilitate nonlinear absorption processes, minimizing thermal effects and resulting in precise material removal and modification at the micro- and nanoscale levels [41]. The femtosecond laser has emerged as an ideal tool for SiC machining and is currently attracting growing attention.

In this comprehensive review, we aim to explore the diverse applications and manufacturing aspects of femtosecond laser processing in SiC. Various processing methods such as direct processing, composite processing, processing environment, and beam shaping were explored. Additionally, the advantages and limitations of femtosecond laser processing in SiC were analyzed, along with highlighting recent advancements in the field. By offering an overview of the current state of the art, challenges, and future prospects, this review aims to enhance the understanding of the potential and opportunities presented by femtosecond laser processing in SiC applications. Researchers and engineers in the field stand to gain valuable insights from this review, empowering them to discover new pathways and advance the frontiers of SiC-based technologies.

2. Processing Methods
2.1. Direct Processing
2.1.1. Conventional Processing Methods

Figure 2 illustrates the femtosecond laser direct processing method. Femtosecond laser direct processing of SiC involves creating different surface features like craters, ripple structures, nanostructures, grooves, and large surface areas.

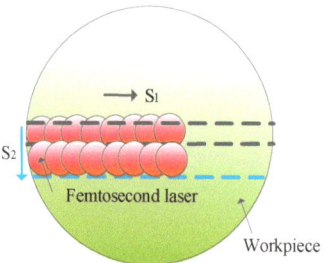

Figure 2. Femtosecond laser direct processing method.

1. Craters

Numerous scholars have conducted research in this field and have obtained the following results. Percussion drilling is accomplished by focusing femtosecond laser pulses onto the material surface, with each laser pulse removing a specific depth and radius of material. This process gradually increases the depth of the micro-craters, leading to the formation of a micro-crater structure [42]. Percussion drilling demonstrates higher efficiency compared to scanning and helical processing methods. However, when the pulse count reaches a certain value, there may be a termination of processing as the number of pulses increases [43,44]. Additionally, when the pulse count reaches a certain value, the micro-craters may exhibit bending [45,46]. The taper and the width of the heat-affected zone are two quality indicators for craters, typically considered better when smaller in size. The experiment demonstrated that optimizing laser fluence, scan speed, and focal position can lead to achieving good crater quality [47]. When the focal point is above the surface of the workpiece, larger diameter ablation craters can be machined, but a conical structure may result due to laser divergence. Conversely, positioning the focal point below the surface of the workpiece can also create larger diameter ablation craters, but a pointed crater bottom may form due to laser beam contraction incidence [48,49].

A comprehensive examination of the surface morphology of SiC following femtosecond laser processing uncovered bond breakage within the carbide ablation craters, the conversion of crystalline SiC to amorphous silicon and amorphous carbon, and the presence of a thin, deformed, and accumulated amorphous layer near the geometric focus of the subsurface in the laser-modified zone [50–52]. The TEM image of femtosecond-laser-modified SiC surface at a peak fluence of 124 J/cm^2 was shown in Figure 3.

Figure 3. TEM image of femtosecond-laser-modified SiC surface at a peak fluence of 124 J/cm^2. (**a**) Low resolution view. (**b**) Magnified TEM image indicated by an arrow in (**a**). (**c**) High-resolution image [50].

Due to the targeted and detailed analysis of the mechanisms involved in the femtosecond laser–SiC interaction in the author's published papers—Refs. [41,43]—the relevant processes induced by femtosecond laser pulses in SiC will be briefly introduced in this paper. The energy transfer of craters in femtosecond laser processing of single crystal SiC can be described as the main role of the free carrier absorption coefficient in the total absorption coefficient [41]. The primary carrier density absorption mechanism is two-photon absorption, followed by three-photon absorption and Auger recombination effect, with collision ionization playing a less significant role. By integrating THz-TDS scattering near-field optical microscopy with ultrafast pump–probe spectroscopy, the dynamics of photo-excited carriers in semiconductors can be examined at femtosecond and nanoscale resolutions, enabling the observation of carrier decay processes up to 100 ps [53]. The damage threshold energy density and non-thermal melting threshold at a wavelength of 800 nm increase with the pulse width, and the damage threshold energy density is inversely proportional to the wavelength. Both non-thermal melting and thermal melting play a role in material removal through melting [43]. Femtosecond laser drilling on SiC can be categorized into single pulse processing and multi-pulse processing. The material removal mechanism during femtosecond laser single pulse machining of SiC involves Coulomb explosion, photomechanical fragmentation, and thermal melting. These physical mechanisms contribute to material removal under low energy density ablation. On the other hand, non-thermal melting, thermal melting, and vaporization are the primary removal mechanisms at high energy densities. Furthermore, at even higher energy densities, phase explosion, thermal melting, vaporization, and non-thermal melting become the dominant removal mechanisms [43,54]. Vaporization may occur in the form of surface evaporation or explosive decomposition (phase explosion) of superheated liquids [55]. The nanoparticles formed during the phase explosion process and through condensation in the ablation plume are closely linked to the thermodynamic state of the irradiated material and the conditions within the laser ablation plume, primarily driven by thermal processes. Multi-pulse ablation demonstrates incubation effects, where the accumulation of pulse numbers can modify the optical properties of the surface being irradiated and the deposition of laser energy [56–58].

2. Ripple Structures, Nanostructures, and Grooves

The evolution of surface morphology in laser ablation areas at different scan speeds is shown in Figure 4. It has shown that as the scan speed decreases, micro/nanostructures can transform from fine ripples, coarse ripples, and nanoparticles into V-shaped grooves. When femtosecond laser lines are applied to SiC, with the increase in energy density and the decrease of scan speed, the processed structure gradually evolves from fine ripples, coarse ripples, and nanoparticles to V-shaped grooves [59,60]. The depth of the groove increases with higher laser energy density, repetition frequency, multi-pass scanning, and numerical aperture. Conversely, it decreases with increased scan speed. The groove width, material removal rate, and heat affected zone increase with higher laser energy density, repetition frequency, and multi-pass scanning but decrease with increased scan speed and numerical aperture. Improving the aspect ratio can be achieved by increasing the number of scans and laser energy density, reducing the scan speed, and minimizing the z-layer feed step [61,62]. The direction of the ripple structure is influenced by the laser polarization angle. For a linearly polarized laser, the ripple structure direction is perpendicular to the laser polarization direction [63–65]. A self-organization model based on laser-induced surface instability can be used to explain this phenomenon [66–68]. The model suggests that changes in the electric field direction disrupt the surface symmetry, causing the ripple direction to rotate with the laser electric field direction, always remaining perpendicular to the laser electric field direction. For elliptically polarized laser, the ripple structure direction is perpendicular to the major axis of the ellipse, but the continuity of the ripple structure is poorer. For circularly polarized laser, well-defined ripple structures are not observed [63].

3. Large surface areas

Due to the limited processing size of the femtosecond laser on SiC, which is typically confined to a single point or a single line segment, its practical application prospects are restricted. However, by optimizing parameters such as energy density, scan speed, and scanning spacing, it becomes possible to fabricate larger periodic structural arrays on SiC using femtosecond laser processing [51,69]. Under high laser energy density, SiC undergoes selective sublimation, leading to unbalanced removal of carbon and silicon elements. With increasing energy density, the number of hanging bonds increases, along with an enhanced ability to capture oxygen elements [70,71]. Residual tensile stress is generated in the preparation of SiC periodic arrays using a femtosecond laser, and the residual tensile stress increases with the increase in scanning distance. Laser energy density can increase lattice disorder [72,73]. Femtosecond lasers can also be applied to polish SiC [74]. The polishing results of the original surface defects of SiC have a significant impact, and a small incident angle is more conducive to obtaining a smooth surface and significantly reducing the degree of oxidation and graphitization in the ablation zone [74]. Due to the different laser absorption rates and energy distribution of SiC at different laser incidence angles, the ablation threshold and morphology of SiC change with the change in the laser incidence angle [75].

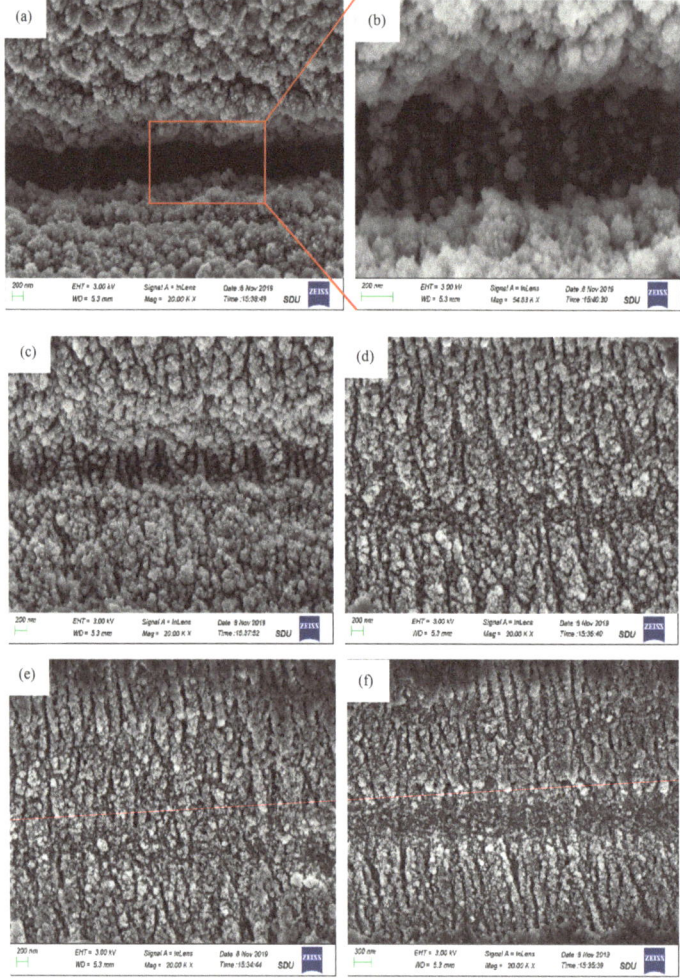

Figure 4. Evolution of surface morphology in laser ablation areas at scan speed S: (**a**,**b**) 50 µm/s, (**c**) 100 µm/s, (**d**) 200 µm/s, (**e**) 300 µm/s, and (**f**) 400 µm/s (fluence, F = 0.76 J/cm^2) [59].

2.1.2. Advanced Processing Methods

1. Processing Environment

Table 1 shows the comparison of processing results under different processing environments. In an air environment, air ionization results in a partial loss of laser energy, leading to a reduction in ablation size and negative effects such as debris eruption [76,77]. In a vacuum environment, the reduction in ambient pressure leads to lower gas molecule density, weakening the ionization effect of air and decreasing laser energy loss. This results in improved surface quality. Conversely, an increase in ambient pressure causes more collisions between plasmas, reducing energy loss and increasing ablation depth [78]. However, this also leads to an increase in the deposition of sediment around the ablation crater. When processing materials in environments with etching gases or nitrogen, it can result in a rough surface quality within the processing area [79]. When comparing corrugated structures processed in air with those generated in an argon environment, the latter shows smaller surface defects and higher ripple heights. X-ray photoelectron spectroscopy analysis reveals a decrease in workpiece surface oxidation during processing in an argon atmosphere [80]. When comparing corrugated structures generated in an argon environment to those processed in air, the former exhibits higher diffraction light intensity, minimal surface defects, and higher amplitude ripples. X-ray photoelectron spectroscopy analysis indicates a reduced ratio of oxygen to metal species on the sample surface in an argon atmosphere, resulting in minimal oxidation [80]. In the case of glass samples subjected to femtosecond laser micromachining in various solutions (including NaCl and KOH aqueous solutions, water, and on the surface of the workpiece), the resulting groove depths and widths were evaluated [81]. At low scanning speeds (50 mm/s), the dissolution of NaCl in distilled water led to an increase in charge density and accelerated multiphoton ionization in the ablation region, resulting in deeper grooves. However, under the same conditions, when NaCl aqueous solution was coated on the surface of the glass sample, the groove depth formed was lower compared to dry samples or those formed in water and KOH aqueous solutions. The study suggests that under multiple scanning glass cutting conditions, the localized temperature increase caused by laser irradiation enhances the corrosion rate of KOH solution, resulting in higher groove depths. The addition of a KOH aqueous solution film improves ablation efficiency, leading to increased groove depth. Furthermore, low-temperature femtosecond laser processing can suppress plasma shielding effects, increase ablation depth, significantly improve surface quality, reduce oxidation and surface roughness, and mitigate thermal accumulation effects [82].

Table 1. Comparison of processing results in different processing environments.

Processing Environments	Advantage	Disadvantage
Air		Partial loss of laser energy
Vacuum	Weakens the ionization effect of air and reduce laser energy loss	
Nitrogen		Rough surface quality
Argon		Small surface defects and high ripple heights
KOH aqueous solutions	Deep grooves	
Low-temperature	Increases ablation depth, improves surface quality, and mitigates thermal accumulation effects	

2. Beam Shaping

Many scholars have also investigated modifying laser beam characteristics to enhance processing outcomes. By utilizing diffraction optical software, the laser beam can be split into several beams, enabling the creation of several modified regions with similar surface morphology in a single scan [83]. In addition to obtaining the same beam of laser in a single scanning process, femtosecond laser pulses can also be divided into several sub pulses that are delayed in time [84]. The delay time of the next sub pulse in two adjacent femtosecond laser sub pulses can be adjusted by the time it reaches the corresponding reflector [85]. The modified regions can then be selectively chemically etched to remove the modified material, resulting in an increase in processing efficiency and consistency. Adjusting parameters like the number of laser beams, beam spacing, and beam energy can optimize the manufacturing of microlens arrays, resulting in superior optical performance [86]. Figure 5 is a schematic representation illustrating the shaping of a single pulse into two pulses. In the process of dual-pulse ablation, when the energy density of the dual pulses is relatively low, the ablation depth decreases as the delay time between the pulses increases [87]. The maximum ablation depth is reached when the delay time is zero, and the ablation depth of the dual pulses is significantly greater than that of a single pulse with the same energy. When the energy density of the dual pulses is high, the delay time has minimal impact on the ablation depth [87]. When the energy of the first pulse is higher than that of the second pulse, the ablation depth noticeably increases, and the recast layer is significantly reduced [88]. Multiphysics layer simulations enable theoretical investigations of the ultrafast thermalization dynamics induced by femtosecond laser double-pulse vortex excitation. Enhanced optical absorption and amplified carrier–phonon coupling dynamics due to double-pulse vortex excitation can further amplify the vortex thermodynamics [89,90]. In addition to temporal shaping, beam shaping can also be performed in space. The Gaussian beam's longitudinal energy distribution is limited due to its propagation attenuation characteristics, leading to reduced ablation size. In contrast, Bessel beams can maintain a constant profile over long distances [91]. By converting a Gaussian beam into a line beam using a cylindrical lens and adjusting the scanning direction perpendicular to the principal axis, the manufacturing efficiency of anti-reflective structures can be significantly enhanced [92]. Additionally, a new method for high-quality fiber Bragg grating fabrication utilizes high refractive index oil to exploit the transverse dispersion effects along the fiber, altering the focal length of the laser beam without requiring additional spatial beam-shaping devices. This approach allows for flexible control over the width and height of the grating plane [93]. Figure 6 shows the effects of different types of refractive index oils on the morphology of the grating and its transmission spectra.

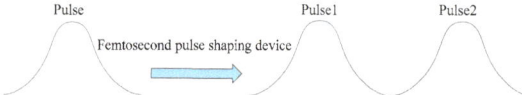

Figure 5. Processing by ultrasonic vibration compounded with femtosecond laser.

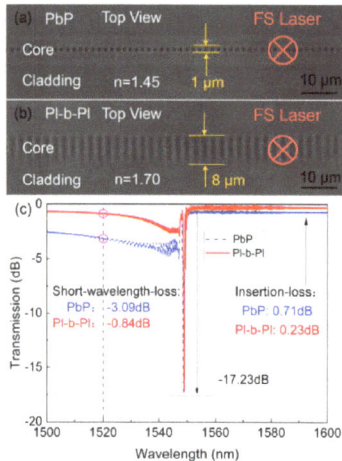

Figure 6. The effects of different types of refractive index oils on the morphology of the grating and its transmission spectra: (**a**) n = 1.45, (**b**) n = 1.7, and (**c**) transmission [93].

2.1.3. Existing Problems and Development Trends

Although the surface morphology and size variation of craters, ripple structures, nanostructures, grooves, and large surface areas resulting from femtosecond laser processing of SiC have been extensively studied, the understanding of the processing mechanisms is currently limited to the formation of ablation craters on SiC via femtosecond laser irradiation. There is a lack of research on the mechanism of femtosecond laser processing of SiC when the laser spot moves both laterally and longitudinally. Further research and exploration are needed to study and expand the processing techniques, processing mechanisms, and applications of processing environments and beam shaping technology. Additionally, the Gaussian beam's longitudinal energy distribution is limited due to its propagation attenuation characteristics, leading to reduced ablation size. In micro-crater machining, it is challenging to generate sufficient intensity of ablation as femtosecond laser pulses are distributed within a specific area.

2.2. Composite Processing

2.2.1. Laser–Water Jet Composite Processing Technology

In laser water jet composite processing technology, as depicted in Figure 7, the material is softened through laser heating, which reduces its yield strength. Subsequently, the high-pressure water jet impacts the heated material, causing shear forces and facilitating its removal [9,94]. Figure 8 shows a schematic diagram of laser–water jet equipment. During the laser–water jet machining of SiC, the maximum wall shear stress decreases as the nozzle target distance increases due to the divergence of the water jet. The turbulence and aeration of the high-pressure water lead to a degradation in the quality of the laser beam once it enters the water [9]. Figure 9 shows the surface morphology of the micro-groove. There are no obvious recast layers, cracks, or debris on the edges of the micro-grooves. As the laser power increases, the bottom of the micro-groove gradually darkens until it is no longer visible [95]. To achieve high-precision drilling, Li et al. [96] proposed a method that involves laser direct drilling followed by secondary refinement of micro-crater shapes using coaxial water-jet-assisted laser drilling. Computational fluid dynamics simulations were utilized to optimize the water channel structure and nozzle shape in the device. Furthermore, response surface methodology was employed to optimize multiple process parameters affecting the crater entrance diameter, exit diameter, and taper.

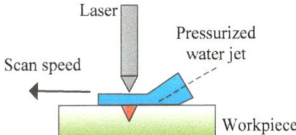

Figure 7. Hybrid laser–water jet processing method.

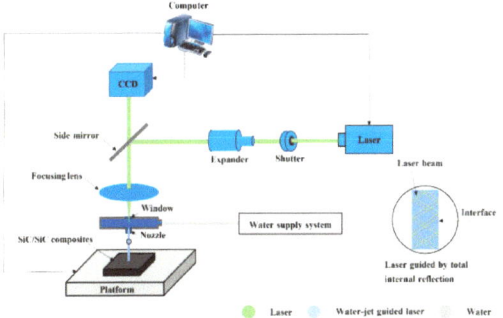

Figure 8. Schematic diagram of laser–water jet equipment [95].

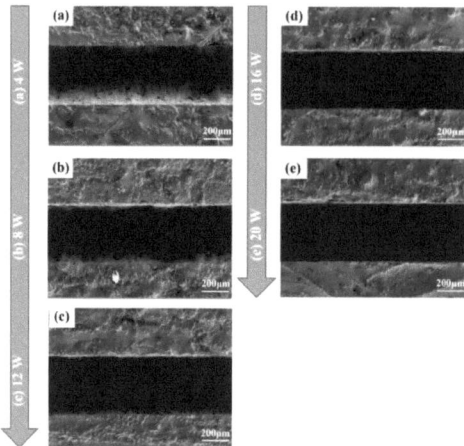

Figure 9. The surface morphology of the micro-groove: (**a**) P = 4 W, (**b**) P = 8 W, (**c**) P = 12 W, (**d**) P = 16 W, and (**e**) P = 20 W [95].

Laser–water jet machining is an advanced processing technology capable of executing complex and precise cuts. It can handle a variety of materials, including those sensitive to high temperatures, as the water jet helps cool the material and prevents heat damage. Unlike traditional mechanical cutting, laser–water jet machining does not require frequent replacement of tools or drill bits, reducing equipment wear. Additionally, this method does not produce harmful gases or significant waste. However, as the energy of the water jet dissipates when penetrating thick materials, it limits the thickness of the materials that can be cut. Compared to pure laser cutting, the processing speed of laser–water jet machining is slower because the water jet takes time to work through the material. The equipment not only has a high initial cost but also consumes a substantial amount of electricity and water during operation. The energy requirements of high-pressure water pumps and laser systems make the overall operational costs relatively high. The operation of high-pressure water jets generates noise and vibration, necessitating additional soundproofing or vibration damping measures. The precision of water jet machining may be affected by the quality of water used. Impurities and hardness in the water can impact the cutting results and normal operation of the machine.

2.2.2. Underwater Laser Composite Processing Technology

To mitigate the recast layer and thermal damage associated with higher energy density in direct femtosecond laser processing of SiC, various methods can be employed. Ethanol, distilled water, and underwater processing have been found to reduce re-deposition and thermal damage caused by ablative materials, effectively minimizing oxygen element incorporation [97–99]. The assistance of liquid media can help reduce the re-deposition of debris, obtain a smooth sidewall surface, and obtain structures with high material removal rates [100,101]. Ren et al. [102] used an underwater femtosecond laser to process craters in a layered circular pattern by adjusting the focus position. This technology can significantly increase crater depth, reduce adverse thermal effects, surface roughness, and recast layers. It was found that the diameter, taper, wall quality, and surface uniformity of holes are affected by the repetition frequency. In order to eliminate micro-crater defects and achieve consistency in the roundness of the inlet and outlet of the craters, copper or aluminum can be coated on the surface of SiC as a protective layer. The protective layer and water are used to assist the femtosecond laser in processing craters on SiC. After processing, the protective layer is removed via chemical corrosion. In the femtosecond laser experimental equipment, a 2× beam expander was used to obtain a smaller focused spot and to improve the roundness of the hole. The λ/4-wave plate was used to convert linearly polarized light into circularly polarized light. In order to filter out stray light around the laser beam, an aperture was used [103]. The equipment diagram for underwater processing is shown in Figure 10. In the water environment, the inhibition of oxidation reaction occurred via a water film during the decomposition process of SiC [104]. The temperature and oxygen content in the ablation area are the main reasons for corrosion mechanisms in both air and water. During processing in a liquid medium, in addition to the periodic diffraction of plasma in air, high-temperature melting of crystals in the liquid also occurs [105]. Therefore, nanoparticles can be observed in Figure 11. However, the presence of laser-induced bubbles and material deposition introduces uncertainty, which significantly impacts the reflection, refraction, and scattering of laser beams.

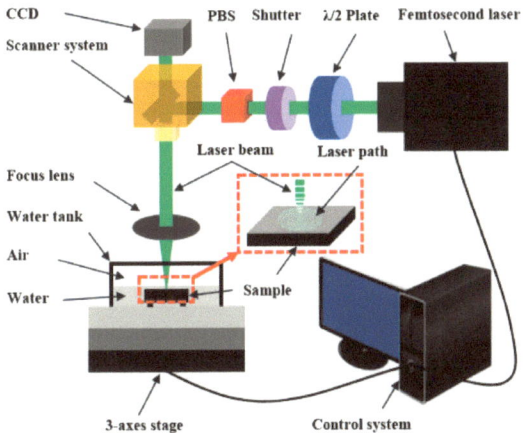

Figure 10. The equipment diagram for underwater processing [104].

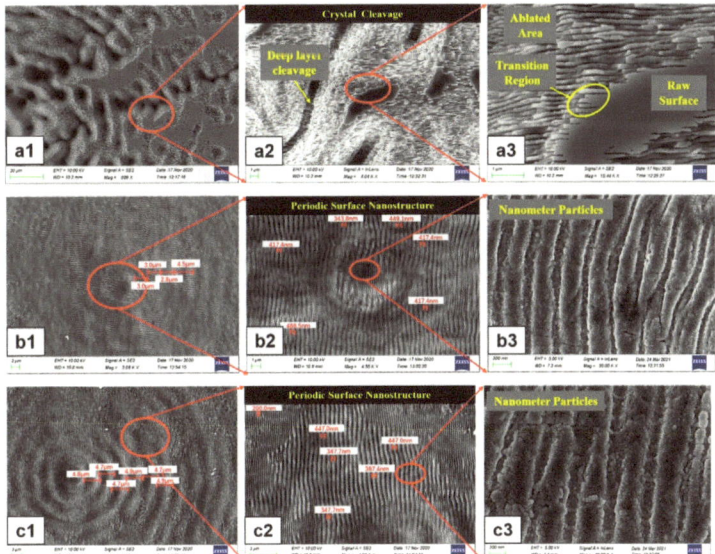

Figure 11. The surface morphology of the craters: (**a1–a3**) in air, (**b1–b3**) in water, and (**c1–c3**) in HF [105].

2.2.3. Laser–Wet Etching Composite Processing Technology

Due to the exceptional chemical stability of SiC, it can undergo selective chemical etching after being modified via femtosecond lasers, as illustrated in Figure 12. Following femtosecond laser processing, crystalline SiC transforms into amorphous SiC, which can undergo chemical reactions with a mixture of hydrofluoric acid and nitric acid [106–108]. This process enables well-defined and clean SiC ripples without introducing impurities, as shown in Figure 13. Raman spectroscopy can be used to measure the crystallinity of corrugated structures before and after etching. It was found that the Raman intensity of amorphous SiC is highly correlated with the etching rate of the ripple structure [109]. However, when the laser fluence is high, significant ablation occurs on the surface of SiC, resulting in poor surface roughness.

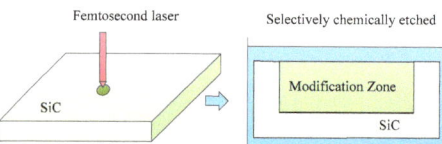

Figure 12. Femtosecond laser combined with selective chemical etching processing method.

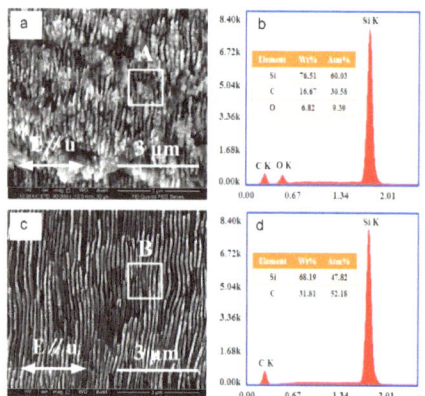

Figure 13. The LISC zone: (**a**) cross-section SEM images before chemical-selective etching; (**b**) chemical composition of area A in (**a**); (**c**) cross-section SEM images after chemical-selective etching; (**d**) chemical composition of area B in (**c**) [108].

In order to improve the surface quality after etching, wet etching was performed in an ultrasonic cleaning instrument with a heating function. Wang et al. [110] found that the etching rate was related to the crystal direction, and the size of the etching pit was related to the etching time. Wet etching of array structures composed of multiple microcraters, groove array structures, and grid array structures can achieve super hydrophilic performance, a high absorption rate, and low light reflectivity on silicon surfaces.

In terms of femtosecond-laser-assisted wet etching, other process methods can also be introduced. In order to achieve efficient and high-quality processing, silicon can be processed using a femtosecond laser in air and deionized water, respectively, followed by wet etching, which can improve the optical properties of silicon [111]. A Bessel beam femtosecond laser can be used to assist in wet etching to prepare small taper, large depth, and high aspect ratio structures [112]. This preparation method is simple and can shorten processing time by about 10 times. The combination of femtosecond laser with dynamic wet etching significantly removes the recast layer from the surfaces of micro-craters in SiC/SiC [113]. This is attributed to the SiO_2 etching induced by NaOH solution in a wet oxygen coupling environment at high temperatures. The experimental equipment diagram is shown in Figure 14. However, the interaction between the laser and the solution causes shock pressure, resulting in ablation pits, fractured SiC fibers, fragmented SiC matrix, and cracks at the bottom of the SiC/SiC holes. Wet etching reactions occur in both the laser-induced oxide layer and localized high-temperature regions. The removal mechanism of SiC/SiC treated with a femtosecond laser combined with dynamic wet etching involves a wet oxygen coupling environment, high-temperature decomposition, localized chemical wet etching, and the influence of the liquid layer.

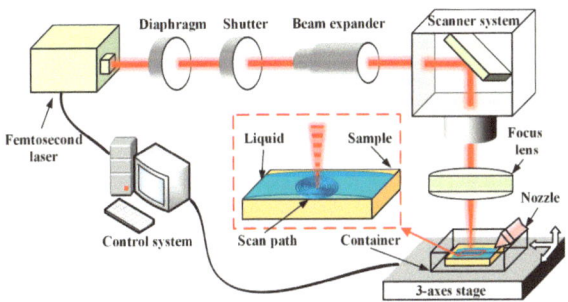

Figure 14. The experimental equipment diagram [113].

2.2.4. Laser–Electrochemical Composite Processing Technology

Laser–electrochemical composite processing technology combines laser and electrochemical machining processes within single equipment for processing purposes [114]. Figure 15 shows the schematic diagram of laser–electrochemical composite processing technology. The temperature rise caused by the laser can enhance kinetic effects in the electrochemical reaction zone, resulting in an increase in current density. In electrochemical machining conditions that promote the formation of passivating electrolytes, a passive layer forms on the workpiece surface. However, the laser weakens the microstructure of the surface passive layer. Figure 16 shows the morphology of laser electrochemical composite processing. In order to measure the sound pressure signal resulting from pulse laser breakdown of electrolytes and study the influence of cavitation on laser–electrochemical composite processing, a hydrophone was employed to capture the sound pressure signal generated when a pulsed laser was focused on the electrolyte [115]. A detection system for laser–electrochemical composite processing was established for this purpose. During laser and electrochemical composite etching, impact cavitation plays a vital role in material removal. As the laser energy increases, both the bubble radius and energy expand. Higher laser energy leads to the formation of plasma shock waves and jet forces, resulting in more efficient material removal. Bubble pulsation facilitates electrolyte flow, ensuring a constant supply of fresh ions and enhancing the electrochemical reaction rate. Moreover, it reduces the heat-affected zone and improves the surface quality of microstructure etching. The combination of laser processing and electrochemical etching creates micro-grooves on the material surface, which helps reduce friction with the surroundings. The width and density of these grooves are crucial in determining the friction reduction effect. Initially, as the width and density of the grooves increase, more contact points form between the material and the external medium, leading to an increase in the surface friction coefficient. However, beyond a certain point, further widening or increasing the density of the grooves may cause a decrease in the surface friction coefficient due to the reduced effective contact area [116]. In laser–electrolytic composite machining, the system incorporates a plunger hydraulic pump and an electric control valve to pressurize the electrolyte and regulate its flow rate. Temperature sensors and ion selective electrodes are utilized for data detection, enabling real-time control of electric control valves, heaters, and other components. This ensures automatic control of the electrolyte circulation control system [117]. In the synchronous processing of jet-assisted laser and electrochemical composite materials, the laser is focused on the processing area through an electrolyte jet. The metal nozzle serves as the cathode, while the workpiece acts as the anode [118]. Synchronous jet-assisted laser and electrochemical composite machining can achieve electrochemical dissolution of workpiece materials, simultaneously remove the recast layer on the laser-processed surface, and reduce thermal impact. It improves mass transfer efficiency in the processing area, enhances electrochemical processing efficiency, reduces laser energy loss caused by processing products, and increases the depth of composite processing. The method also reduces surface sputtering during laser processing, provides timely cooling to the laser processing area, and minimizes thermal impact. Laser–electrolytic sequential composite machining eliminates the need for a laser stable coupling device. The material removal mechanism in each individual process is well studied, and electrolytic machining improves surface quality while allowing control over machining accuracy. However, challenges arise in the process of laser–electrolytic sequential composite machining. The processing head requires frequent replacement, and accurate positioning of the wire or tube-shaped micro tool electrode at the center of the laser-preformed hole is crucial. Misalignment between the tool electrode and the laser preformed hole, as well as uneven distribution of the recast layer from laser processing, can result in variations in the electrolysis rate of the crater wall and fluctuations in the crater wall itself. This makes it difficult to maintain precision consistency in micro-crater processing. Moreover, the technology has limitations in deep processing due to the restricted depth of laser drilling [118]. However, the laser and electrochemical composite machining process is complex, involving the intricate coupling of multiple physical fields such as the light field, flow field, electrochemical dissolution field, thermal field, etc. The distribution patterns of processing products, bubbles, plasma, and other phenomena within the machining gap are not yet well understood, making it challenging to model the laser and electrochemical composite machining process accurately. Additionally, some experimental process parameter data are difficult to monitor. Therefore, modeling and simulation optimization of the process are crucial for providing support in intelligent control of the machining process, enhancing its controllability, and improving predictability.

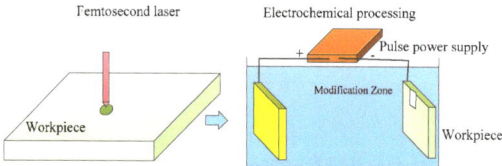

Figure 15. Laser–electrochemical composite processing technology.

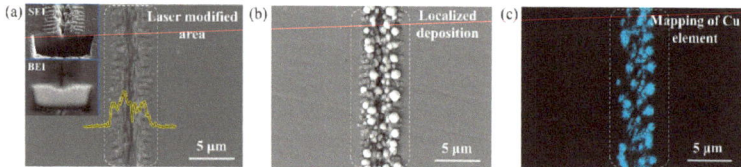

Figure 16. (**a**) Surface morphology diagram of laser-structured steel surface. (**b**) Surface morphology diagram of steel surface after electrochemical deposition. (**c**) Distribution of copper elements in sedimentary areas [114].

2.2.5. Laser–Ultrasonic Vibration Composite Processing Technology

By combining ultrasonic vibration with ultrafast laser technology, the efficiency of ultrafast laser micro-crater manufacturing can be further enhanced. This combination improves the depth-to-diameter ratio of micro-craters and enhances the quality of micro-crater morphology [119]. Figure 17 shows the schematic diagram of ultrasonic vibration compounded with femtosecond laser composite processing technology. Figure 18 depicts the surface morphology of C/SiC composite materials processed under different ultrasonic amplitudes. Compared to femtosecond laser polishing alone, ultrasonic-vibration-assisted femtosecond laser polishing enhances the material removal rate and reduces particles, debris, and surface oxidation, while also increasing residual compressive stress in carbon fibers [120]. Currently, there are two main forms of ultrasonic-vibration-assisted laser drilling: one involves applying ultrasonic vibration to the laser lens, while the other involves applying ultrasonic vibration directly to the workpiece. When ultrasonic vibration acts on the laser lens, it can be controlled to change the energy distribution at the laser focal point, resulting in better processing quality of micro holes compared to traditional laser drilling methods. Ultrasonic vibration applied to the workpiece effectively enhances the absorption of the laser beam via the material inside the hole, further promoting beam reflection and improving the processing depth and efficiency of micro holes. When vertical ultrasonic vibration and coaxial low-pressure assistance are used, additional kinetic energy is provided for the flow and injection of the molten material, enhancing the material removal rate. This approach also reduces re-solidification of splashed melt on the crater wall and decreases the thickness of the recast layer. During laser processing, the material undergoes heating, melting, and evaporation. The sound pressure gradient generated via ultrasonic vibration within the molten pool promotes continuous circulation and flow of the molten material. Additionally, high-frequency ultrasonic vibration facilitates the flow of plasma above the processing area, which helps improve the quality of micro-crater processing. The mixing and stirring effects induced by ultrasound enhance the flow and convection of molten materials within the pores. This improves the homogenization of composition and microstructure, reduces component segregation, and minimizes the formation of micro-cracks [121]. The composite processing technology that combines ultrasonic vibration, laser processing, and water jet processing incorporates the advantages of these three methods [122]. In this technology, the ultrasonic module utilizes a direct connection between the ultrasonic transducer, vibration plate (where the workpiece is placed), and the multifunctional, ultra-precision fine tuning platform. This setup eliminates system errors caused by laser vibration and enhances the stability of the processing process. The amplitude is measured using a laser displacement sensor. The jet system is powered by an electric-motor driven plunger pump, and the jet parameters are controlled by adjusting the overflow valve and throttle valve [122]. By incorporating ultrasonic vibration into the process using the same laser parameters, the depth of laser-etched grooves gradually increases in conjunction with an elevation in laser current. The cavitation effect induced by high-frequency ultrasonic vibration enhances the kinetic energy of the molten pool and promotes vertical agitation of the slag. During laser cutting, part of the molten material evaporates, while another part is removed from the ablation groove due to the impact of the water jet. The combination of ultrasonic water jet and laser processing significantly reduces the surface recast layer in the groove. Additionally, there is a small amount of crystallization at the groove boundaries, and the groove width increases, resulting in relatively good surface quality. However, due to increased energy loss caused by the water jet's impact during laser etching, the groove depth decreases.

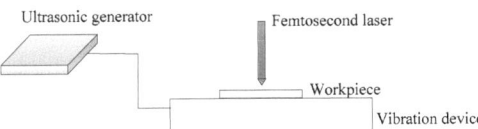

Figure 17. Processing via ultrasonic vibration compounded with femtosecond laser.

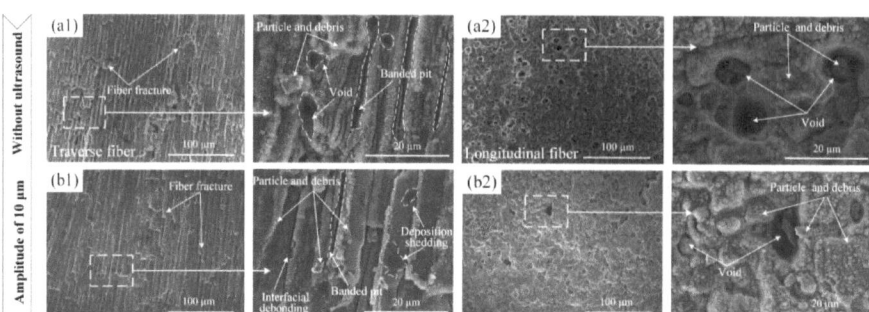

Figure 18. Surface morphology of C/SiC composites processed at different ultrasonic amplitudes: (**a1,a2**) without ultrasound; (**b1,b2**) 10 μm [120].

2.2.6. Laser–Chemical Mechanical Polishing Processing Technology

The femtosecond laser–chemical mechanical polishing (CMP) hybrid processing technique involves the initial use of a femtosecond laser for material pre-treatment, followed by the application of colloidal silica slurry to polish the workpiece surface under a polishing pressure of 120 kPa and a rotational speed of 50 min^{-1} [123]. The ripple structures generated on the workpiece surface by the femtosecond laser facilitate increased the contact area between the CMP slurry and the substrate surface. Processing in ambient air introduces oxygen elements into the laser-modified zone and generates an amorphous layer, both of which enhance the CMP process [124]. In the femtosecond laser–CMP hybrid processing technique, the material removal rates of the C-face and Si-face of silicon carbide samples increased by 77% and 207%, respectively, compared to untreated samples [125]. Additionally, the research found that the surface roughness (Ra) of the workpiece polished with alumina slurry was reduced to as low as 0.081 nm [126]. However, the oxide coating on the SiC surface reduced the likelihood of abrasive contact with the SiC-Si surface, resulting in minor scratching after polishing.

The femtosecond laser–ultrasonic vibration polishing hybrid method can be used to improve micro-cracks generated during femtosecond laser processing [127]. Diamond particles move up and down at an extremely high speed in the vertical direction relative to the workpiece. In the laser-modified zone, the raised peaks fracture under the intense collision of the diamond particles and are carried away by the diamond slurry, resulting in lower surface roughness.

However, further research is needed to assess the processing efficiency of femtosecond-laser-assisted chemical–mechanical polishing, as well as the evaluation of subsurface stress layers and crystal defects post-processing.

2.2.7. Laser–Inductively Coupled Plasma Etching

The femtosecond laser and inductively coupled plasma (ICP) etching hybrid processing technology can accelerate the etching rate of silicon carbide and achieve a better quality processed surface [128]. Following femtosecond laser irradiation, the ICP etching rate of silicon carbide can be increased by up to 117.18% compared to processing without femtosecond laser treatment [128]. The induced silica and rough surface on silicon carbide via femtosecond-laser-facilitated ICP etching is effectively removed after etching [129]. The femtosecond laser and inductively coupled plasma etching hybrid processing technology can significantly improve the surface morphology at the bottom of the modified area, reduce surface roughness, eliminate amorphous carbon, and enhance its deformation capability.

Laser–inductively coupled plasma etching is an advanced micro- and nano-fabrication technique known for its high precision and efficiency. It enables precise machining at the micro- and even nano-meter scale, boasting exceptional machining resolution and the capability to handle intricate structures and multi-layered devices. Compared to conventional plasma etching techniques, it offers higher processing rates and enhanced production efficiency while minimizing device damage during processing. However, the acquisition and maintenance costs of this equipment are relatively high, and the operation and optimization of process parameters can be complex. Moreover, it demands stringent environmental conditions for operation.

2.2.8. Existing Problems and Development Trends

Femtosecond laser composite processing involves multiple machining processes, necessitating precise control and optimization of various process parameters. Femtosecond laser composite processing systems typically require multiple devices and complex control systems, resulting in high costs. The mechanism of femtosecond laser composite processing remains to be further investigated, and the composite interaction field between femtosecond laser and other processing methods needs to be further determined. Additionally, the surface characteristics after femtosecond laser composite processing need to be further established. Future advancements in laser composite processing will move towards intelligence and automation, leveraging machine learning and artificial intelligence technologies to optimize machining processes and parameter control, thereby enhancing production efficiency and quality stability. The integration of multiple machining methods into a single system will enable multifunctional machining to meet diverse material and processing requirements. Enhancing adaptability and flexibility will be emphasized for different types of materials to achieve a wider range of applications. Furthermore, in order to improve processing efficiency, achieve clean manufacturing, enhance surface quality, and ensure sidewall inclination, new approaches for femtosecond laser composite processing need to be further studied.

3. Applications

Depending on the structure of femtosecond laser processing, the processed surface exhibits different applications. Femtosecond laser machining of SiC micro-grooves is primarily used in applications such as micro-cavities, micro-sensors, and micro-mechanical systems. Micro-groove texturing on non-contact single crystal SiC mechanical sealing surfaces can reduce the friction coefficient; improve lubricant flowability; store a certain amount of lubricant, thereby reducing friction, and expanding the range of lubrication applications [130]. Micro-grooves on SiC ceramic surfaces can reduce friction losses in rolling bearings and improve rotational performance [131]. Femtosecond-laser-machined SiC micro-grooves are also commonly applied in droplet-based microfluidic devices [132], microfluidic sensors [133], and biosensors [134], where high aspect ratio micro-grooves help reduce friction in microfluidic devices and improve flow properties [135]. The micro–nano structural morphology induced by femtosecond lasers has potential applications in various fields, including bio-molecular labeling [136], surface texturing [137], self-cleaning surfaces and hydrophobic materials [138], super-wetting surfaces [139], wide-field imaging [140], improved cutting performance [141], solar cells [142], enhanced conductivity of laser-modified regions [143], light absorption enhancement, photocurrent enhancement [144], and MEMS [145]. The structure of SiC can be further extended using femtosecond lasers to create a series of parallel micro-grooves on the surface of SiC. By utilizing the highly graded surface capillary structure of the micro-grooves, water has a very high diffusion function in the inertial capillary flow state, resulting in excellent micro core and nano core suction functions and cooling performance on the surface of SiC, and this function has long-term stability [146]. Therefore, it can be applied to M-cycle air conditioning systems to reduce energy loss, and can also be applied to improve power generation efficiency, waste heat recovery, and thermal management [146]. In the final analysis, the fabrication of micro–nano structures on SiC holds significant practical application significance.

3.1. Microelectromechanical Systems

SiC is highly suitable for applications in microelectromechanical systems due to its high mechanical strength, chemical inertness, and electrical stability. It finds uses in actuators, pressure sensors, temperature sensors, transverse resonance devices, optoelectronics, and sensors [147,148]. For large-sized manufactured SiC structures, many scholars are currently committed to applying this structure to sensors. Due to its excellent high-temperature stability, chemical inertness, and mechanical and electrical properties, SiC is an ideal material to replace silicon for manufacturing high-temperature pressure sensors operating at temperatures above 500 °C [149]. Micro-sensing devices made of SiC can achieve remote sensing predictions of high temperatures and maintain high sensitivity [150]. However, SiC is a typical difficult-to-machine material with high hardness and chemical inertness, making it challenging to meet the manufacturing requirements of sensitive diaphragms for high-temperature pressure sensors using traditional mechanical machining and chemical etching methods.

To meet the performance requirements of sensors, femtosecond laser can be utilized for deep etching on SiC diaphragms [151,152]. The sensitivity of pressure sensors manufactured using femtosecond lasers is twice that of other SiC sensors [153]. Femtosecond lasers can be used to fabricate square-shaped sensitive diaphragms on SiC, which exhibit good output characteristics at high and low temperatures [153,154]. Femtosecond lasers can also be used to fabricate circular sensitive diaphragms on SiC, which exhibit good operational capabilities under high temperatures and pressures [155–157]. A femtosecond laser was employed to create blind holes with a diameter of 1200 μm and a depth of 270 μm on a 350 μm thick 4H-SiC wafer to form an 80 μm thick circular membrane as the pressure sensor's diaphragm [151]. The excess SiC material was gradually removed layer by layer using the femtosecond laser until the desired depth of the blind hole was achieved, with the remaining membrane at the bottom serving as the diaphragm for the pressure sensor. The smoothness, flatness, sidewall inclination angle, and surface roughness of the diaphragm are important factors that influence the performance of the sensor [158–160]. The sensor was connected to a single-arm bridge for pressure sensor characterization experiments. As pressure was applied and released, the output voltage of the sensor varied accordingly. The nonlinearity error can reach up to 0.13%, and the repeatability error is 1.49% [161,162]. Under extreme temperatures and corrosive environments, the sensitivity of the sensor can reach 76.0 μV/V/MPa, with an average daily drift of only 1.61% [162].

Femtosecond lasers can be used to fabricate accelerometers on SiC [163]. Compared to pressure sensors, accelerometers have a more complex structure and require higher machining precision. To ensure the sensor's shock resistance and longevity, the morphology of the sensitive beam also needs to meet high standards.

Compared to other machining methods for SiC sensor membranes, using a combination of femtosecond lasers and inductively coupled ICP etching has several advantages, including high machining dimensional accuracy, low surface roughness, and relatively high processing efficiency [164]. A detailed simulation and optimization analysis of sensor size was conducted based on the anisotropic elastic modulus parameters of SiC. Each layer scan differs by 30° from the previous scan path to ensure uniformity of surface removal [165]. After femtosecond laser machining, the processed SiC can undergo wet cleaning and ICP etching to significantly improve the surface morphology of the membrane's bottom side.

3.2. Existing Problems and Development Trends

However, the efficiency of femtosecond laser deep etching on SiC is relatively low, processing quality is difficult to ensure, elliptical and excessive etching are prone to occur, and it is challenging to ensure sidewall inclination. Further research is needed to understand the mechanism of large-area removal of SiC using femtosecond lasers. The next step would involve further exploration and research into efficient and high-quality methods for removing SiC, as well as understanding the underlying mechanisms. The specific applications of femtosecond laser processing on SiC can be further expanded based on the needs of businesses and industries.

4. Conclusions

This article provides a detailed discussion on the machining methods and applications of femtosecond laser processing on SiC. The methods for femtosecond laser processing on SiC are categorized into direct processing, hybrid processing, processing environment modification, and beam shaping methods. The application of femtosecond laser processing on SiC in MEMS is also described.

Femtosecond lasers can directly process craters, ripple structures, nanoscale structures, grooves, and large-area surfaces on SiC. After femtosecond laser processing of SiC, the unbalanced removal of silicon and carbon elements in SiC occurs, and oxygen is introduced due to the capturing effect of dangling bonds. The region modified via femtosecond lasers exhibits residual tensile stress and strain, with the residual tensile stress increasing with the scan spacing increment. The modified region also exhibits lattice disorder, which increases with the increase in energy density. During femtosecond laser ablation of SiC, both thermal melting and non-thermal melting coexist. The dominant mechanism for carrier absorption is multiphoton absorption, while collisional ionization has negligible effects.

Laser–water jet hybrid processing, underwater processing, and femtosecond laser and chemical selective etching hybrid processing techniques can be used for SiC machining. Hybrid laser processing methods can effectively increase the ablation depth and reduce the re-deposition of ablated material. However, laser–water jet and underwater processing techniques may affect the beam quality due to fluid turbulence and bubbles. Selective chemical etching using a mixture of hydrofluoric acid and nitric acid is not easy to handle in practical experiments and carries certain risks. Processing in a vacuum environment can reduce energy losses and achieve better surface quality. Beam shaping can be used to obtain uniform processing dimensions or larger aspect ratios, resulting in better surface quality of the material.

Femtosecond laser processing on SiC offers various benefits for the fabrication of sensors, micro-cavities, and micro-mechanical systems. Micro-groove texturing on SiC surfaces offers benefits like reduced friction coefficient, improved lubricant flowability, and enhanced rotational performance. Femtosecond laser technology enables the precise fabrication of highly accurate diaphragms for sensors, including pressure and accelerometer devices. These diaphragms can be created in square or circular shapes with exceptional precision. They exhibit excellent output characteristics even under challenging conditions, such as high temperatures, pressures, and extreme environments. However, challenges remain in terms of low efficiency and sidewall inclination during deep etching, necessitating further research to develop more efficient and high-quality removal methods and mechanisms. Additionally, exploring specific applications of femtosecond laser processing on SiC can be tailored to meet the needs of different industries. An alternative approach using a combination of femtosecond laser and ICP etching shows promise in achieving high machining accuracy and low surface roughness for sensor membranes. Further improvements have been observed through post-processing steps like wet cleaning and ICP etching, significantly enhancing the surface morphology of the processed SiC.

In summary, this review emphasizes on different femtosecond laser composite processing methods and the application of SiC structures processed by femtosecond laser in MEMS. And it is expected to promote the development and application of femtosecond laser composite processing technology.

Despite the contributions of femtosecond laser processing, there are still limitations that need to be addressed in the context of SiC processing. The processing mechanisms for micro-grooving or milling of SiC using femtosecond laser are not yet clear and require further investigation. The mechanisms of femtosecond laser composite processing also need to be studied in more depth. Surface characteristics after femtosecond laser composite processing are still subject to further research. Additionally, new methods for femtosecond laser composite material processing need to be explored to improve processing efficiency, achieve clean manufacturing, enhance surface quality, and ensure sidewall taper. The effects of processing environment and beam shaping on composite processing can be further investigated in the future. Furthermore, the application scope of femtosecond-laser-processed SiC structures needs to be expanded, and process optimization should be performed based on specific application requirements.

Author Contributions: Conceptualization, Q.W.; supervision and validation, R.Z.; investigation and writing, R.Z.; investigation and visualization, Q.W.; validation, Q.C. and R.D. All authors have read and agreed to the published version of the manuscript.

Funding: This work was supported in part by the Jinan City School Integration Development Strategy Engineering Project: Intelligent Manufacturing "VR+" Online Education Technology Innovation Research Institute under Grant JNSX2021050, in part by the Construction of Case Library for the Course of Intelligent Manufacturing Technology Based on VR Technology under Grant ALK202108, and in part by Shandong Natural Science Foundation under Grant ZR202211260018.

Conflicts of Interest: The authors declare no conflicts of interest.

References

1. Li, J.; Yang, G.; Liu, X.; Luo, H.; Xu, L.; Zhang, Y.; Cui, C.; Pi, X.; Yang, D.; Wang, R. Dislocations in 4H silicon carbide. *J. Phys. D Appl. Phys.* **2022**, *55*, 463001. [CrossRef]
2. Tian, R.; Ma, C.; Wu, J.; Guo, Z.; Yang, X.; Fan, Z. A review of manufacturing technologies for silicon carbide super junction devices. *J. Semicond.* **2021**, *42*, 061801. [CrossRef]
3. Sheng, K.; Ren, N.; Xu, H.Y. A recent review on silicon carbide power devices technologies. *Proc. CSEE* **2020**, *40*, 1741–1748.
4. Palani, I.A.; Paul, C.P. A review on laser assisted deposition and micro-etching of silicon carbide (SiC) coating for MEMs device applications. *Int. J. Des. Manuf. Technol.* **2013**, *7*, 7–16.
5. Sarro, P.M. Silicon carbide as a new MEMS technology. *Sens. Actuators* **2000**, *82*, 210–218. [CrossRef]
6. Fisher, G.R.; Barnes, P. Towards a unified view of polytypism in silicon carbide. *Philos. Mag. B* **1990**, *61*, 217–236. [CrossRef]
7. Casadt, J.B.; Johnson, R.W. Status of silicon carbide (SiC) as a wide-Bandgap Semiconductor for High-Temperature Applications: A Review. *Solid State Electron.* **1996**, *39*, 1409–1422. [CrossRef]
8. Wu, J. When group-III nitrides go infrared: New properties and perspectives. *J. Appl. Phys.* **2009**, *106*, 011101. [CrossRef]
9. Feng, S.C. Study on Near-Damage-Free Micromachining Mechanisms of Silicon Carbide Wafer Using Hybrid Laser-Waterjet. Ph.D. Thesis, Department of Electrical Engineering, Shandong University, Jinan, China, 2018.
10. Wang, X.; Yu, X.; Song, J.; Huang, W.; Xiang, Y.; Dai, X.; Zhang, H. Two-dimensional semiconducting antimonene in nanophotonic applications—A review. *Chem. Eng. J.* **2021**, *406*, 126876. [CrossRef]
11. Wang, M.; Pu, J.; Hu, Y.; Zi, Y.; Wu, Z.; Huang, W. Functional graphdiyne for emerging applications: Recent advances and future challenges. *Adv. Funct. Mater.* **2024**, *34*, 2308601. [CrossRef]
12. Mohajeri, A.; Shahsavar, A. Tailoring the optoelectronic properties of graphyne and graphdiyne: Nitrogen/sulfur dual doping versus oxygen containing functional groups. *J. Mater. Sci.* **2017**, *52*, 5366–5379. [CrossRef]
13. Wang, X.H.; Zhang, Z.C.; Wang, J.J.; Chen, X.D.; Yao, B.W.; Hou, Y.X.; Yu, M.X.; Li, Y.; Lu, T.B. Synthesis of wafer-scale molayer pyrenyl graphdiyne on ultrathin hexagonal boron nitride for multibit optoelectronic memory. *ACS Appl. Mater. Interfaces* **2020**, *12*, 33069–33075. [CrossRef] [PubMed]
14. Hu, Y.; Wang, M.; Hu, L.; Hu, Y.; Guo, J.; Xie, Z.; Wei, S.; Wang, Y.; Zi, Y.; Zhang, H.; et al. Recent advances in two-dimensional graphdiyne for nanophotonic applications. *Chem. Eng. J.* **2022**, *450*, 138228. [CrossRef]
15. Huang, W.; Wang, M.; Hu, L.; Wang, C.; Xie, Z.; Zhang, H. Recent Advances in semiconducting monoelemental selenium nanostructures for device applications. *Adv. Func. Mater.* **2020**, *30*, 2003301. [CrossRef]
16. Hu, H.; Shi, Z.; Khan, K.; Cao, R.; Liang, W.; Tareen, A.K.; Zhang, Y.; Huang, W.; Guo, Z.; Luo, X.; et al. Recent advances in doping engineering of black phosphorus. *J. Mater. Chem. A* **2020**, *8*, 5421–5441. [CrossRef]
17. Wang, M.; Zhu, J.; Zi, Y.; Wu, Z.-G.; Hu, H.; Xie, Z.; Zhang, Y.; Hu, L.; Huang, W. Functional two-dimensional black phosphorus nanostructures towards next-generation devices. *J. Mater. Chem. A* **2021**, *9*, 12433–12473. [CrossRef]
18. Wang, M.; Zhu, J.; Zi, Y.; Wu, Z.-G.; Hu, H.; Xie, Z.; Zhang, Y.; Hu, L.; Huang, W. Emerging mono-elemental bismuth nanostructures: Controlled synthesis and their versatile applications. *Adv. Funct. Mater.* **2020**, *31*, 2007584. [CrossRef]
19. Wang, M.; Hu, Y.; Zi, Y.; Huang, W. Functionalized hybridization of bismuth nanostructures for highly improved nanophotonics. *APL Mater.* **2022**, *104*, 050901. [CrossRef]
20. Shi, Z.; Cao, R.; Khan, K.; Tareen, A.K.; Liu, X.; Liang, W.; Zhang, Y.; Ma, C.; Guo, Z.; Luo, X.; et al. Two-Dimensional Tellurium: Progress. Challenges, and Prospects. *Nano Micro Lett.* **2020**, *12*, 99. [CrossRef]
21. Chen, Z.; Yang, Q.; Mo, F.; Li, N.; Liang, G.; Li, X.; Huang, Z.; Wang, D.; Huang, W.; Fan, J.; et al. Aqueous zinc–tellurium batteries with ultraflat discharge plateau and high volumetric capacity. *Adv. Mater.* **2020**, *32*, 20014699. [CrossRef]

22. Wang, P.Z. Study on the Surface Layer Crack Damage in Diamond Wire Sawing of Single Crystal Silicon Carbide. Ph.D. Thesis, Department of Electrical Engineering, Shandong University, Jinan, China, 2020.
23. Wang, P.; Ge, P.; Ge, M.; Bi, W.; Meng, J. Material removal mechanism and crack propagation in single scratch and double scratch tests of single-crystal silicon carbide by abrasives on wire saw. *Ceram. Int.* **2019**, *45*, 384–393. [CrossRef]
24. Ge, M.; Zhang, C.; Wang, P.; Li, Z.; Ge, P. Modeling of electroplated diamond wire and its application towards precision slicing of semiconductors. *J. Manuf. Process.* **2023**, *87*, 141–149. [CrossRef]
25. Zhang, P. Research on the Key Technologies of Ultra-Precision Polishing of Silicon Carbide Single Crystal Substrate. Ph.D. Thesis, Mechatronics Engineering, Shandong University, Jinan, China, 2017.
26. Xu, H.M.; Wang, J.B.; Li, Q.A. Research progress in chemical mechanical polishing technology for silicon carbide wafers. *Mod. Manuf. Eng.* **2022**, *6*, 153–159.
27. Zhang, X.; Wang, R.; Zhang, X.; Yang, D.; Pi, X. Research status and development trend of silicon carbide single crystal substrate machining technology. *J. MUC (Nat. Sci. Ed.)* **2021**, *30*, 5–12.
28. Yang, X.; Yang, X.; Gu, H.; Kawai, K.; Arima, K.; Yamamura, K. Efficient and slurry less ultrasonic vibration assisted electrochemical mechanical polishing for 4H–SiC wafers. *Ceram. Int.* **2022**, *48*, 7570–7583. [CrossRef]
29. Dinh, T.; Nguyen, N.T.; Dao, D.V.; Dinh, T.; Nguyen, N.T.; Dao, D.V. Fabrication of SiC MEMS Sensors. In *Thermoelectrically Effect in SiC for High-Temperature MEMS Sensors*; Springer: Singapore, 2018; pp. 55–74.
30. Racka-Szmidt, K.; Stonio, B.; Żelazko, J.; Filipiak, M.; Sochacki, M. A review: Inductively coupled plasma reactive ion etching of silicon carbide. *Materials* **2021**, *15*, 123. [CrossRef]
31. Zhao, G.; Zhao, B.; Ding, W.; Xin, L.; Nian, Z.; Peng, J.; He, N.; Xu, J. Nontraditional energy-assisted mechanical machining of difficult-to-cut materials and components in aerospace community: A comparative analysis. *Int. J. Extrem. Manuf.* **2024**, *6*, 022007. [CrossRef]
32. Kishawy, H.A.; Hosseini, A. *Machining Difficult-to-Cut Materials: Basic Principles and Challenges*; Springer: Cham, Switzerland, 2019.
33. An, Q.; Chen, J.; Ming, W.; Chen, M. Machining of SiC ceramic matrix composites: A review. *Chin. J. Aeronaut.* **2021**, *34*, 540–567. [CrossRef]
34. Zhao, S.; Yu, M.; Yan, R.; Zhai, Q.; Liu, J.; Dai, X.; Lei, C.; Wang, D. Surface roughness and micro hardness improvement of laser cladding stainless-steel 316 by laser polishing based on multiple remelting. *Opt. Laser Technol.* **2024**, *176*, 110903. [CrossRef]
35. Feng, S.; Huang, C.; Wang, J.; Jia, Z. Surface quality evaluation of single crystal 4H-SiC wafer machined by hybrid laser-waterjet: Comparing with laser machining. *Mater. Sci. Semicond. Process.* **2019**, *93*, 238–251. [CrossRef]
36. Zhao, G.; Xin, L.; Li, L.; Zhang, Y.; He, N.; Hansen, H.N. Cutting force model and damage formation mechanism in milling of 70wt% Si/Al composite. *Chin. J. Aeronaut.* **2023**, *36*, 114–128. [CrossRef]
37. Zhao, G.; Nian, Z.; Zhang, Z.; Li, L.; He, N. Enhancing the machinability of C_f/SiC composite with the assistance of laser-induced oxidation during milling. *J. Mater. Res. Technol.* **2023**, *22*, 1651–1663. [CrossRef]
38. Zhang, R.; Huang, C.; Wang, J.; Zhu, H.; Yao, P.; Feng, S. Micromachining of 4H-SiC using femtosecond laser. *Ceram. Int.* **2018**, *44*, 17775–17783. [CrossRef]
39. Lu, Y.C.; Jia, J.C. A Precision Micro Hole Machining System Based on Femtosecond Laser. CN202011627090.9, 16 April 2021.
40. Lu, Y.C.; Jia, J.C. A Femtosecond Laser Processing System. CN214518289U, 29 October 2021.
41. Zhang, R.; Wang, Q.; Huang, C.; Wang, J.; Tang, A.; Zhao, W. Energy transfer between femtosecond laser and silicon carbide. *JOM* **2023**, *75*, 4047–4058. [CrossRef]
42. Xia, B. Mechanism and Online Observation of High-Aspect-Ratio, High-Quality Micro Holes Drilling with Femtosecond Laser. Ph.D. Thesis, Department of Electrical Engineering, Beijing Institute of Technology, Beijing, China, 2016.
43. Zhang, R. Study on the Machining Performance and Material Removal Mechanism of Single-Crystal Silicon Carbide Substrate by Femtosecond Lasers. Ph.D. Thesis, Department of Electrical Engineering, Shandong University, Jinan, China, 2021.
44. Li, W.B. Research on Femtosecond Laser Purse Polishing of Silicon Carbide Ceramic Material. Ph.D. Thesis, Department of Electrical Engineering, Harbin Institute of Technology, Harbin, China, 2011.
45. Xia, B.; Jiang, L.; Li, X.; Yan, X.; Lu, Y. Mechanism and elimination of bending effect in femtosecond laser deep-hole drilling. *Opt. Express* **2015**, *21*, 27853. [CrossRef] [PubMed]
46. Xia, B.; Jiang, L.; Wang, S.; Yan, X.; Liu, P. Femtosecond laser drilling of micro-holes. *Chin. J. Lasers* **2013**, *40*, 0201001. [CrossRef]
47. Liu, C.; Zhang, X.; Gao, L.; Jiang, X.; Li, C.; Yang, T. Feasibility of micro-hole machining in fiber laser trepan drilling of 2.5D C/SiC composite: Experimental investigation and optimization. *Opt. Int. J. Light. Electron. Opt.* **2021**, *242*, 167186. [CrossRef]
48. Wang, Q.W. Generation and Application of Micro Structures on Monocrystalline Silicon Fabricated by Femtosecond Laser Irradiation and Wet Etching. Master's Thesis, Department of Electrical Engineering, Shandong University, Jinan, China, 2021.
49. Wei, H.L. Study on Femtosecond Laser Machining Mechanism and Surface Integrity of Single-Crystal Gallium Nitride. Master's Thesis, Department of Electrical Engineering, Shandong University, Jinan, China, 2023.
50. Rehman, Z.U.; Janulewicz, K.A. Structural transformations in femtosecond laser-processed n-type 4H-SiC. *Appl. Surf. Sci.* **2016**, *385*, 1–8. [CrossRef]
51. Zhang, R.; Huang, C.; Wang, J.; Wang, Q.; Feng, S.; Zhao, W.; Tang, A. Analysis of the microscopic characteristics of periodic structure arrays on silicon carbide fabricated by femtosecond laser. *Ceram. Int.* **2024**, *50*, 1193–1204. [CrossRef]

52. Chen, X.; Zhou, W.; Feng, Q.; Zheng, J.; Liu, X.; Tang, B.; Li, J.; Xue, J.; Peng, S. Irradiation effects in 6H-SiC induced by neutron and heavy ions: Raman spectroscopy and high-resolution XRD analysis. *J. Nucal Mater.* **2016**, *478*, 215–221. [CrossRef]
53. Gokus, T.; Albert, J.; Danilov, A.; Paul, S.; Huber, A.J. Time-resolved THz-TDS nanoscopy for probing carrier dynamics with femtosecond temporal and nanometer spatial resolution. In Proceedings of the International Conference on Infrared Millimeter and Terahertz Waves, Montreal, QC, Canada, 17–22 September 2023; pp. 1–2.
54. Li, L. An Investigation into the Micro/Nano Machining Process for Germanium Substrates Using Femtosecond Lasers. Ph.D. Thesis, Department of Electrical Engineering, The University of New South Wales, Sydney, Australia, 2019.
55. Shugaev, M.V.; He, M.; Levy, Y. *Handbook of Laser Micro-and Nano-Engineering*; Springer: Berlin, Germany, 2021; pp. 83–149.
56. Bonse, J.; Baudach, S.; Krüger, J.; Kautek, W.; Lenzner, M. Femtosecond laser ablation of silicon modification thresholds and morphology. *Appl. Phys. A* **2002**, *74*, 19–25. [CrossRef]
57. Wan, D.P.; Wang, J.; Mathew, P. Energy deposition and non-thermal ablation in femtosecond laser grooving of silicon. *Mach. Sci. Technol.* **2011**, *15*, 263–283. [CrossRef]
58. Tsibidis, G.D.; Stratakis, E.; Loukakos, P.A.; Fotakis, C. Controlled ultrashort-pulse laser-induced ripple formation on semiconductors. *Appl. Phys. A* **2014**, *114*, 57–68. [CrossRef]
59. Zhang, R.; Huang, C.; Wang, J.; Feng, S.; Zhu, H. Evolution of micro/nano-structural arrays on crystalline silicon carbide by femtosecond laser ablation. *Mater. Sci. Semicond. Process.* **2021**, *121*, 105299. [CrossRef]
60. Gai, X.C. Research on Experimental System and Technique of Femtosecond Laser Micromachining. Master's Thesis, Department of Electrical Engineering, Harbin Institute of Technology, Harbin, China, 2013.
61. Zhang, R.; Huang, C.; Wang, J.; Chu, D.; Liu, D.; Feng, S. Experimental investigation and optimization of femtosecond laser processing parameters of silicon carbide–based on response surface methodology. *Ceram. Int.* **2022**, *48*, 14507–14517. [CrossRef]
62. Zhang, R.; Huang, C.; Wang, J.; Zhu, H.; Liu, H. Fabrication of high-aspect-ratio grooves with high surface quality by using femtosecond laser. *Ind. Lubr. Tribol.* **2021**, *73*, 718–726. [CrossRef]
63. Li, L.; Wang, J. Direct writing of large-area micro/nano-structural arrays on single crystalline germanium substrates using femtosecond lasers. *Appl. Phys. Lett.* **2017**, *110*, 251901. [CrossRef]
64. Gemini, L.; Hashida, M.; Shimizu, M.; Miyasaka, Y.; Inoue, S.; Tokita, S.; Limpouch, J.; Mocek, T.; Sakabe, S. Metal-like self-organization of periodic nanostructures on silicon and silicon carbide under femtosecond laser pulses. *J. Appl. Phys.* **2013**, *114*, 194903. [CrossRef]
65. Tan, B.; Venkatakrishnan, K. A femtosecond laser-induced periodical surface structure on crystalline silicon. *J. Micromech. Microeng.* **2006**, *16*, 1080. [CrossRef]
66. Varlamova, O.; Reif, J.; Varlamov, S.; Bestehorn, M. The laser polarization as control parameter in the formation of laser-induced periodic surface structures: Comparison of numerical and experimental results. *Appl. Surf. Sci.* **2011**, *257*, 5465–5469. [CrossRef]
67. Hernández, A.; Kudriavtsev, Y.; Gallardo, S.; Avendaño, M.; Asomoza, R. Formation of self-organized nano-surfaces on III–V semiconductors by low energy oxygen ion bombardment. *Mater. Sci. Semicond. Process.* **2015**, *37*, 190–198. [CrossRef]
68. Sipe, J.E.; Young, J.F.; Preston, J.S.; Van Driel, H.M. Laser-induced periodic surface structure. *Phys. Rev. B* **1983**, *27*, 1141–1154. [CrossRef]
69. Ji, X. Geometrical Morphologies Adjustment of Micro/Nanostructures on the Silicon Surface Induced by Femtosecond Laser. Ph.D. Thesis, Department of Electrical Engineering, Beijing Institute of Technology, Beijing, China, 2015.
70. Tangwarodomnukun, V.; Eng, B.; Eng, M. Towards Damage-Free Micro-Fabrication of Silicon Substrates Using a Hybrid Laser-Water Jet Technology. Ph.D. Thesis, Department of Electrical Engineering, The University of New South Wales, Sydney, Australia, 2012.
71. Roma, G. Linear response calculation of first-order Raman spectra of point defects in silicon carbide. *Phys. Status Solidi A* **2016**, *213*, 2995–2999. [CrossRef]
72. Xie, X.; Hu, X.; Chen, X.; Liu, F.; Yang, X.; Xu, X.; Wang, H.; Li, J.; Yu, P.; Wang, R. Characterization of the three-dimensional residual stress distribution in SiC bulk crystals by neutron diffraction. *CrystEngComm* **2017**, *19*, 6527–6532. [CrossRef]
73. Menzel, R.; Gärtner, K.; Wesch, W.; Hobert, H. Damage production in semiconductor materials by a focused Ga+ ion beam. *J. Appl. Phys.* **2000**, *88*, 5658–5661. [CrossRef]
74. Zheng, Q.; Mei, X.; Jiang, G.; Yan, Z.; Fan, Z.; Wang, W.; Pan, A.; Cui, J. Influence of surface morphology and processing parameters on polishing of silicon carbide ceramics using femtosecond laser pulses. *Surf. Interfaces* **2023**, *36*, 102528. [CrossRef]
75. Chen, J.; Chen, X.; Zhang, X.; Zhang, W. Effect of laser incidence angle on the femtosecond laser ablation characteristics of silicon carbide ceramics. *Opt. Laser Eng.* **2024**, *36*, 107849. [CrossRef]
76. Qi, L.; Nishii, K.; Yasui, M.; Aoki, H.; Namba, Y. Femtosecond laser ablation of sapphire on different crystallographic facet planes by single and multiple laser pulses irradiation. *Opt. Laser Eng.* **2010**, *48*, 1000–1007. [CrossRef]
77. Manickam, S.; Wang, J.; Huang, C. Laser-material interaction and grooving performance in ultrafast laser ablation of crystalline germanium under ambient conditions. *J. Eng. Manuf.* **2013**, *227*, 1714–1723. [CrossRef]
78. Qi, Y. Experimental Studies on Femtosecond Laser Ablation of Semiconductor and Metals. Ph.D. Thesis, Department of Electrical Engineering, Jilin University, Changchun, China, 2014.
79. Yang, M. Femtosecond Laser Induced Microstructures and Nanostructures on Silicon Surface. Ph.D. Thesis, Department of Electrical Engineering, Nankai University, Tianjin, China, 2014.

80. Karkantonis, T.; Gaddam, A.; Tao, X.; See, T.L.; Dimov, S. The influence of processing environment on laser-induced periodic surface structures generated with green nanosecond laser. *Surf. Interfaces* **2022**, *31*, 102096. [CrossRef]
81. Mačernytė, L.; Skruibis, J.; Vaičaitis, V.; Sirutkaitis, R.; Balachninaitė, O. Femtosecond laser micromachining of soda–lime glass in ambient air and under various aqueous solutions. *Micromachines* **2019**, *10*, 354. [CrossRef]
82. Yu, X.; Jiang, L.; Luan, Q.; Cai, Y.; Song, Q.; Wang, B.; Liu, Z. Investigation of mechanism and surface morphology on the femtosecond laser ablation of silicon nitride under different auxiliary processing environments. *Ceram. Int.* **2023**, *49*, 13425–13434. [CrossRef]
83. Li, Y.; Chen, T.; Pan, A.; Li, C.; Tang, L. Parallel fabrication of high-aspect-ratio all-silicon grooves using femtosecond laser irradiation and wet etching. *J. Micromech. Microeng.* **2015**, *25*, 115001. [CrossRef]
84. Liu, C.; Wang, R. A Device and Measurement Method for Measuring Femtosecond Laser Contrast. 202010193927.7, 1 May 2020.
85. Zhang, J.Y.; Yan, Z.; Gao, J.C. A Device and Method for Reducing the Time Required for Introducing Femtosecond Lasers into Structures. 202010962160.X, 14 September 2020.
86. Wang, W.; Qi, D.; Lei, P.; Shi, W.; Li, Z.; Zhang, J.; Ho, W.; Zheng, H. Chemical etching-assisted femtosecond laser multi-beam rapid preparation of As_2Se_3 microlens arrays. *J. Manuf. Process.* **2024**, *120*, 460–466. [CrossRef]
87. Roberts, D.; du Plessis, A.; Botha, L. Femtosecond laser ablation of silver foil with single and double pulses. *Appl. Surf. Sci.* **2010**, *256*, 1784–1792. [CrossRef]
88. Le Harzic, R.; Breitling, D.; Sommer, S.; Föhl, C.; König, K.; Dausinger, F.; Audouard, E. Processing of metals by double pulses with short laser pulses. *Appl. Phys. A* **2005**, *81*, 1121–1125. [CrossRef]
89. Du, G.; Yu, F.; Waqas, A.; Chen, F. Ultrafast thermalization dynamics in silicon wafer excited by femtosecond laser double-pulse vortex beam. *Opt. Laser Technol.* **2024**, *174*, 110619. [CrossRef]
90. Du, G.; Yu, F.; Lu, Y.; Kai, L.; Yang, Q.; Hou, X.; Chen, F. Ultrafast thermalization dynamics in Au/Ni film excited by femtosecond laser double-pulse vortex beam. *Int. J. Therm. Sci.* **2023**, *187*, 108208. [CrossRef]
91. Zhao, L.L.; Wang, F.; Xie, J.; Zhao, W.W. Fabrication of high-aspect-ratio structural change microregions in silicon carbide by femtosecond Bessel beams. *Adv. Mater. Res.* **2015**, *1102*, 143–147. [CrossRef]
92. Zhou, K.; Yuan, Y.; Wang, C.; Zhang, K.; Chen, J.; He, H. Rapid fabrication of antireflective structures on ZnS surface by spatial shaping femtosecond laser. *Opt. Laser Technol.* **2024**, *171*, 110393. [CrossRef]
93. Li, X.; Chen, F.; Bao, W.; Wang, R.; Qiao, X. Beam-shaping device-free femtosecond laser plane-by-plane inscription of high-quality FBGs. *Opt. Laser Technol.* **2023**, *161*, 109226. [CrossRef]
94. Zhu, H. An Investigation into The Micro Grooving Process for Germanium Substrates Using a Hybrid Laser-Waterjet Technology. Ph.D. Thesis, Department of Electrical Engineering, The University of New South Wales, Sydney, Australia, 2016.
95. Hu, T.; Yuan, S.; Wei, J.; Zhou, N.; Zhang, Z.; Zhang, J.; Li, X. Water jet guided laser grooving of SiC_f/SiC ceramic matrix composites. *Opt. Laser Technol.* **2024**, *168*, 109991. [CrossRef]
96. Li, Z.; Duan, L.; Zhao, R.; Zhang, Y.; Wang, X. Experimental investigation and optimization of modification during coaxial waterjet-assisted femtosecond laser drilling. *Opt. Laser Technol.* **2024**, *177*, 111072. [CrossRef]
97. Li, C.; Shi, X.; Si, J.; Chen, T.; Chen, F.; Liang, S.; Wu, Z.; Hou, X. Alcohol-assisted photoetching of silicon carbide with a femtosecond laser. *Opt. Commun.* **2009**, *282*, 78–80. [CrossRef]
98. Ma, Y.; Shi, H.; Si, J.; Ren, H.; Chen, T.; Chen, F.; Hou, X. High-aspect-ratio grooves fabricated in silicon by a single pass of femtosecond laser pulses. *J. Appl. Phys.* **2012**, *111*, 093102. [CrossRef]
99. Zheng, Q.; Fan, Z.; Jiang, G.; Pan, A.; Yan, Z.; Lin, Q.; Cui, J.; Wang, W.; Mei, X. Mechanism and morphology control of underwater femtosecond laser micro grooving of silicon carbide ceramics. *Opt. Express* **2019**, *27*, 2–18. [CrossRef] [PubMed]
100. Charee, W.; Tangwarodomnukun, V.; Dumkum, C. Laser ablation of silicon in water under different flow rates. *Int. J. Adv. Manuf. Technol.* **2015**, *78*, 19–29. [CrossRef]
101. Wang, W.; Song, H.; Liao, K.; Mei, X. Water-assisted femtosecond laser drilling of 4H-SiC to eliminate cracks and surface material shedding. *Int. J. Adv. Manuf. Technol.* **2020**, *112*, 553–562. [CrossRef]
102. Ren, N.; Gao, F.; Wang, H.; Xia, K.; Song, S.; Yang, H. Water-induced effect on femtosecond laser layered ring trepanning in silicon carbide ceramic sheets using low-to-high pulse repetition rate. *Opt. Commun.* **2021**, *496*, 127040. [CrossRef]
103. Liu, B.; Fan, P.; Song, H.; Liao, K.; Wang, W. Fabrication of 4H–SiC microvias using a femtosecond laser assisted by a protective layer. *Opt. Mater.* **2022**, *123*, 111695. [CrossRef]
104. Wei, J.; Yuan, S.; Zhang, J.; Zhou, N.; Zhang, W.; Li, J.; An, W.; Gao, M.; Fu, Y. Femtosecond laser ablation behavior of SiC/SiC composites in air and water environment. *Corros. Sci.* **2022**, *208*, 110671. [CrossRef]
105. Wu, C.; Fang, X.; Kang, Q.; Sun, H.; Zhao, L.; Tian, B.; Fang, Z.; Pan, M.; Maeda, R.; Jiang, Z. Crystal cleavage, periodic nanostructure and surface modification of SiC ablated by femtosecond laser in different media. *Surf. Coat. Technol.* **2021**, *424*, 127652. [CrossRef]
106. Khuat, V.; Ma, Y.; Si, J.; Chen, T.; Chen, F.; Hou, X. Fabrication of through holes in silicon carbide using femtosecond laser irradiation and acid etching. *Appl. Surf. Sci.* **2014**, *289*, 529–532. [CrossRef]
107. Gao, B.G.B.; Chen, T.C.T.; Khuat, V.K.V.; Si, J.S.J.; Hou, X.H.A.X. Fabrication of grating structures on silicon carbide by femtosecond laser irradiation and wet etching. *Chin. Opt. Lett.* **2016**, *14*, 021407. [CrossRef]
108. Ma, Y.; Khuat, V.; Pan, A. A simple method for well-defined and clean all-SiC nano-ripples in ambient air. *Opt. Lasers Eng.* **2016**, *82*, 141–147. [CrossRef]

109. Liang, Y.C.; Li, Y.E.; Liu, Y.H.; Kuo, J.F.; Cheng, C.W.; Lee, A.C. High-quality structures on 4H-SiC fabricated by femtosecond laser LIPSS and chemical etching. *Opt. Laser Technol.* **2023**, *163*, 109437. [CrossRef]
110. Wang, Q.; Yao, P.; Chu, D.; Qu, S.; He, W.; Xu, X.; Zhu, H.; Zou, B.; Liu, H.; Huang, C. Array structure of monocrystalline silicon surface processed by femtosecond laser machining assisted with anisotropic chemical etching. *Opt. Laser Technol.* **2024**, *169*, 110165. [CrossRef]
111. Wang, Q.; Yao, P.; Li, Y.; Jiang, L.; Xu, J.; Liang, S.; Chu, D.; He, W.; Huang, C.; Zhu, H.; et al. Inverted pyramid structure on monocrystalline silicon processed by wet etching after femtosecond laser machining in air and deionized water. *Opt. Laser Technol.* **2023**, *157*, 108647. [CrossRef]
112. Song, Y.; Xu, J.; Liu, Z.; Zhang, A.; Yu, J.; Qi, J.; Chen, W.; Cheng, Y. Fabrication of high-aspect-ratio fused silica microstructures with large depths using Bessel-beam femtosecond laser-assisted etching. *Opt. Laser Technol.* **2024**, *170*, 110305. [CrossRef]
113. Li, X.; Yuan, S.; Zhou, N.; Wei, J.; Gao, M.; Hu, T.; Shi, X. Investigation on femtosecond laser combined with dynamic wet etching machining of SiC/SiC. *J. Eur. Ceram. Soc.* **2024**, 1–15. [CrossRef]
114. Xu, J.; Zhang, G.; Lv, J.; Guo, Z.; Wang, F.; Zhang, Y.; Xu, J.; Cheng, G. High-efficiency localized electrochemical deposition based on ultrafast laser surface modification. *Surf. Coat. Technol.* **2023**, *471*, 129923. [CrossRef]
115. Mao, W.P.; Ding, W.; Zhang, C.Y. Detection and experiment of impact cavitation in laser electrochemical composite processing. *Laser Technol.* **2014**, *38*, 753–758.
116. Yao, Y.H. Study on the friction reduction effect of laser electrochemical composite micromachining groove texture. *Wirel. Internet Technol.* **2020**, *38*, 130–133.
117. Chang, Y.B.; He, L.; Song, Y.Z. Research on electrolyte circulation control system for laser electrolytic composite processing. *Group Technol. Prod. Modif.* **2016**, *33*, 55–59.
118. Wang, Y.F.; Yang, Y.; Zhang, W.W. Research status and prospects of laser and electrolytic composite processing technology. *Electr. Mach. Molds* **2022**, *4*, 1–13.
119. Wu, X.L. Research on the Mechanism of Micro-Hole Processing by Ultrasonic Vibration Compounded with Femtosecond Laser. Master's Thesis, Department of Electrical Engineering, Anhui Jianzhu University, Hefei, China, 2022.
120. Zheng, Q.; Mei, X.; Jiang, G.; Cui, J.; Fan, Z.; Wang, W.; Yan, Z.; Guo, H.; Pan, A. Investigation on ultrasonic vibration-assisted femtosecond laser polishing of C/SiC composites. *J. Eur. Ceram. Soc.* **2023**, *43*, 4656–4672. [CrossRef]
121. Wang, H.; Zhu, S.; Asundi, A.; Xu, Y. Experimental characterization of laser trepanning performance enhanced by water-based ultrasonic assistance. *Opt. Laser Technol.* **2019**, *109*, 547–560. [CrossRef]
122. Yuan, Z.Z. Design of Ultrasonic Vibration and Jet-Assisted Laser Composite Machining System. Master's Thesis, Department of Electrical Engineering, Anhui Jianzhu University, Hefei, China, 2018.
123. Wang, C.; Kurokawa, S.; Doi, T.; Yuan, J.; Sano, Y.; Aida, H.; Zhang, K.; Deng, Q. The polishing effect of SiC substrates in femtosecond laser irradiation assisted chemical mechanical polishing (CMP). *ECS J. Solid State Sci. Technol.* **2017**, *6*, 105–112. [CrossRef]
124. Xie, X.; Peng, Q.; Chen, G.; Li, J.; Long, J.; Pan, G. Femtosecond laser modification of silicon carbide substrates and its influence on CMP process. *Ceram. Int.* **2021**, *47*, 13322–13330. [CrossRef]
125. Chen, Y.D.; Liu, H.Y.; Cheng, C.Y.; Chen, C. Comparison of C face (000-1) and Si face (0001) of silicon carbide wafers in femtosecond laser irradiation assisted chemical–mechanical polishing process. *Appl. Phys. A* **2022**, *128*, 1094. [CrossRef]
126. Chen, G.; Li, J.; Long, J.; Luo, H.; Zhou, Y.; Xie, X.; Pan, G. Surface modulation to enhance chemical mechanical polishing performance of sliced silicon carbide Si-face. *Appl. Surf. Sci.* **2021**, *536*, 147963. [CrossRef]
127. Wang, H.; Guan, Y.; Zheng, H. Smooth polishing of femtosecond laser induced craters on cemented carbide by ultrasonic vibration method. *Appl. Surf. Sci.* **2017**, *426*, 399–405. [CrossRef]
128. Huang, Y.; Tang, F.; Guo, Z.; Wang, X. Accelerated ICP etching of 6H-SiC by femtosecond laser modification. *Appl. Surf. Sci.* **2019**, *488*, 853–864. [CrossRef]
129. Wu, C.; Fang, X.; Liu, F.; Guo, X.; Maeda, R.; Jiang, Z. High speed and low roughness micromachining of silicon carbide by plasma etching aided femtosecond laser processing. *Ceram. Int.* **2020**, *46*, 17896–17902. [CrossRef]
130. Chen, C.-Y.; Chung, C.-J.; Wu, B.-H.; Li, W.-L.; Chien, C.-W.; Wu, P.-H.; Cheng, C.-W. Microstructure and lubricating property of ultra-fast laser pulse textured silicon carbide seals. *Appl. Phys. A* **2012**, *107*, 345–350. [CrossRef]
131. Murzin, S.P.; Balyakin, V.B. Micro structuring the surface of silicon carbide ceramic by laser action for reducing friction losses in rolling bearings. *Opt. Laser Technol.* **2017**, *88*, 96–98. [CrossRef]
132. Huang, L.; Maerkl, S.J.; Martin, O.J. Integration of plasmonic trapping in a microfluidic environment. *Opt. Express* **2009**, *17*, 6018–6024. [CrossRef] [PubMed]
133. Ong, W.-L.; Kee, J.-S.; Ajay, A.; Ranganathan, N.; Tang, K.-C.; Yobas, L. Buried microfluidic channel for integrated patch-clamping assay. *Appl. Phys. Lett.* **2006**, *89*, 093902. [CrossRef]
134. Jugessur, A.S.; Dou, J.; Aitchison, J.S.; De La Rue, R.M.; Gnan, M. A photonic nano-Bragg grating device integrated with microfluidic channels for bio-sensing applications. *Microelectron. Eng.* **2009**, *86*, 1488–1490. [CrossRef]
135. Zhao, G.; Xia, H.; Zhang, Y.; Li, L.; He, N.; Hansen, H.N. Laser-induced oxidation assisted micro milling of high aspect ratio micro-groove on WC-Co cemented carbide. *Chin. J. Aeronaut.* **2020**, *34*, 465–475. [CrossRef]
136. Liu, N.; Sun, Y.; Wang, H.; Liang, C. Femtosecond laser-induced nanostructures on Fe-30Mn surfaces for biomedical applications. *Opt. Laser Technol.* **2021**, *139*, 106986. [CrossRef]

137. Kodama, S.; Suzuki, S.; Hayashibe, K.; Shimada, K.; Mizutani, M.; Kuriyagawa, T. Control of short-pulsed laser induced periodic surface structures with machining picosecond laser micro/nanotexturing with ultraprecision cutting. *Precis. Eng.* **2019**, *55*, 433–438. [CrossRef]
138. Vorobyev, A.Y.; Guo, C. Multifunctional surfaces produced by femtosecond laser pulses. *J. Appl. Phys.* **2015**, *117*, 033103. [CrossRef]
139. Yin, K.; Duan, J.; Sun, X.; Wang, C.; Luo, Z. Formation of superwetting surface with line-patterned nanostructure on sapphire induced by femtosecond laser. *Appl. Phys. A* **2015**, *119*, 69–74. [CrossRef]
140. Song, Y.M.; Xie, Y.; Malyarchuk, V.; Xiao, J.; Jung, I.; Choi, K.-J.; Liu, Z.; Park, H.; Lu, C.; Kim, R.-H.; et al. Digital cameras with designs inspired by the arthropod eye. *Nature* **2013**, *497*, 95–99. [CrossRef] [PubMed]
141. Meng, B.; Yuan, D.; Zheng, J.; Xu, S. Molecular dynamics study on femtosecond laser aided machining of monocrystalline silicon carbide. *Mater. Sci. Semicond. Process.* **2019**, *101*, 1–9. [CrossRef]
142. Sánchez, M.I.; Delaporte, P.; Spiegel, Y.; Franta, B.; Mazur, E.; Sarnet, T. A laser-processed silicon solar cell with photovoltaic efficiency in the infrared. *Phys. Status Solidi (A)* **2021**, *218*, 2000550. [CrossRef]
143. Deki, M.; Yamamoto, M.; Ito, T.; Tomita, T.; Matsuo, S.; Hashimoto, S.; Kitada, T.; Isu, T.; Onoda, S.; Ohshima, T.; et al. Femtosecond laser modification aiming at the enhancement of local electric conductivities in SiC. *AIP Conf. Proc.* **2011**, *1399*, 119–120.
144. Zhao, Q.Z.; Ciobanu, F.; Malzer, S.; Wang, L.J. Enhancement of optical absorption and photocurrent of 6H-SiC by laser surface nanostructuring. *Appl. Phys. Lett.* **2007**, *91*, 121107. [CrossRef]
145. Pecholt, B.; Vendan, M.; Dong, Y.; Molian, P. Ultrafast laser micromachining of 3C-SiC thin films for MEMS device fabrication. *Int. J. Adv. Manuf. Technol.* **2008**, *39*, 239–250. [CrossRef]
146. Fang, R.; Zhang, H.; Zheng, J.; Li, R.; Wang, X.; Luo, C.; Yang, S.; Li, S.; Li, C.; Chen, Y.; et al. High-temperature silicon carbide material with wicking and evaporative cooling functionalities fabricated by femtosecond laser surface nano/microstructuring. *Ceram. Int.* **2023**, *49*, 20138–20147. [CrossRef]
147. Papanasam, E.; Kumar, P.; Chanthini, B.; Manikandan, E.; Agarwal, L. A Comprehensive Review of Recent Progress, Prospect and Challenges of Silicon Carbide and its Applications. *Silicon* **2022**, *14*, 12887–12900.
148. Monica, V.; Paul, M. Femtosecond pulsed laser microfabrication of SiC MEMS micro gripper. *J. Laser Appl.* **2007**, *19*, 149–154.
149. Marsi, N.; Majlis, B.Y.; Hamzah, A.A.; Mohd-Yasin, F. High Reliability of MEMS Packaged Capacitive Pressure Sensor Employing 3C-SiC for High Temperature. *Energy Procedia* **2015**, *68*, 471–479. [CrossRef]
150. Suster, M.; Ko, W.; Young, D. An optically powered wireless telemetry module for high-temperature MEMS sensing and communication. *IEEE J. Microelectromech. Syst.* **2004**, *13*, 536–541. [CrossRef]
151. Zhao, Y.; Zhao, Y.L.; Wang, L.K. Application of femtosecond laser micromachining in silicon carbide deep etching for fabricating sensitive diaphragm of high temperature pressure sensor. *Sens. Actuators A Phys.* **2020**, *309*, 112017. [CrossRef]
152. Zheng, J.X.; Tian, K.S.; Qi, J.Y.; Guo, M.R.; Liu, X.Q. Advances in fabrication of micro-optical components by femtosecond laser with etching technology. *Opt. Laser Technol.* **2023**, *167*, 109793. [CrossRef]
153. Nguyen, T.K.; Phan, H.P.; Dinh, T.; Dowling, K.M.; Foisal, A.R.; Senesky, D.G.; Nguyen, N.T.; Dao, D.V. Highly sensitive 4H-SiC pressure sensor at cryogenic and elevated temperatures. *Mater. Des.* **2018**, *156*, 441–445. [CrossRef]
154. Gupta, S.; Molian, P. Design of laser micro machined single crystal 6H–Sic diaphragms for high-temperature micro-electro-mechanical-system pressure sensors. *Mater. Des.* **2011**, *32*, 127–132. [CrossRef]
155. Zehetner, J.; Vanko, G.; Dzuba, J.; Ryger, I.; Lalinsky, T.; Benkler, M.; Lucki, M. Laser ablation for membrane processing of AlGaN/GaN- and micro structured ferroelectric thin film MEMS and SiC pressure sensors for extreme conditions, Barcelona, Spain. *Proc. SPIE* **2015**, *9517*, 951721.
156. Zehetner, J.; Kraus, S.; Lucki, M.; Vanko, G.; Dzuba, J.; Lalinsky, T. Manufacturing of membranes by laser ablation in SiC, sapphire, glass and ceramic for GaN/ferroelectric thin film MEMS and pressure sensors. *Microsyst. Technol. Micro Nano Syst. Inf. Storage Process. Syst.* **2016**, *22*, 1883–1892. [CrossRef]
157. Dong, Y.; Nair, R.; Molian, R.; Molian, P. Femtosecond-pulsed laser micromachining of a 4H–SiC wafer for MEMS pressure sensor diaphragms and via holes. *J. Micromech. Microeng.* **2008**, *18*, 035022. [CrossRef]
158. Wang, L.; Zhao, Y.; Yang, Y.; Zhao, Y. Two-step femtosecond laser etching for bulk micromachining of 4H-SiC membrane applied in pressure sensing. *Ceram. Int.* **2022**, *48*, 12359–12367. [CrossRef]
159. Wang, L.; Zhao, Y.; Zhao, Y.; Yang, Y.; Li, B.; Gong, T. Mass fabrication of 4H-SiC high temperature pressure sensors by femtosecond laser etching. In Proceedings of the 16th Annual IEEE International Conference on Nano/Micro Engineered and Molecular Systems, Xiamen, China, 25–29 April 2021; pp. 1478–1481.
160. Wang, L.; Zhao, Y.; Yang, Y.; Pang, X.; Hao, L.; Zhao, Y. Piezoresistive 4H-SiC pressure sensor with diaphragm realized by femtosecond laser. *IEEE Sens. J.* **2022**, *22*, 11535–11542. [CrossRef]
161. Osipov, A.A.; Iankevich, G.A.; Osipov, A.A.; Speshilova, A.B.; Karakchieva, A.A.; Endiiarova, E.V.; Levina, S.N.; Karakchiev, S.V.; Alexandrov, S.E. Silicon carbide dry etching technique for pressure sensors design. *J. Manuf. Process.* **2022**, *73*, 316–325. [CrossRef]
162. Wang, L.; Zhao, Y.; Yang, Y.; Pang, X.; Hao, L.; Zhao, Y.; Liu, J. Development of laser-micromachined 4H-SiC MEMS piezoresistive pressure sensors for corrosive environments. *IEEE Trans. Electron Devices* **2022**, *69*, 2009–2014. [CrossRef]
163. Yang, Y.; Zhao, Y.; Wang, L.; Zhao, Y. Application of femtosecond laser etching in the fabrication of bulk SiC accelerometer. *J. Mater. Res. Technol.* **2022**, *17*, 2577–2586. [CrossRef]

164. Wang, L.; Zhao, Y.; Yang, Z.; Zhao, Y.; Yang, X.; Gong, T.; Li, C. Femtosecond laser micromachining in combination with ICP etching for 4H–SiC pressure sensor membranes. *Ceram. Int.* **2021**, *47*, 6397–6408. [CrossRef]
165. Wu, C.; Fang, X.; Fang, Z.; Sun, H.; Li, S.; Zhao, L.; Tian, B.; Zhong, M.; Maeda, R.; Jiang, Z. Fabrication of 4H-SiC piezoresistive pressure sensor for high temperature using an integrated femtosecond laser-assisted plasma etching method. *Ceram. Int.* **2023**, *49*, 29467–29476. [CrossRef]

Disclaimer/Publisher's Note: The statements, opinions and data contained in all publications are solely those of the individual author(s) and contributor(s) and not of MDPI and/or the editor(s). MDPI and/or the editor(s) disclaim responsibility for any injury to people or property resulting from any ideas, methods, instructions or products referred to in the content.

Article

Understanding the Effect of Dispersant Rheology and Binder Decomposition on 3D Printing of a Solid Oxide Fuel Cell

Man Yang [1], Santosh Kumar Parupelli [1,2], Zhigang Xu [3] and Salil Desai [1,2,*]

1. Industrial and Systems Engineering, North Carolina A & T State University, Greensboro, NC 27411, USA; amandayang2005@gmail.com (M.Y.); sparupel@ncat.edu (S.K.P.)
2. Center of Excellence in Product Design and Advanced Manufacturing, North Carolina A & T State University, Greensboro, NC 27411, USA
3. Mechanical Engineering, North Carolina A & T State University, Greensboro, NC 27411, USA; zhigang@ncat.edu
* Correspondence: sdesai@ncat.edu; Tel.: +1-(336)-285-3725

Abstract: Solid oxide fuel cells (SOFCs) are a green energy technology that offers a cleaner and more efficient alternative to fossil fuels. The efficiency and utility of SOFCs can be enhanced by fabricating miniaturized component structures within the fuel cell footprint. In this research work, the parallel-connected inter-digitized design of micro-single-chamber SOFCs (μ-SC-SOFCs) was fabricated by a direct-write microfabrication technique. To understand and optimize the direct-write process, the cathode electrode slurry was investigated. Initially, the effects of dispersant Triton X-100 on LSCF (La0.6Sr0.2Fe0.8Co0.2O3-δ) slurry rheology was investigated. The effect of binder decomposition on the cathode electrode lines was evaluated, and further, the optimum sintering profile was determined. Results illustrate that the optimum concentration of Triton X-100 for different slurries was around 0.2–0.4% of the LSCF solid loading. A total of 60% of solid loading slurries had high viscosities and attained stability after 300 s. In addition, 40–50% solid loading slurries had relatively lower viscosity and attainted stability after 200 s. Solid loading and binder affected not only the slurry's viscosity but also its rheology behavior. Based on the findings of this research, a slurry with 50% solid loading, 12% binder, and 0.2% dispersant was determined to be the optimal value for the fabricating of SOFCs using the direct-write method. This research work establishes guidelines for fabricating the micro-single-chamber solid oxide fuel cells by optimizing the direct-write slurry deposition process with high accuracy.

Keywords: 3D printing; binder concentration; dispersant; rheology behavior; solid-oxide fuel cells

Citation: Yang, M.; Parupelli, S.K.; Xu, Z.; Desai, S. Understanding the Effect of Dispersant Rheology and Binder Decomposition on 3D Printing of a Solid Oxide Fuel Cell. *Micromachines* 2024, 15, 636. https://doi.org/10.3390/mi15050636

Academic Editors: Yao Liu and Jinjie Zhou

Received: 8 April 2024
Revised: 5 May 2024
Accepted: 8 May 2024
Published: 9 May 2024

Copyright: © 2024 by the authors. Licensee MDPI, Basel, Switzerland. This article is an open access article distributed under the terms and conditions of the Creative Commons Attribution (CC BY) license (https://creativecommons.org/licenses/by/4.0/).

1. Introduction

The dramatic convergence of global economic evolution, technological advancements, a fast-growing population, and social improvements have escalated the demand for energy resources [1–3]. Currently, there exist several green methods [4,5] to produce power which include solar power [6], hydropower [7], wind power [8], nuclear power [9], geothermal power [10], biodiesel [11], etc. Nevertheless, the above-mentioned power generation techniques have distinct limitations and drawbacks. Therefore, there is a significant need for the development of novel ways of generating power. Fuel cells, an advanced prospective green energy solution, is obtaining eminence importance for producing power [12,13]. A fuel cell consists of an unglazed earthenware electrolyte and gold/platinum foil electrodes [14,15]. Fuel cells, a green energy technology, have the inherent ability to utilize minimal fuel more energy efficiently than the above-mentioned competing technologies with minimal or negligible pollutant discharges into the environment [16–20]. Generally, a fuel cell consists of three components which include two electrodes—an anode (positive) and a cathode (negative), as well as an electrolyte.

Currently, the highest-temperature fuel cells in development are solid oxide fuel cells which can be operated over a wider temperature range from 600 °C to 1000 °C, allowing for several fuels to be used. Solid oxide, or solid ceramic material, is the electrolyte for high-temperature activities because it conducts oxygen ions (O_2). Along with an operational efficiency of over 60%, this fuel cell has one of the highest rates of energy generation compared to other energy conversion devices [21,22]. This is enabled due to the high operational temperature of SOFCs enabling cogeneration systems to create high-pressure steam, which can be utilized in multiple applications. The development of portable electronics and micro-electromechanical systems (MEMSs) has made miniature fuel cells desirable options for power sources. Micro-solid oxide fuel cells provide shorter recharge times and greater energy densities than traditional electrochemical batteries [23–28].

Researchers are investigating various microfabrication procedures to fabricate μ-SOFCs [29–31] with enhanced efficiency [32–35]. To reduce the ohmic losses in the electrolyte, micropatterning technologies can be used, where electrodes are placed closer together. Moreover, by applying microfabrication, the single-chamber solid oxide fuel cell (SC-SOFC) architecture can be coupled in series or parallel arrays to enhance the overall voltage, current, and power of the assembly. Hibino (1996), the pioneer of SC-SOFC, created a new cell design with an interpenetrating comb-like electrode pattern on the same surface of an electrolytic material with feature sizes in the millimeter range [36]. Chung and Chung applied finite element modeling to simulate a micro-single-chamber solid oxide fuel cells (μ-SC-SOFCs) with a side-by-side architecture [37]. The research aimed to enhance the efficiency and lower the operating temperature of μ-SC-SOFCs. Most studies have used a co-planar configuration to investigate the properties, operating parameters, and output power of various electrodes and electrolyte materials [38,39]. In the direct writing process [40–42], the suspension filament is extruded onto the substrate, which is appropriate for mass production. The suspension can be solidly loaded up to 60% using this approach. The suspension has a higher viscosity than the soft lithography solution. The initiation of the direct ink writing technique for the fabrication of three-dimensional ceramic pieces without depending on costly tooling dies or lithographic masks was spearheaded by Prof. Jennifer A. Lewis [43,44]. Employing this technology enables the fabrication of intricate three-dimensional shapes with custom design patterns without expensive and skilled tooling and masking. For applications in the structural, functional, and biological domains, three-dimensional (3D) ceramic material patterning is crucial [45,46]. Yong Bum Kim [47] effectively used the direct write method in 2005 to build integrated planar solid oxide fuel that was serially coupled. The anode (YSZ + NiO) and cathode (LSCF) were directly written on the side of a zirconia substrate that was moderately stabilized with Yttria. Sung-Jin Ahn et al. [48] extruded the cathode (LSM) and anode (NiO-SDC-Pd) using a syringe nozzle onto the YSZ substrate in the year 2006 using the direct write technique. Anode and cathode widths were approximately 500 μm and 450 μm, individually. The direct write technique was utilized to fabricate parallel-connected SC-SOFC on an electrolyte pie, involving a method where the electrolyte width exceeded 500 μm, as reported by Melanie Kuhn's research group [49–52] in 2007. With a mean electrode gap of about 1 mm, the experiment determined that both cells maintained stable open circuit voltage at 0.9 V and attained a peak power density of 2.3 mWcm^{-2}.

There exist several slurry compositions with a good level of rheological analysis; however, our research focuses on understanding the composition and rheology of a specific slurry composition for LCSF-based cathode electrodes. Ali et al. [53] research study focused on evaluating LSCF ink composition for powder and electrochemical characterization in screen printing applications. It is important to note that our work focuses on 3D printing of an LSCF cathode electrode which has different rheological requirements compared to screen printing applications. Moreover, the authors in the above-mentioned publication did not evaluate any rheological behavior of slurry which is critical in our case given the 3D printing application. Finally, 3D printing permits high solid loading of 40 to 60% and thus controlled porosity of microstructures as compared to very low viscosity and solid loading content in screen printing inks. Thus, though there are several slurry products and similar compositions in the literature, they do not address important rheological properties required in 3D printing of an LSCF cathode electrode, which our research comprehensively addresses.

It is very crucial to understand the effect of the dispersant, slurry loading, and binder on the direct-write process as it has a significant impact on the fabricated SOFCs. In our research, the focus is on fabricating parallel-connected inter-digitized design µ-SC-SOFCs. The cathode electrode slurry was investigated to understand and optimize the direct-write technique. Initially, the effects of dispersant on LSCF (La0.6Sr0.2Fe0.8Co0.2O3-δ) slurry rheology were investigated. Further, the binder removal process was investigated, and the optimum sintering profile was determined as it has a substantial impact on fuel cell microstructure.

2. Materials and Methods

The direct-writing process of the SC-SOFC in coplanar design is illustrated in Figure 1 as reported in the previous research work [54]. The direct writing system consists of a micro extrusion system, a sample holder, a pressure regulator, and a motion controller. The micro extrusion system incorporates an air cylinder, a reservoir retainer, piston, and a micronozzle. This technique consists of various sequential steps. Initially, the electrolyte substrate is prepared; then, the anode slurry is loaded into the reservoir retainer, which is held in an air cylinder. The air pressure for the piston loaded on the end of the retainer is regulated. By altering the pressure, the slurry is extruded from the micronozzle. Concurrently, the velocity, acceleration, and deceleration of the motion stage are regulated by the motion controller. Consecutive lines of the anode are created on the electrolyte substrate. After that, the platinum wire is attached to the anode slurry, and the annealing of anode lines is completed. Similarly, the cathode slurry is loaded in the reservoir and handled in the above-mentioned manner to build the cathode lines. Finally, the fuel cell is examined to determine its performance [54].

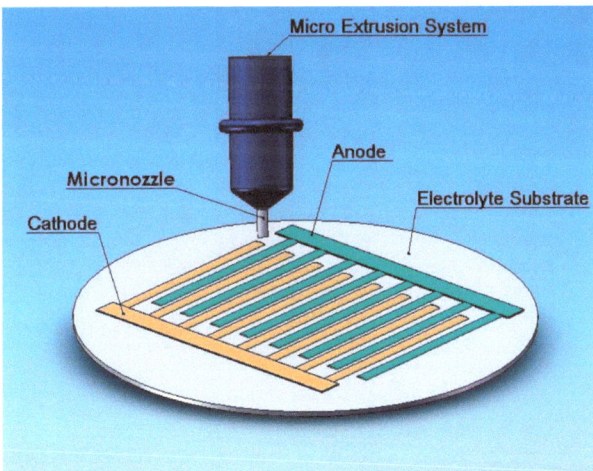

Figure 1. Schematic illustration of a SC-SOFC in coplanar design and process.

The custom microstructure of the anode and cathode can be fabricated by controlling the air pressure, velocity, the mixture ratio of the slurry, and the distance between the nozzle and the substrate. After annealing, the structures of printed electrodes are maintained.

2.1. Requirements of the Slurry for Direct Writing

Nano- or submicron-sized powders exhibit excellent potential for SOFC applications because of their low sintering temperature and high surface area with the capability of microstructure tuning when the attractive forces are substantial [55]. This study's inclination for nano- or submicron powder sizes is corroborated by findings showcasing micro-size fuel cell production employing comparable powder sizes [56]. To accomplish relatively isotropic properties, the powder's particle size should be notably smaller, specifically about an order of magnitude lesser than the micro component's smallest internal dimension [57].

Need for Dispersant

The simultaneous presence of fine particles and high solid loading causes strong particle–particle interactions, which in turn hinders slurry management in the direct writing process. These outcomes enhance slurry viscosity. Particle size, concentration, and viscosity, the entirety of which are impacted by the addition of binder, have a greater impact on the stability and sedimentation of the slurry. Ceramic powders clump due to attractive Van der Waals forces; nevertheless, this can be prevented by employing appropriate dispersants that alter the powder's surface properties. This permits repulsive forces (derived from steric hindrance from large-molecule adsorption or electrostatic repulsion through electrical double-layer overlap) to overcome attractive forces and maintain the dispersion of the particles [56]. A high solid content slurry's viscosity can be considerably reduced by adding dispersants. To stabilize oxide powder slurries, these dispersants are essential in the ceramics sector. Stable, high-solid-content slurries that produce faultless, superior goods demand careful consideration of the appropriate dispersant and the right quantity for a certain ceramic powder. Lacking a dispersant, a lower-energy liquid typically instantly wets and submerges a high-energy ceramic powder surface. These changes have an abrupt effect and cannot be reversed. In contrast, the processes of dispersion and stability are more complex and take different amounts of time to unfold depending on particle density in the suspension.

2.2. Experiments

2.2.1. Cathode Slurry Materials

The cathode material of choice was LSCF powder, which has a surface area of 6.3 m^2/g and was supplied by the Fuel cell material company. The LSCF powders were mixed with organic solvent terpineol, dispersant: Triton X-100 (Sigma-Aldrich, St. Louis, MO, USA), and binder: Polyvinyl butyral (Sigma-Aldrich, St. Louis, MO, USA) for the formulation of the cathode slurry. Between 40 wt%, 50 wt%, and 60 wt%, the solid proportions were altered. The choice of Triton X-100 as the dispersant for stabilizing Yttrium-stabilized zirconia (YSZ) submicron powder was supported by its effective use, as registered in the literature [58]. Triton X-100 ($C_{14}H_{22}O(C_2H_4O)_{10}$) connects a hydrophilic polyethylene oxide chain with a lipophilic or hydrophobic hydrocarbon group. It is a nonionic surfactant that normally contains 9.5 units of ethylene oxide. This compound is employed in the distribution of carbon atoms in pliable composite materials.

2.2.2. Binder Concentration

It is crucial that the fuel cell electrodes incessantly retain their porous structure throughout the sintering procedure. Polyvinyl butyral (PVB) was employed as the binder for this purpose. Since PVB has a larger molecular size, combining it as a binder improves the slurry's viscosity and stability. This advancement in viscosity helps the direct writing method achieve its envisioned purposes by progressing precision. Reaching temperatures of up to 400 °C can eliminate PVB fully [59]. In this dispersant exploration, the binder concentration was held constant at 12 wt% of the solid powder to examine the effects of the dispersant on the rheological properties of the slurry.

2.2.3. Variations in Dispersant and Solid Loading Concentrations

Three distinct slurries with solid contents of 40 wt%, 50 wt%, and 60 wt% were employed. Dispersant concentrations differed between 0.2, 0.4, 0.6, 1, and 1.5 wt % in relation to the solid content. Determining the slurry's pumpability and significance of dipping or coating operations demands a detailed understanding of its rheological properties. An essential metric for evaluating the flow characteristics of various phases of matter, such as gases, liquids, and semi-solids, is viscosity [60]. In this scrutiny, measurements of viscosity were carried out in conjunction with considerations of flow behavior, the effectiveness of the direct writing method, and the caliber of the sintered lines.

The study investigated the rheological behavior of mixtures with varied concentrations, as presented in Table 1, to evaluate the effect of dispersant on slurries with diverse solid contents. Utilizing a SC4-15 concentric cylinder geometry and a rotational viscometer manufactured by Brookfield LVDV-III Ultra Rheometers (AMETEK Brookfield, Middleborough, MA, USA), rheological measurements were taken at 22.4 °C. To guarantee reproducibility, slurry viscosity was measured at least three times ($n = 3$) at shear rates between 0.05 and 7 s^{-1}.

Table 1. Slurry concentration combinations.

Solid Loading (wt%/v)	Dispersant (% of Solid Loading wt)	Binder (% of Solid Loading wt)
40%	0.0	12%
40%	0.2	12%
40%	0.4	12%
40%	0.6	12%
40%	1.0	12%
40%	1.5	12%
50%	0.0	12%
50%	0.2	12%
50%	0.4	12%
50%	0.6	12%
50%	1.0	12%
50%	1.5	12%
60%	0.0	12%
60%	0.2	12%
60%	0.4	12%
60%	0.6	12%
60%	1.0	12%
60%	1.5	12%

2.2.4. Slurry Preparation Process

Two discrete milling steps were employed in slurry preparation. The nano-sized LSCF powder was initially moistened by mixing it with a solvent (50% volume) involving terpineol and Triton dispersion. This mixture was ball-milled for 20 min in a Spex Mixer/Mill (Spex800 M- Antylia Scientific, Vernon Hills, IL, USA). In addition, a mixture of 50% solvent and 12% binder weight was heated to 100 °C for ball milling. Then, for a total of 90 min, the LSCF powder and binder fusion were combined and ball milled.

2.2.5. YSZ Substrate Preparation

In this investigation, 34% of the YSZ particles in the base layer had a size range of 5–10 nm, while 66% of the particles had a particle size of 55 nm. A die with a 25 mm diameter was filled with the YSZ powder blend. The electrolyte base was established by sintering at 1350 °C for ten hours after a pressure of 10,000 lb was applied for five minutes. The sample surfaces were polished after sintering to positively affect cathode trace expansion. A sequence of sandpapers (180-, 400-, 600-, and 800-grain) were handled in this polishing method, with a minimum 15 min interval between each one under a 12-pound weight.

2.2.6. Testing Procedures

The fluid's internal resistance to flow is reflected in the viscosity of the cathode slurry. In Newtonian and non-Newtonian fluids, the correlation between shear stress and shear rate is unlike. When the temperature is fixed, the relationship for Newtonian fluids is linear and has a constant slope throughout a range of shear rates. On the other hand, non-Newtonian fluids display a fluctuating relationship in which viscosity either enhances at higher shear rates (shear-thickening) or drops at higher shear rates (shear thinning) [61]. This phenomenon was observed in the rheological properties of several slurries; the exact way in which they influenced direct writing was also recorded. Essential details about the slurry's rheological behavior are presented below.

Stability of Slurry Viscosity with Respect to Time

The critical idea behind a rotational viscometer is that shear stress or torque is essential while rotating an item in a fluid. It estimates the torque expected to turn a spindle in the fluid at a set speed. During operation, this apparatus produces torque in the slurry. For slurries with differing compositions, different periods ought to achieve the constant viscosity level at a given torque. Consequently, it is imperative to establish the amount of time required to achieve a steady viscosity measurement prior to conducting rheological analyses. In this study, until a consistent viscosity was recorded, examinations were taken at 30 s intervals to ascertain the times desired for the viscosity to stabilize at a constant shear rate of 22.4 °C.

Effect of Shear Rate and Shear Stress Variation on Slurry Viscosity

Once stable times for various slurries were determined, viscosity was evaluated at discrete spindle speeds to examine the rheological characteristics of probable cathode slurries. With this process, the viscosity of the slurry was recorded by declining the shear rate from its greatest value until stability was reached. RheocalcT software (https://store.brookfieldengineering.com/rheocalct-software-standard-edition/) was used for the plotting and analysis of the shear rate against viscosity as well as the analysis of power law and other fluid models, in addition to Herschel–Bulkley and others.

2.2.7. Direct Writing Cathode Lines on the YSZ Substrate

A 100 μm nozzle was utilized to dispense cathode slurries. For individual slurry variants, cathode traces of 10 mm length—four lines per slurry type—were generated on the YSZ substrate under the following conditions: nozzle-to-substrate gap of 100 μm; nozzle velocity of 0.5 mm/s; and extrusion pressures of 200 kPa, 100 kPa, and 30 kPa. Based on the lowest possible pressure essential to extrude a given slurry composition—200 kPa for slurries with 60% solids, 100 kPa for those with 50% solids, and 30 and 15 kPa for those with 40% solids, respectively—direct-writing pressures were preferred. Moreover, 50% solid content slurries were extruded at 200 kPa to compare the direct-writing efficacy of the two mixtures. Similar to this, 40% solid slurries were extruded at 100 kPa to compare the contents with 50% and 40% solids. However, 40% solid slurries were not dispensed at 200 kPa while the resulting line width was considered very high for the manufacturing of cathode electrodes.

Evaluating Width and Height of the Line

Using a Zeiss microscope, cathode trace observations were fabricated. The cathode lines were photographed from both the top and side profiles. Image-Pro Plus software version 11 was employed to determine the height and width of these lines both before and after they were sintered. As the average height of the profile peaked, the line height was documented.

3. Results

3.1. Stability of Slurries with Respect to Time

Due to its high viscosity, the slurry with a 60% solid load stabilized in 300 s. On the other hand, due to their diminished viscosities, the slurries encompassing 50% and 40% of the solid load attained stability after 200 s. The times required to reach stable viscosity rates for slurries holding 60%, 50%, and 40% solids, respectively, are shown in Figure 2. As all these measurements were obtained at the constant temperature of 22.4 °C, it is obvious that the viscosity of each slurry remained constant throughout the experiment.

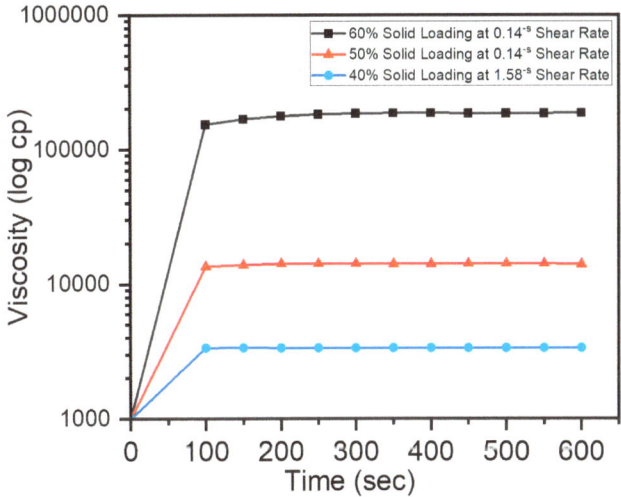

Figure 2. Viscosity plots of different solid loading slurries with a 0.2% dispersant and a 12% binder.

3.2. Rheological Characteristics of Slurries

The rheological behavior of slurries was assessed by observing the apparent viscosity obtained from the shear rate versus shear stress curves at 22.4 °C. Based on various shear rates for different slurries, the shear rate varied from $0.01~\text{s}^{-1}$ to $7~\text{s}^{-1}$. The rheological data were evaluated using the Herschel–Bulkley model, a generalized model for non-Newtonian fluids. The mathematical representation for this three-parameter model is presented in Equation (1), where τ = shear stress; D = shear rate; k = consistency index; n = flow index; and τ^o = yield stress.

$$\tau = \tau^o + kD^n \tag{1}$$

- When $\tau^o = 0$ and $n = 1$, the slurry is a Newtonian fluid.
- When $\tau^o = 0$ and $n < 1$, the slurry is a pseudoplastic fluid.
- When $\tau^o = 0$ and $n > 1$, the slurry is a dilatant fluid (shear-thickening).
- When $\tau^o \neq 0$ and $n = 1$, the slurry is a Bingham fluid.
- When $\tau^o \neq 0$ and $n < 1$, the slurry is a viscoplastic fluid.

In a static state, a Bingham or a viscoplastic fluid behaves like a solid. It takes a specified amount of force—referred to as yield stress—to initiate a flow. The flow starts as soon as this yield threshold is surpassed. Plastic fluids may then demonstrate Newtonian, pseudoplastic, or dilatant flow characteristics. Figures 3–5 illustrate the viscosity–shear rate relationship as well as shear rate and shear stress plots (made possible by Hershel–Bulkley analysis) for slurries with different concentrations which highlight the inimitable flow characteristics of each slurry.

Figure 3a–c show the rheological characteristics of 60% solid loading with 0% and 0.2% dispersants, respectively. The slurry with a 0% dispersant displayed significantly higher viscosity (400,000 cP) and initial yield stress (9.64 D/cm^2) as seen from the Hershel–Bulkley plots. However, an increase in the shear rate resulted in a lowering of viscosity indicating shear thinning (viscoplastic) fluid behavior, whereas the slurry with a 0.2% dispersant had lower viscosity (260,000 cP), initial yield stress (5.14 D/cm^2) and marginal reduction in viscosity with increasing shear rate, indicating a Bingham fluid behavior. This can inferred from the lower flow index of 0.76 for a 0% dispersant (Figure 3b) versus a 1.0 flow index for 0.2% dispersant (Figure 3c) slurries, respectively. The characterization of high solid loading slurries has implications in selecting high extrusion pressures and relatively wider nozzle sizes. In contrast, higher solid loading can benefit from denser LSCF cathode traces.

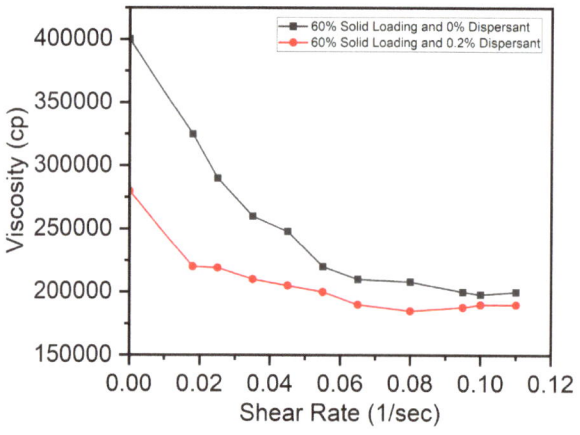

(a)

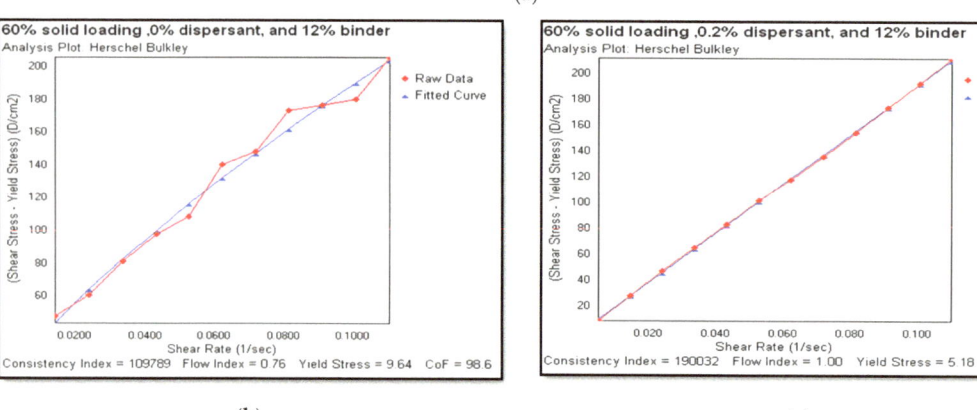

(b) (c)

Figure 3. (a) Viscosity and shear rate plot for 60% solid loadings and a shear rate and shear stress plot (Hershel–Bulkley analysis plot) for slurries with a 60% solid loading and a 12% binder with (b) a 0% dispersant (viscoplastic) and (c) a 0.2% dispersant (Bingham).

Figure 4a–c show the rheological characteristics of a 50% solid loading with 0% and 0.4% dispersants, respectively. The slurry with a 0% dispersant (Figure 4b) displayed relatively higher viscosity (50,000cP) and initial yield stress (3.20 D/cm^2) as seen from the Hershel–Bulkley plots. However, an increase in the shear rate did not result in changes to the flow index (1.04), indicating a Bingham fluid behavior, whereas the slurry with a 0.4% dispersant (Figure 4c) had lower viscosity (26,500cP) and varying viscosity with increasing shear rate, indicating a pseudoplastic fluid behavior. The 50% slurries provide an option for usage of both finer and relatively larger extrusion nozzle sizes, thus providing process parameter choices for 3D printing practitioners while depositing LSCF cathode traces.

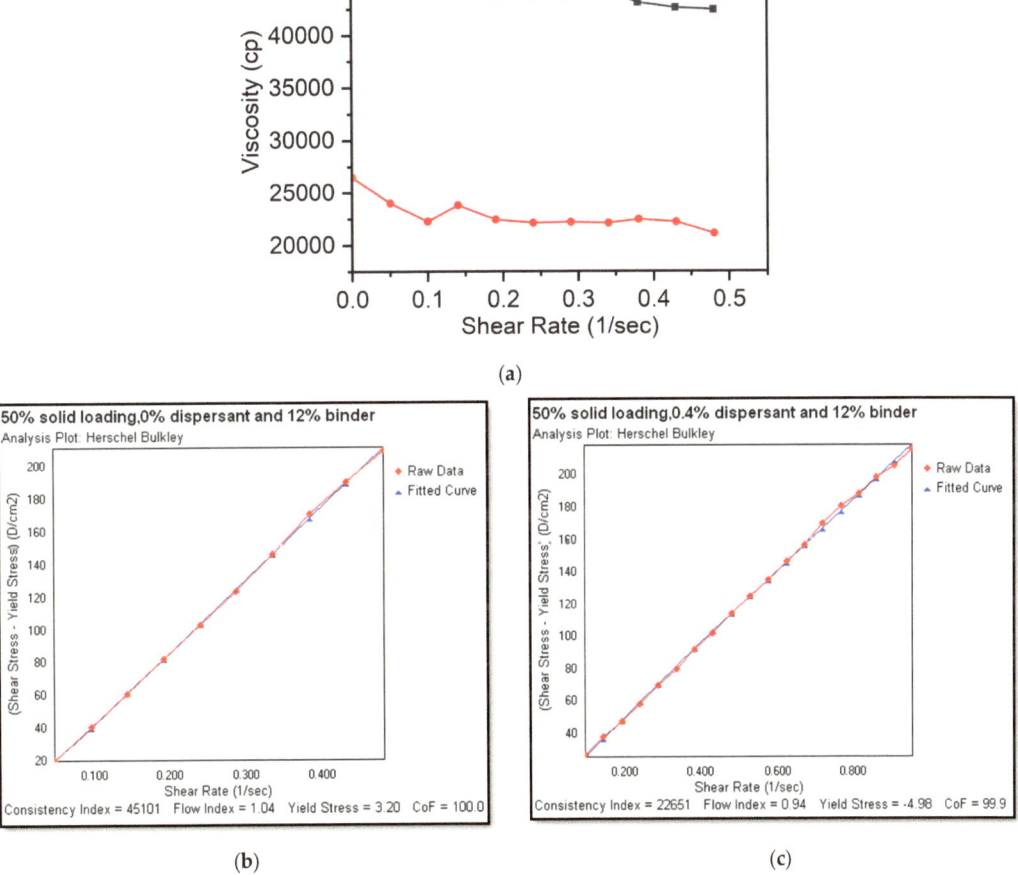

Figure 4. (**a**) Viscosity and shear rate plot for 50% solid loadings and a shear rate and shear stress plot (Hershel–Bulkley analysis plot) for slurries with a 50% solid loading and a 12% binder with (**b**) a 0% dispersant (Bingham) and (**c**) a 0.4% dispersant (pseudoplastic).

Figure 5a–c show the rheological characteristics of a 40% solid loading with 0% (Figure 5b) and 0.2% dispersants, respectively. Both the slurries displayed Newtonian fluid behavior. However, the addition of a 0.2% dispersant (Figure 5c) had a significant influence on the reduction in slurry viscosity from 11,000 cP to 3600 cP. Lower viscosities can permit deposition at lower extrusion pressures. Moreover, finer nozzle dimensions can be employed to deposit thinner cathode electrodes. However, extrusion pressure and line deposition speeds need to be modified to ensure consistent trace width and height.

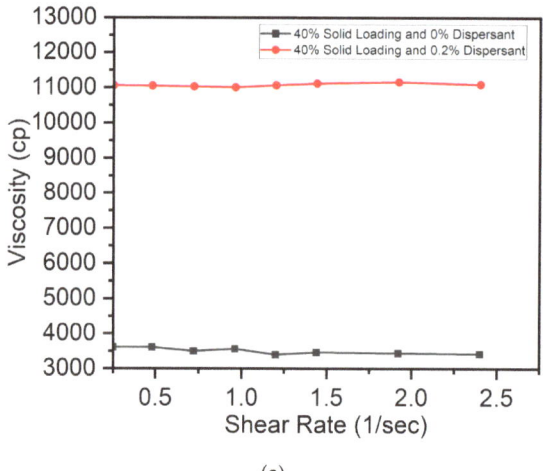

(a)

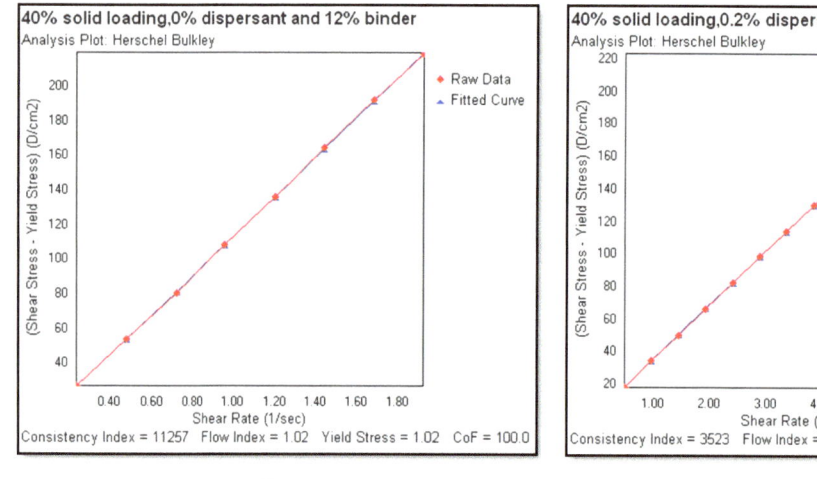

(b)

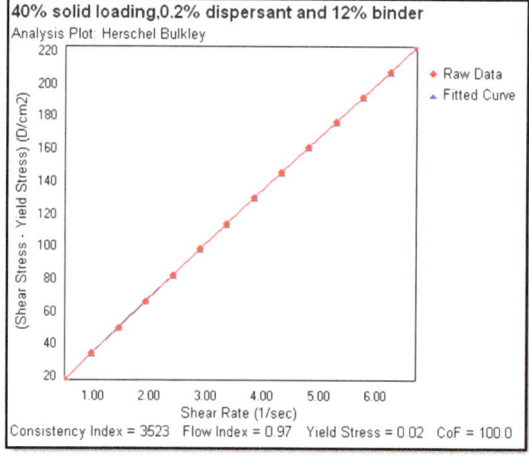

(c)

Figure 5. (**a**) Viscosity and shear rate plot for 40% solid loadings and a shear rate and shear stress plot (Hershel–Bulkley analysis plot) for slurries with a 40% solid loading and a 12% binder with (**b**) a 0% dispersant (Newtonian) and (**c**) a 0.2% dispersant (Newtonian).

The usage of different slurry compositions mentioned above provides rheological guidance for selecting process parameters for depositing LSCF cathode electrodes. These include extrusion pressure, line speed, and nozzle dimensions, which can significantly impact the trace dimensions and subsequent SOFC operation.

3.3. Evaluating the Optimal Dispersant Concentration

To evaluate how dispersants influenced the flow properties of slurries, rheology measurements were employed. Slurries that were not treated with dispersants had their viscosity evaluated in order to create a baseline. Shear stress was roughly 158 D/cm^2, and apparent viscosities were measured when torque equaled 70%. The viscosity of the candidate slurries is plotted against the amount of dispersant in wt% of the LSCF powder employed. With a 0.4 wt% dispersant concentration, the Triton dispersant attained the lowest viscosity for slurries with a 60% solid content, decreasing it by approximately 70% when compared to the dispersion without any dispersant (Figure 6).

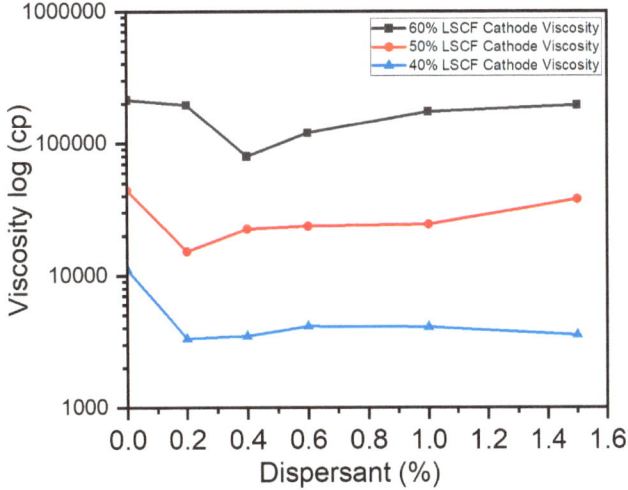

Figure 6. Viscosities of different LSCF solid loading slurries vs. dispersant concentration.

The optimum dispersion level for slurries including 60% solids is 0.4 weight percent. Due to its enormous adsorption on the slurry powder surfaces, the dispersant substantially reduced the Hamaker constant and improved the characteristics of the surfaces, which is why it was so effective. A dispersant concentration ranging from 0.2 to 0.4 weight percent was acquired to produce the least viscosity for slurries with 50% and 40% solid contents (Figure 6). But when the dispersant increased over this threshold, the extra dispersant molecules that stayed liberated in the mixture caused viscosity to increase. This occurred due to interactions between the excess dispersant molecules and the solvent [58].

3.4. Evaluating Width and Height of Lines

The three cathode lines were deposited with the slurries on the YSZ substrate. For the 60% solid content slurry, 200 kPa was the bare minimum extrusion pressure. It was feasible to extrude slurries with 40% and 50% solid contents at 100 kPa and slurries with a 40% solid content at as low as 15 kPa. Nevertheless, operating the extrusion at these low pressures caused obstacles during the direct-writing process. As a result, the measurements recorded in the table relate to slurries with a 40% solid content at 100 kPa, a 50% solid content at 200 kPa, and a 60% solid content at 200 kPa.

For 60% solid loading slurries, under 200 kPa, the range of line width was 190 μm −330 μm. For 50% solid loading slurries, under 200 kPa, the range of line width was 360 μm–670 μm. For 40% solid loading slurries, under 100 kPa, the range of line width was 550 μm–1250 μm. The line width increased with the increase in the solid loading significantly. Table 2 shows the summary of viscosity and dimension without slurry composition as slurries with different solid loadings, binder concentrations, and dispersants can provide varying viscosities. This outline can be used to choose the direct writing pressure in general, based on the viscosity of the slurry.

Table 2. Summary of viscosity and line dimensions.

Pressure (kPa)	Viscosity (kcp)	Line Width before Sintering (μm)	Line Height before Sintering (μm)
200	150–220	180–260	20–36
200	80–150	280–330	30–40
200	20–45	360–500	23–40
200	10–20	>600	Around 100
100	10–15	Around 600	Around 190
100	3.2–5	>800	40–100

3.5. Binder Removal Process

The sintering temperature curve was determined by the binder removal research study. Figure 7 illustrates the Thermo Gravimetric Analysis (TGA) and Differential Scanning Calorimetry (DSC) data for PVB drying at a 5 °C/min heating rate from room temperature to 800 °C. The TGA shows the changes in the weight of a specimen while its temperature is increased. The black curves represent the heating cycle from room temperature to 800 °C and a corresponding loss in binder mass at each temperature data point. DSC is a thermodynamical tool for direct assessment of the heat energy uptake, which occurs in a sample within a regulated increase or decrease in temperature. The markings in DSC graph (blue line) show both exothermic (heat-releasing) and endothermic (heat-absorbing) reactions occurring within the slurry mixture at different temperatures showing phase change in the materials and evaporation of the dispersant. An endothermic process was liable for the notable 87% mass loss that was seen at temperatures between 150 °C and 200 °C. Both endothermic and exothermic reactions were associated with the binder removal process between 200 °C and 600 °C, which led to the residue-free breakdown of the final 10% of the mass. The mass did not vary between 600 °C and 800 °C, indicating that the binder was wholly eliminated. These outcomes concur with Kim's group research study [22].

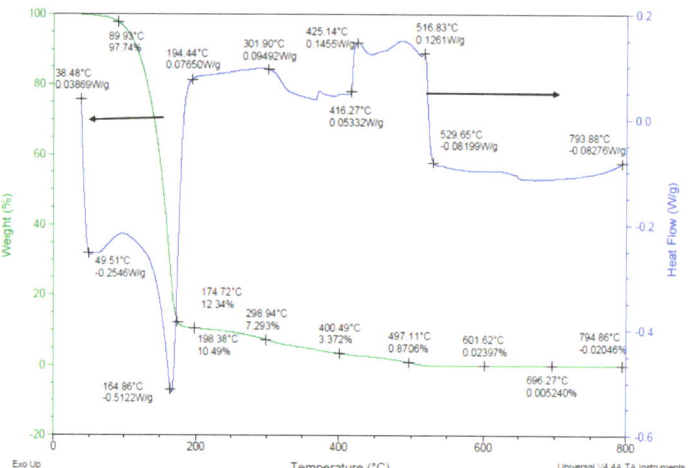

Figure 7. Binder decomposition process.

The experimental results of the drying process using TGA and DSC investigations are illustrated in Figure 8. The LSCF slurry, which has a 60% solid content, a 12% binder, and a 1.5% dispersant, was dried at temperatures as high as 800 °C. Between 50 °C and 150 °C, a significant amount of the overall mass, roughly 83%, vanished. Exothermic reactions caused a continuous mass loss of approximately 20% between temperatures of 150 °C and 350 °C. The mass remained stable above 350 °C and an endothermic process occurred. There are noticeable differences in the way the binder is eliminated when blended with LSCF powder when comparing Figures 7 and 8. To be more specific, Figure 8 illustrates that the slurry released a significant quantity of heat in the 150–300 °C region. With a boiling point of about 219 °C at 100 kPa, the solvent (α-terpineol) may have vaporized, which could be the cause of this event.

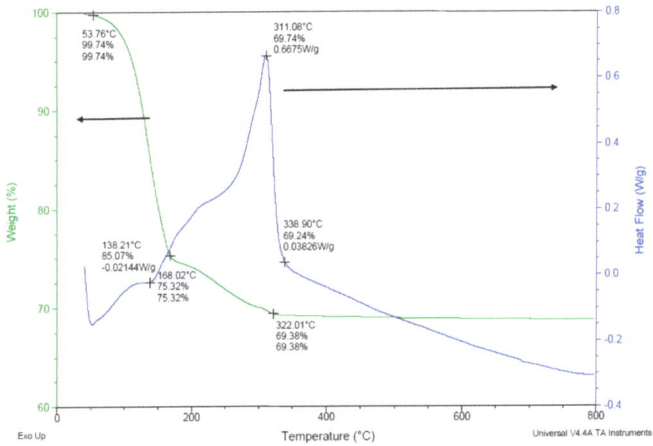

Figure 8. Drying process of a 60 wt% LSCF slurry with a 12% PVB, a 1.5% Triton.

There were substantial cracks on the surface of the cathode electrode, as illustrated in Figure 9b, which were triggered by the fast-drying rates based on the slurry drying temperature profile that was previously described. The subsequent experiment attempted to address this by slowing down the heating rate to 3 °C per minute while holding the sintering temperature at a constant 150 °C for an hour. The revised drying regimen had modified mass loss and heat flow profile, and at 150 °C, an exothermic reaction with extra mass loss was perceived. The mass loss was broken down as follows: The mass stayed constant from 350 °C to 800 °C, losing only 17% between 150 °C and 350 °C, roughly 8% during the dwell period, and 73% between 50 and 150 °C.

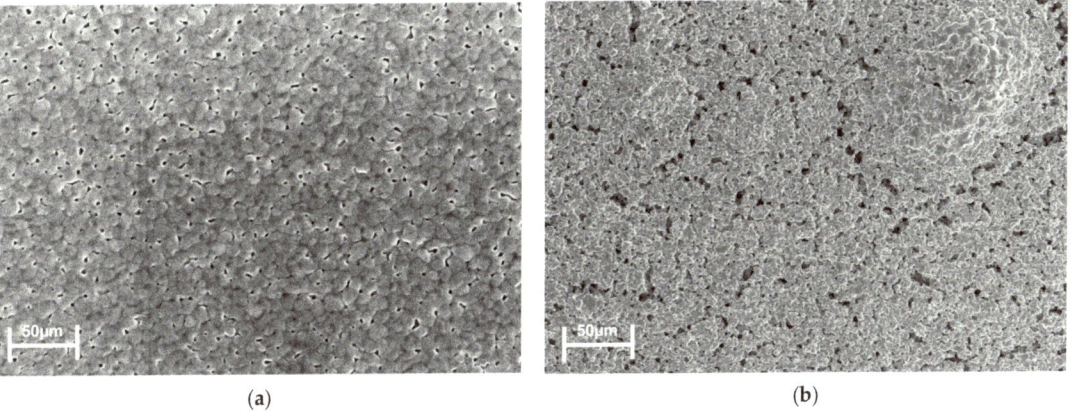

(a) (b)

Figure 9. Scanning electron microscopy (SEM) images of the cathode electrode. (a) No cracks. (b) With cracks.

Figure 10 illustrates the drying process trajectory of the lines which were deposited by a 60 wt% LSCF, a 12% PVB, and a 1.5% dispersant slurry on the substrate. The mass loss was found to be 46.3% in the temperature range of 50 °C–150 °C; throughout the dwell period, it was 7.4% lower; and from 150 °C to 300 °C, it was 20.4% lower. After a 26% drop from 375 °C to 600 °C, the mass stabilized between 600 °C and 800 °C. The mass remained persistent between 300 °C and 375 °C. There was an exothermic reaction at 150 °C that led to more mass loss. Between 350 °C and 400 °C, there was a more substantial mass decrease. In this case, the fabrication of lengthy surface fractures was avoided by reducing the drying speed.

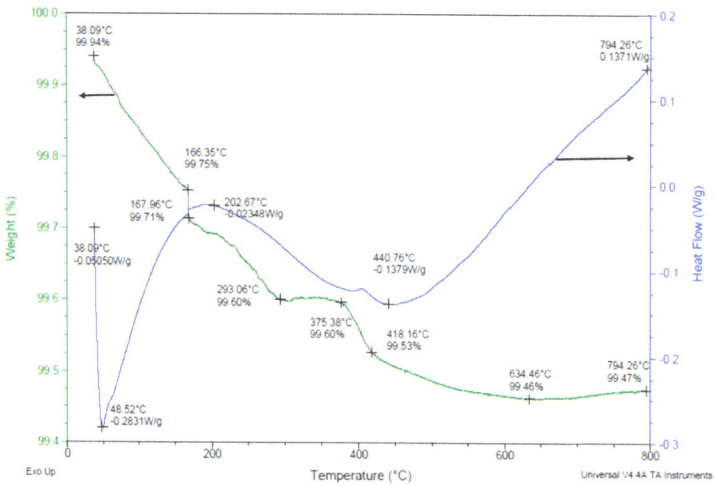

Figure 10. The drying process of the lines which were written by a 60 wt% LSCF, a 12% PVB, and a 1.5% dispersant slurry with dwell and ramp temperature cycling.

By varying the dwell, heat ramp-up, and cooling ramp-down cycles for curing the electrode lines a suitable drying process of the LSCF line was accomplished. Based on the dispersant evaporation and binder decomposition findings, an optimal sintering profile was determined as follows. The heating cycle began at room temperature to 150 °C at a rate of 3 °C/min. This was followed by a dwell period of 1 h. The heating was further initiated to 800 °C at a rate of 3 °C/min followed by a subsequent dwell period of 1 h. The final heating cycle was initiated through 1300 °C at a rate of 5 °C/min. The cathode trace was held at dwell for 3 h at the highest temperature. Finally, a gradual cooling cycle was performed to 50 °C at a rate of 5 °C/min. Thus, a stepwise heating and a dwell profile served as an optimal profile for crack-free and porous cathode electrodes for the SOFC.

4. Conclusions

In this research, a parallel-connected inter-digitized design of a micro-single-chamber SOFC (μ-SC-SOFCs) was fabricated with a direct-writing process. The cathode electrode slurry was investigated to understand and optimize the direct-write process [62]. The effects of dispersant and binder concentration on the cathode line dimensions were evaluated. Analysis of various amounts of the dispersion (Triton) was essential to observe its influence on the LSCF slurry. Higher viscosity slurries with 60% solid loading resembled the traits of Bingham and Viscoelastic fluids, which had a primitive yield stress. While the viscosity of the 50% solid loading slurries was slightly lower than that of the 60% solid loading ones, they demonstrated properties comparable to those of pseudoplastic and shear-thickening fluids. Of all the slurries explored, those with 40% solid loading had the minimum viscosity and mostly posed Newtonian and pseudoplastic characteristics. Triton was found to be an exceptional dispersant for cathode slurries; the optimal concentration for advancing the dispersion in various slurries was roughly 0.2–0.4% of the LSCF solid content. Further, the binder removal process was investigated because it has a significant impact on fuel cell microstructure. The sintering temperature profile was determined from the binder removal process in this study. Based on the binder decomposition results, the sintering profile was determined as follows: Temperature: room to 150 °C; rate: 3 °C/min; dwelling: 1 h; heating: to 800 °C; rate: 3 °C/min; dwelling: 1 h; heating: to 1300 °C; rate: 5 °C/min; dwelling: 3 h; and cooling: to 50 °C; rate: 5 °C/min. This research lays the foundation for developing optimal process parameters and sintering cycles for the deposition of high-fidelity cathode electrodes for micro single-chamber solid oxide fuel cells for green energy.

Author Contributions: M.Y., S.K.P., Z.X. and S.D. contributed to the conceptualization and methodology of this original research. M.Y. conducted all the experiments, ran the statistical analysis, wrote the original results, and carried out the initial paper draft preparation. S.K.P. contributed to the review and editing of the draft. S.D. contributed to the review and editing, supervision, and fund acquisition. All authors have read and agreed to the published version of the manuscript.

Funding: The authors would like to express their gratitude for funding support from the National Science Foundation Grant (NSF CMMI Award #2100739, #2100850), North Carolina Defense Manufacturing Community Support Program, and the Center of Excellence in Product Design and Advanced Manufacturing at North Carolina A&T State University.

Data Availability Statement: The original contributions presented in the study are included in the article, further inquiries can be directed to the corresponding author.

Conflicts of Interest: The authors declare no conflicts of interest.

References

1. Wang, J.; Bentley, Y. Modelling world natural gas production. *Energy Rep.* **2020**, *6*, 1363–1372. [CrossRef]
2. Cichon, K.; Kliks, A.; Bogucka, H. Energy-efficient cooperative spectrum sensing: A survey. *IEEE Commun. Surv. Tutor.* **2016**, *18*, 1861–1886. [CrossRef]
3. Midilli, A.; Dincer, I.; Ay, M. Green energy strategies for sustainable development. *Energy Policy* **2006**, *34*, 3623–3633. [CrossRef]
4. Shafiq, R.R.G.; ElHalawany, B.M.; Ali, N.; Elsherbini, M.M. A power source for E-devices based on green energy. *Energy Harvest. Syst.* **2024**, *11*, 20230078. [CrossRef]
5. Omer, A.M. Green energies and the environment. *Renew. Sustain. Energy Rev.* **2008**, *12*, 1789–1821. [CrossRef]
6. Hayat, M.B.; Ali, D.; Monyake, K.C.; Alagha, L.; Ahmed, N. Solar energy—A look into power generation, challenges, and a solar-powered future. *Int. J. Energy Res.* **2019**, *43*, 1049–1067. [CrossRef]
7. Bartle, A. Hydropower potential and development activities. *Energy Policy* **2002**, *30*, 1231–1239. [CrossRef]
8. Gipe, P. Wind power. *Wind Eng.* **2004**, *28*, 629–631. [CrossRef]
9. Toth, F.L.; Rogner, H.-H. Oil and nuclear power: Past, present, and future. *Energy Econ.* **2006**, *28*, 1–25. [CrossRef]
10. Williamson, K.; Gunderson, R.; Hamblin, G.; Gallup, D.; Kitz, K. Geothermal power technology. *Proc. IEEE* **2001**, *89*, 1783–1792. [CrossRef]
11. Ma, F.; Hanna, M.A. Biodiesel production: A review. *Bioresour. Technol.* **1999**, *70*, 1–15. [CrossRef]
12. Grove, W. XXIV. On voltaic series and the combination of gases by platinum. *Lond. Edinb. Dublin Philos. Mag. J. Sci.* **1839**, *14*, 127–130. [CrossRef]
13. Bacon, F.T. Fuel cells, past, present and future. *Electrochim. Acta* **1969**, *14*, 569–585. [CrossRef]
14. Choi, W. *New Approaches to Improve the Performance of the PEM Based Fuel Cell Power Systems*; Texas A&M University: College Station, TX, USA, 2004.
15. Vielstich, W.; Lamm, A.; Gasteiger, H. *Handbook of Fuel Cells. Fundamentals, Technology, Applications*; PBD: Gateshead, UK, 2003.
16. Carrette, L.; Friedrich, K.A.; Stimming, U. Fuel cells: Principles, types, fuels, and applications. *ChemPhysChem* **2000**, *1*, 162–193. [CrossRef] [PubMed]
17. Meda, U.S.; Rajyaguru, Y.V.; Pandey, A. Generation of green hydrogen using self-sustained regenerative fuel cells: Opportunities and challenges. *Int. J. Hydrogen Energy* **2023**, *48*, 28289–28314. [CrossRef]
18. Stambouli, A.B.; Traversa, E. Fuel cells, an alternative to standard sources of energy. *Renew. Sustain. Energy Rev.* **2002**, *6*, 295–304. [CrossRef]
19. Minh, N.Q. Ceramic fuel cells. *J. Am. Ceram. Soc.* **1993**, *76*, 563–588. [CrossRef]
20. O'Hayre, R.; Colella, W.; Cha, S.-W.; Prinz, F.B. *Fuel Cell Fundamentals*, 2nd ed.; Wiley: Hoboken, NJ, USA, 2009.
21. Kobayashi, Y.; Ando, Y.; Kabata, T.; Nishiura, M.; Tomida, K.; Matake, N. Extremely high-efficiency thermal power system-solid oxide fuel cell (SOFC) triple combined-cycle system. *Mitsubishi Heavy Ind. Tech. Rev.* **2011**, *48*, 9–15.

22. Stambouli, A.; Traversa, E. Solid oxide fuel cells (SOFCs): A review of an environmentally clean and efficient source of energy. *Renew. Sustain. Energy Rev.* **2002**, *6*, 433–455. [CrossRef]
23. Evans, A.; Bieberle-Hütter, A.; Rupp, J.L.; Gauckler, L.J. Review on microfabricated micro-solid oxide fuel cell membranes. *J. Power Sources* **2009**, *194*, 119–129. [CrossRef]
24. Beckel, D.; Bieberle-Hütter, A.; Harvey, A.; Infortuna, A.; Muecke, U.; Prestat, M.; Rupp, J.; Gauckler, L. Thin films for micro solid oxide fuel cells. *J. Power Sources* **2007**, *173*, 325–345. [CrossRef]
25. Evans, A.; Bieberle-Hütter, A.; Galinski, H.; Rupp, J.L.; Ryll, T.; Scherrer, B.; Tölke, R.; Gauckler, L.J. Micro-solid oxide fuel cells: Status, challenges, and chances. *Monatshefte für Chemie-Chem. Mon.* **2009**, *140*, 975–983. [CrossRef]
26. Kwon, C.; Son, J.; Lee, J.; Kim, H.; Lee, H.; Kim, K. High-performance micro-solid oxide fuel cells fabricated on nanoporous anodic aluminum oxide templates. *Adv. Funct. Mater.* **2011**, *21*, 1154–1159. [CrossRef]
27. Jasinski, P. Micro solid oxide fuel cells and their fabrication methods. *Microelectron. Int.* **2008**, *25*, 42–48. [CrossRef]
28. Bieberle-Hütter, A.; Beckel, D.; Infortuna, A.; Muecke, U.P.; Rupp, J.L.; Gauckler, L.J.; Rey-Mermet, S.; Muralt, P.; Bieri, N.R.; Hotz, N.; et al. A micro-solid oxide fuel cell system as battery replacement. *J. Power Sources* **2008**, *177*, 123–130. [CrossRef]
29. Patil, T.C.; Duttagupta, S.P. Micro-Solid Oxide Fuel Cell: A multi-fuel approach for portable applications. *Appl. Energy* **2016**, *168*, 534–543. [CrossRef]
30. Zoupalis, K.; Amiri, A.; Sugden, K.; Kendall, M.; Kendall, K. Micro solid oxide fuel cell thermal dynamics: Incorporation of experimental measurements and model-based estimations for a multidimensional thermal analysis. *Energy Convers. Manag.* **2023**, *277*, 116650. [CrossRef]
31. Corigliano, O.; Pagnotta, L.; Fragiacomo, P. On the technology of solid oxide fuel cell (SOFC) energy systems for stationary power generation: A review. *Sustainability* **2022**, *14*, 15276. [CrossRef]
32. Parupelli, S.K.; Desai, S. Understanding Hybrid Additive Manufacturing of Functional Devices. *Am. J. Eng. Appl. Sci.* **2017**, *10*, 264–271. [CrossRef]
33. Kuhn, M.; Napporn, T.; Meunier, M.; Vengallatore, S.; Therriault, D. Direct-write microfabrication of single-chamber micro solid oxide fuel cells. *J. Micromech. Microeng.* **2007**, *18*, 015005. [CrossRef]
34. Chiabrera, F.; Garbayo, I.; Alayo, N.; Tarancón, A. Micro solid oxide fuel cells: A new generation of micro-power sources for portable applications. *Smart Sens. Actuators MEMS VIII* **2017**, *10246*, 188–198.
35. Evans, A.; Karalić, S.; Martynczuk, J.; Prestat, M.; Tölke, R.; Yáng, Z.; Gauckler, L.J. La0.6Sr0.4CoO3-δ Thin Films Prepared by Pulsed Laser Deposition as Cathodes for Micro-Solid Oxide Fuel Cells. *ECS Trans.* **2012**, *45*, 333. [CrossRef]
36. Hibino, T.; Ushiki, K.; Kuwahara, Y. New concept for simplifying SOFC system. *Solid State Ionics* **1996**, *91*, 69–74. [CrossRef]
37. Chung, C.-Y.; Chung, Y.-C. Performance characteristics of micro single-chamber solid oxide fuel cell: Computational analysis. *J. Power Sources* **2006**, *154*, 35–41. [CrossRef]
38. Hibino, T.; Hashimoto, A.; Inoue, T.; Tokuno, J.-I.; Yoshida, S.-I.; Sano, M. Single-chamber solid oxide fuel cells at intermediate temperatures with various hydrocarbon-air mixtures. *J. Electrochem. Soc.* **2000**, *147*, 2888–2892. [CrossRef]
39. Shao, Z.; Mederos, J.; Chueh, W.C.; Haile, S. High power-density single-chamber fuel cells operated on methane. *J. Power Sources* **2006**, *162*, 589–596. [CrossRef]
40. McKenzie, J.; Parupelli, S.; Martin, D.; Desai, S. Additive manufacturing of multiphase materials for electronics. In *IIE Annual Conference. Proceedings*; Institute of Industrial and Systems Engineers (IISE): Norcross, GA, USA, 2017; pp. 1133–1138.
41. Parupelli, S.K.; Desai, S. Hybrid additive manufacturing (3D printing) and characterization of functionally gradient materials via in situ laser curing. *Int. J. Adv. Manuf. Technol.* **2020**, *110*, 543–556. [CrossRef]
42. Rajaram, G.; Desai, S.; Xu, Z.; Pai, D.M.; Sankar, J. Systematic studies on NI-YSZ anode material for Solid Oxide Fuel Cell (SOFCs) applications. *Int. J. Manuf. Res.* **2008**, *3*, 350–359. [CrossRef]
43. Tohver, V.; Morissette, S.L.; Lewis, J.A.; Tuttle, B.A.; Voigt, J.A.; Dimos, D.B. Direct-Write Fabrication of Zinc Oxide Varistors. *J. Am. Ceram. Soc.* **2002**, *85*, 123–128. [CrossRef]
44. Lewis, J.A.; Smay, J.E.; Stuecker, J.; Cesarano, J. Direct ink writing of three-dimensional ceramic structures. *J. Am. Ceram. Soc.* **2006**, *89*, 3599–3609. [CrossRef]
45. Zhao, X.; Evans, J.R.G.; Edirisinghe, M.J.; Song, J. Direct ink-jet printing of vertical walls. *J. Am. Ceram. Soc.* **2002**, *85*, 2113–2115. [CrossRef]
46. Parupelli, S.; Desai, S. A comprehensive review of additive manufacturing (3D printing): Processes, applications and future potential. *Am. J. Appl. Sci.* **2019**, *16*, 244–272. [CrossRef]
47. Kim, Y.-B.; Ahn, S.-J.; Moon, J.; Kim, J.; Lee, H.-W. Direct-write fabrication of integrated planar solid oxide fuel cells. *J. Electroceram.* **2006**, *17*, 683–687. [CrossRef]
48. Ahn, S.-J.; Kim, Y.-B.; Moon, J.; Lee, J.-H.; Kim, J. Co-planar type single chamber solid oxide fuel cell with micro-patterned electrodes. *J. Electroceram.* **2006**, *17*, 689–693. [CrossRef]
49. Kuhn, M.; Napporn, T.; Meunier, M.; Therriault, D.; Vengallatore, S. Fabrication and testing of coplanar single-chamber micro solid oxide fuel cells with geometrically complex electrodes. *J. Power Sources* **2008**, *177*, 148–153. [CrossRef]
50. Desai, S.; Yang, M.; Xu, Z.; Sankar, J. Direct write manufacturing of solid oxide fuel cells for green energy. *J. Environ. Res. Dev.* **2014**, *8*, 477.
51. Yang, M.; Xu, Z.; Desai, S.; Kumar, D.; Sankar, J. Fabrication of novel single-chamber solid oxide fuel cells towards green technology. *ASME Int. Mech. Eng. Congr. Expo.* **2009**, *43871*, 61–66.

52. Rajaram, G.; Xu, Z.; Jiang, X.S.; Pai, D.M.; Desai, S.; Sankar, J. A statistical approach to the design and fabrication of anode material for solid oxide fuel cells-A case study. *Int. J. Ind. Eng. Theory Appl. Pract.* **2006**, *13*, 349–356.
53. Ali, S.A.M.; Anwar, M.; Baharuddin, N.A.; Somalu, M.R.; Muchtar, A. Enhanced electrochemical performance of LSCF cathode through selection of optimum fabrication parameters. *J. Solid State Electrochem.* **2017**, *22*, 263–273.
54. Yang, M.; Xu, Z.; Desai, S.; Kumar, D.; Sankar, J. Fabrication of micro single chamber solid oxide fuel cell using photolithography and pulsed laser deposition. *J. Fuel Cell Sci. Technol.* **2015**, *12*, 021004. [CrossRef]
55. Renger, C.; Kuschel, P.; Kristoffersson, A.; Clauss, B.; Oppermann, W.; Sigmund, W. Rheology studies on highly filled nano-zirconia suspensions. *J. Eur. Ceram. Soc.* **2007**, *27*, 2361–2367. [CrossRef]
56. Loh, N.H.; Tor, S.B.; Tay, B.Y.; Murakoshi, Y.; Maeda, R. Fabrication of micro gear by micro powder injection molding. *Microsyst. Technol.* **2007**, *14*, 43–50. [CrossRef]
57. Çetinel, F.A.; Bauer, W.; Müller, M.; Knitter, R.; Haußelt, J. Influence of dispersant, storage time and temperature on the rheological properties of zirconia–paraffin feedstocks for LPIM. *J. Eur. Ceram. Soc.* **2010**, *30*, 1391–1400. [CrossRef]
58. Zürcher, S.; Graule, T. Influence of dispersant structure on the rheological properties of highly-concentrated zirconia dispersions. *J. Eur. Ceram. Soc.* **2005**, *25*, 863–873. [CrossRef]
59. Seo, J.; Kuk, S.; Kim, K. Thermal decomposition of PVB (polyvinyl butyral) binder in the matrix and electrolyte of molten carbonate fuel cells. *J. Power Sources* **1997**, *69*, 61–68. [CrossRef]
60. Lee, E.; Kim, B.; Choi, S. Hand-held, automatic capillary viscometer for analysis of Newtonian and non-Newtonian fluids. *Sens. Actuators A Phys.* **2020**, *313*, 112176. [CrossRef]
61. Chinyoka, T. Comparative response of Newtonian and Non-Newtonian fluids subjected to exothermic reactions in shear flow. *Int. J. Appl. Comput. Math.* **2021**, *7*, 75. [CrossRef]
62. Yang, M. Investigating Fabrication Methods for Micro Single-Chamber Solid Oxide Fuel Cells. Ph.D. Thesis, North Carolina Agricultural and Technical State University, Greensboro, NC, USA, 2010.

Disclaimer/Publisher's Note: The statements, opinions and data contained in all publications are solely those of the individual author(s) and contributor(s) and not of MDPI and/or the editor(s). MDPI and/or the editor(s) disclaim responsibility for any injury to people or property resulting from any ideas, methods, instructions or products referred to in the content.

Article

Micromirror Array with Adjustable Reflection Characteristics Based on Different Microstructures and Its Application

Hao Cao [1,*], Zhishuang Xue [2], Hongfeng Deng [1], Shuo Chen [1], Deming Wang [1] and Chengqun Gui [1,*]

[1] The Institute of Technological Sciences, Wuhan University, Wuhan 430072, China
[2] Research Center of Graphic Communication, Printing and Packaging, Wuhan University, Wuhan 430072, China
* Correspondence: 2021106520024@whu.edu.cn (H.C.); cheng.gui@whu.edu.cn (C.G.)

Abstract: The conventional reflective optical surface with adjustable reflection characteristics requires a complex external power source. The complicated structure and preparation process of the power system leads to the limited modulation of the reflective properties and difficulty of use in large-scale applications. Inspired by the biological compound eye, different microstructures are utilized to modulate the optical performance. Convex aspheric micromirror arrays (MMAs) can increase the luminance gain while expanding the field of view, with a luminance gain wide angle > 90° and a field-of-view wide angle close to 180°, which has the reflective characteristics of a large gain wide angle and a large field-of-view wide angle. Concave aspheric micromirror arrays can increase the luminance gain by a relatively large amount of up to 2.66, which has the reflective characteristics of high gain. Industrial-level production and practical applications in the projection display segment were carried out. The results confirmed that convex MMAs are able to realize luminance gain over a wide spectrum and a wide range of angles, and concave MMAs are able to substantially enhance luminance gain, which may provide new opportunities in developing advanced reflective optical surfaces.

Keywords: micromirror array; 3D lithography; reflection characteristics; adjustable

Citation: Cao, H.; Xue, Z.; Deng, H.; Chen, S.; Wang, D.; Gui, C. Micromirror Array with Adjustable Reflection Characteristics Based on Different Microstructures and Its Application. *Micromachines* **2024**, *15*, 506. https://doi.org/10.3390/mi15040506

Academic Editor: Jaeyoun Kim

Received: 12 March 2024
Revised: 31 March 2024
Accepted: 3 April 2024
Published: 8 April 2024

Copyright: © 2024 by the authors. Licensee MDPI, Basel, Switzerland. This article is an open access article distributed under the terms and conditions of the Creative Commons Attribution (CC BY) license (https://creativecommons.org/licenses/by/4.0/).

1. Introduction

Reflective optical surfaces have unique advantages such as a flexible system design and adjustable surface reflection characteristics [1–5], being extensively utilized in multiple applications such as in holographic displays, signal detection, beam shaping, communication systems, and microelectronics [6–11]. Hopkins JB et al. [4,12] fabricated a 1 mm square hexagonal planar micromirror array (MMA) and individually controlled each square hexagonal mirror unit with different mechanical designs to rapidly and stably generate angular tilts, thus controlling the position of the reflected light. Li Z et al. [13] obtained antireflective transparent surfaces consisting of silica nanocaps by the simple heat treatment of silica-coated monolayer colloidal crystal templates, which provided an effective reduction in the reflectivity of the reflected light. Inspired by the moth eye, Yue gang Fu and team [14–17] prepared microstructural reflective surfaces of cylinder, cone, and circular hole shapes by using reactive ion etching to regulate the reflectivity of such surfaces through microstructures for anti-reflection effects, or through other bionic micro–nano structures to modulate the reflected light properties.

However, multiple studies focused on the increase/reduction of reflectivity for a certain direction and the realization of additional driving forces [18–20]. Currently, no desirable solution is available to regulate the reflection characteristics of reflective optical surfaces, such as working ranges, luminance gain intervals, and gain values, by using microstructures. And existing means of preparing reflective optical surfaces, such as precision machining, reactive ion etching, the machining degree of freedom, precision, efficiency, and costs of these microstructure preparation methods cannot be satisfied simultaneously, and the processable format is small each time, making industrial applications difficult [21–23].

Herein, different microstructures prepared by means of 3D lithography are used to modulate reflective properties. We successfully obtained reflective optical surfaces with diverse characteristics without the need for complex drive systems, and they have excellent color performance and image quality. Convex MMAs have a large gain wide angle and field of view wide angle, while concave MMAs have a high gain peak. Finally, we attempted industrial-level production and tested the actual application in the projection display segment, and results showed that we obtained products with excellent performance with lower production costs.

2. Design and Preparation of MMA

Light can "disappear" out of the void after it enters moth eyes. Studies indicate that the surface hexagonal convex microstructure produces absorptive effects on the incident light. Inspired by this, an aspherical micromirror, a microstructure, was prepared as a reflected surface, consequently creating a large-format MMA. The caliber of the aspherical micromirror is 80 μm and its height is 10 μm; the surface aluminum layer thickness is approximately 60 nm. After incident light is reflected by the MMA, the reflected light is regulated by the reflected surface for diverse effects, such as large signal gain, wide coverage, large color gamut, and small chromatic differences.

Figure 1 shows the optical simulation model and results for different MMAs. The simulation model is composed of a light source, a collimating plane, an MMA, and a detector. The object distance between the light source and the micromirror is 0.2 m, while the interval between the detector and the micromirror is 0.1 m. The simulation result is as shown in the right side of Figure 1. It can be observed that the collimating beam can maintain high illuminance, at 9×10^5 lm/cm^2 and 5×10^5 lm/cm^2, respectively, after the beam is regulated by two MMAs, while the illuminance of an ordinary planar mirror can only reach 2×10^5 lm/cm^2 under the same conditions. In addition, the concave micromirror is superior to the convex micromirror in terms of optical gain, while the convex micromirror is superior to the concave micromirror in terms of the visual angle width as it can reach a luminance gain wide angle above 100°. The optical simulation results verify that the microstructural array is able to modulate the characteristics of the reflective optical system, in which the concave micromirror is able to significantly enhance the luminance gain and the convex micromirror is able to expand the gain range (the process of simulation is shown in the Supporting Information).

The entire MMA preparation process was mainly composed of 3D lithography for the microstructure, UV transfer, and plating a reflective coating. As shown in Figure 2, the laser beam was focused on the photoresist surface for exposure; the photoresist of various depths was exposed by adjusting the exposure dose of the laser beam, and the microstructures of various heights were obtained after development. During the exposure, the laser beam passed through the integrated optical system. In the system, the attenuator and the diaphragm regulated the light intensity and numerical aperture; the regulated laser beam was focused on the photoresist via the collecting mirror as a light source to complete the exposure of the photoresist. Gray-level photoresist is sensitive to exposure power; diverse exposure powers result in different exposure depths of the photoresist. Therefore, different photoresist depths can be exposed by regulating the exposure power to obtain a 3D structure with an ideal shape. Electrical modulation was adopted for the device used in this study to regulate the exposure power; if the gray-level exposure of a light beam reaches 4096, the result is that 4096 microstructures with diverse heights and gradients can be obtained.

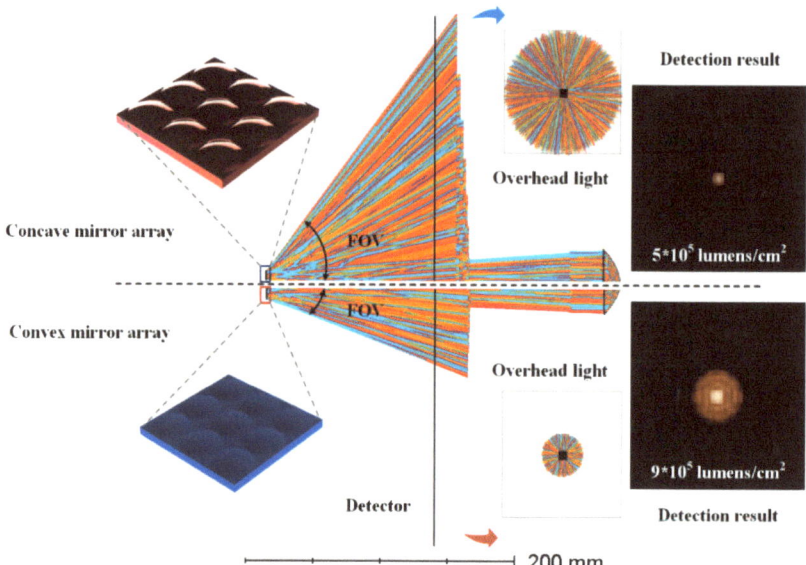

Figure 1. Optical simulation model and results.

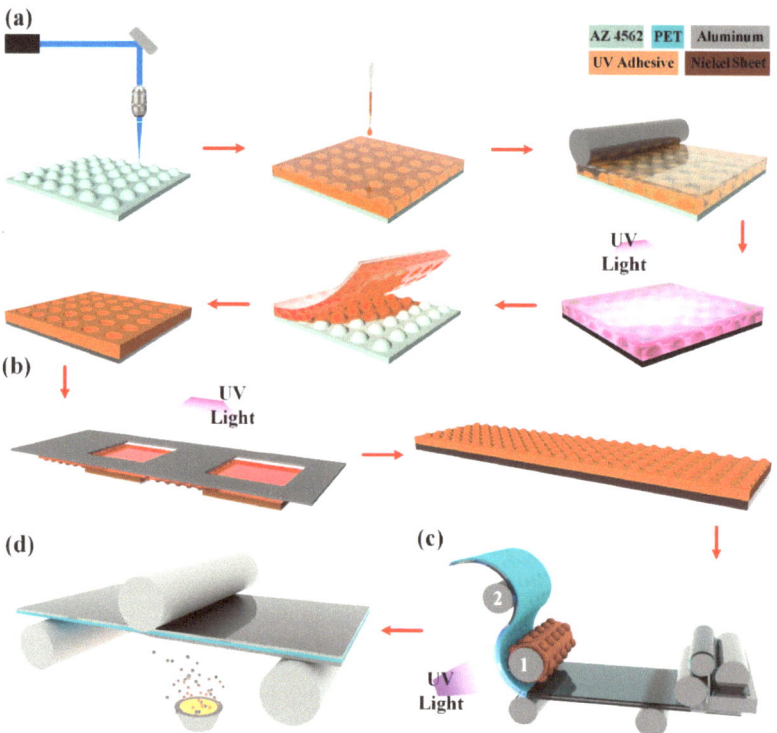

Figure 2. Manufacturing process of MMA: (**a**) 3D lithography; (**b**) preparation of templates; (**c**) UV transfer; (**d**) aluminized film.

After 3D lithography, UV transfer was performed to obtain a template. First of all, UV adhesive was dropped onto the prepared microstructural slab; the UV adhesive was solidified through ultraviolet light polymerization, and a UV slab with microstructures was obtained after demolding; afterwards, UV exposure was performed for different locations to make large-format UV slabs. Through large-format lithography, a 100-inch MMA could be made by two splicing, not only to improve the production efficiency, but also to reduce the structural losses during splicing. A roll-to-roll approach was employed for the UV compression molding transfer: the template was placed on Cylinder 1 and the PET/PMMA flexible film was placed on Cylinder 2. When the PET flexible film rotated, the UV adhesive was applied to the film; microstructures took shape after the film passed the template, and ultraviolet light polymerization was performed. Finally, thermal evaporation was employed for plating the aluminum film as a reflecting layer. Figure 3a,b show the MMAs and the SEM image of the surface aluminum layers. Figure 3c,d demonstrate the height variation in the center region (30 μm × 30 μm) with a surface roughness of 8.5 nm, which meets the requirements for the use of optical devices.

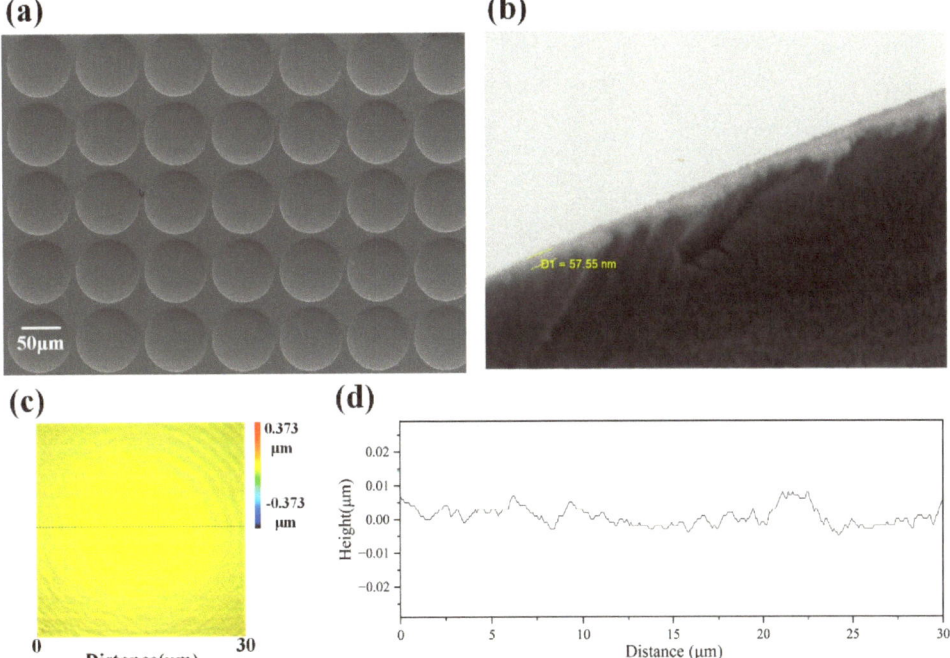

Figure 3. Characterization result of MMA: (**a**,**b**) are the SEM images of the MMAs and the surface aluminum layers; (**c**,**d**) are the height change in the central area.

3. Results and Discussion

3.1. Analysis of Reflection Characteristics

As shown in Figure 4a, the light source D65 was adopted to illuminate the screen from left and right sides of the screen from an angle of 45°, and a PR-705 spectrophotometer (hereinafter referred to as "PR-705") was utilized to collect the reflected light radiance at 0°, 15°, 30°, 45°, 60°, and 75° respectively. D65 and PR-705 were both positioned 2 m from the MMA. The D65 intensity was adjusted to third gear. PR-705 receives reflectance spectra in the range of 380–780 nm. The results are shown in Figure 4b–f.

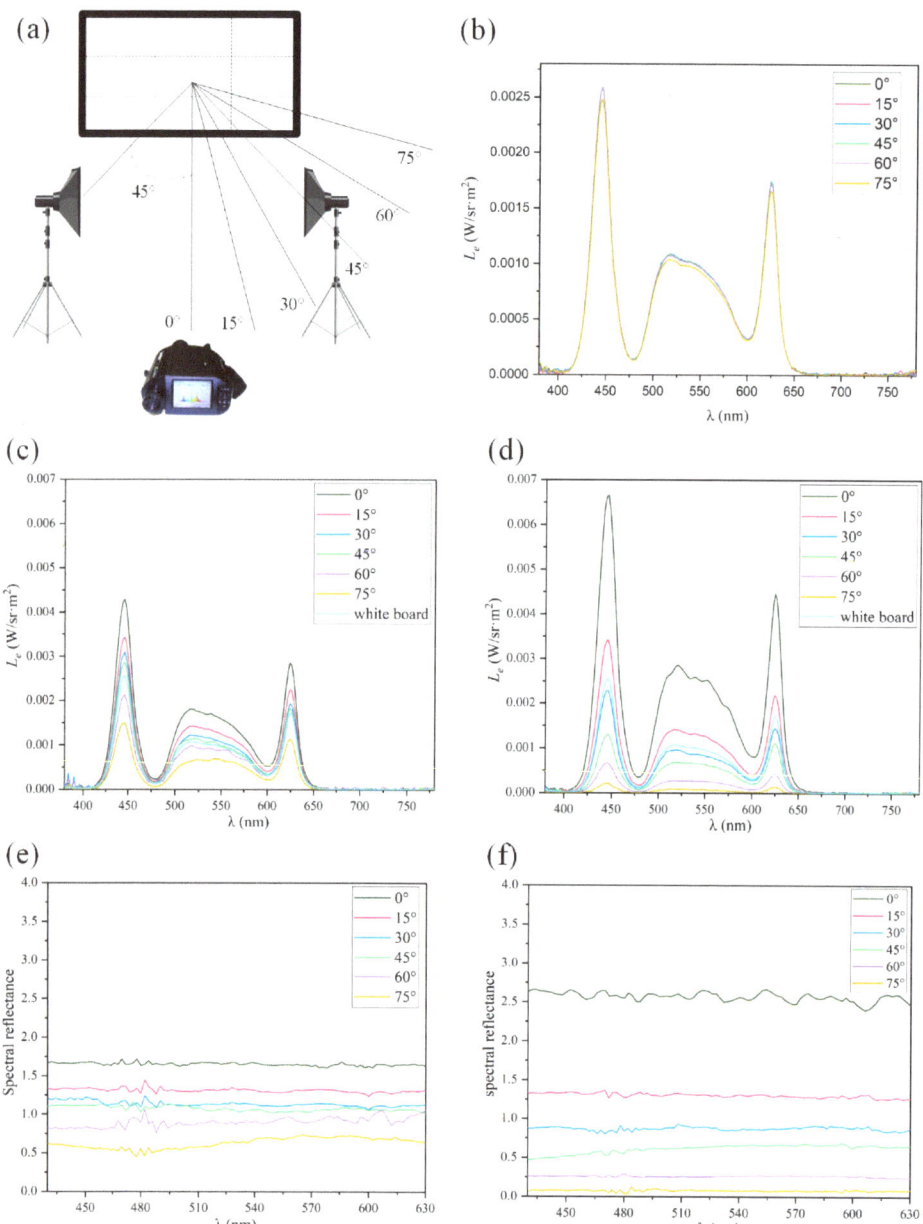

Figure 4. (**a**) is the schematic diagram of the testing conditions; (**b**) shows the reflected light radiance of the standard reflective white board at various angles; (**c**) demonstrates the reflected light radiance of the convex MMA at different angles; (**d**) shows the reflected light radiance of the concave MMA at diverse angles; (**e**) demonstrates the spectral reflectivity of the convex MMA at various angles; and (**f**) is the spectral reflectivity of the concave MMA at different angles.

Figure 4b shows the results of a standard white board. It can be observed that the white board has nearly constant reflected light: both curves have three maximum values and two minimum values; the three maximum values appear at (446 nm, 0.0026 W/sr·m^2), (518 nm, 0.0011 W/sr·m^2), and (624 nm, 0.0017 W/sr·m^2), respectively, while the two minimum values appear at (480 nm, 0.00014 W/sr·m^2) and (600 nm, 0.00032 W/sr·m^2). The curve shapes are the same as the spectral distribution of light source D65 because the standard white board is a typical diffuse reflection surface and the reflectivity is approximate to 1 in all directions to reflect the incident light from the light source to all directions in a lossless manner. The experimental results indicate the Lambertian characteristics of standard reflective white boards.

Figure 4c,e demonstrate the results for the convex MMA. It can be found that, within the scope of 380–780 nm visible light, the array has the highest radiance at 0°; the curve shape is consistent with the light source spectrum, being smooth and reaching the maximum value of 0.0043 W/sr·m^2 at 446 nm. The array starts declining at 15° and the curve shape is consistent with the light source spectrum, being smooth; at 30°, it continues to decline with a slow amplitude, and the curve shape is consistent with the light source spectrum, being smooth. It declines slightly at 45°, and the reflected light radiance at each waveband is slightly higher than that of the standard white board. The curve shape is consistent with the light source spectrum, but there is a small fluctuation in the middle waveband; it continues to decline at 60°, and the reflected light radiance at each waveband is slightly lower than that of the standard white board. It has the lowest radiance at 75°, reaching the minimum value of 0.000041 W/sr·m^2 at 480 nm. This is because, at 0°, the convex aspheric micromirror reflects incident light to a large range, but the structureless area has the strongest reflection; therefore, the radiance is highest at 0°, and the overall radiance has large gains compared with those of the white board. After it deflects to 15°, the convex aspheric micromirror can reflect incident light into the scope. The reflected light is relatively strong, while the reflectivity of the structureless area declines and the radiance is lower, but it still has gains. After it deflects to 30° and 45°, the convex aspheric micromirror can reflect incident light into the scope and the reflected light is strong, so the radiance declines in turn, but it still has gains. After it deflects to 60° and 75°, the convex aspheric micromirror can still reflect incident light into the scope, but the reflected light is not as strong as before and the structureless area can barely reflect, so the radiance is lower than that of the white board, but it can still obtain easily identifiable reflected light. The experimental results indicate that the gain wide angle of the convex MMA is >90° and its working interval is approximate to 180°.

Figure 4d,f show the results for the concave MMA. It can be observed that, within the scope of 380–780 nm visible light, the array has the highest radiance at 0°; the curve shape is consistent with the light source spectrum, reaching the maximum value of 0.0067 W/sr·m^2 at 446 nm. The array starts declining at 15°, and the curve shape is consistent with the light source spectrum, being smooth. At 30°, it continues to decline with a slow amplitude; the reflected light radiance of the wavebands has small differences with that of the standard white board at this moment, and the curve shape is consistent with the light source spectrum. It declines at 45° and continues declining at 60°, reaching the minimum value of 0.000039 W/sr·m^2 at 480 nm. It declines to the lowest level at 75°; the curve shape is nearly a straight line and the values of all of the wavebands are approximate to 0. The reason is that, at 0°, the concave aspheric micromirror has a convergence function whereby it converges incident light to the signal receiver and the structureless area is a mirror reflection, and it has the highest radiance because of the strongest reflection at that moment. The overall radiance has large gains compared with those of the white board, but the spectrum curve has some fluctuations; after it deflects to 15°, the reflected light of the concave MMA can still reach the scope, and the light is strong with a large amplitude of declination, so the radiance is lower, but it still has some gains. After it deflects to 30°, the reflected light converged by the concave MMA can still reach the scope, but the light declines, so the radiance declines to less than the white board. After it deflects to 45° and 60°, the array acts as a shelter; very little reflected light from the array can reach the scope, so the radiance is reduced dramatically and is smaller than that of the white board. After it deflects to 75°, the shelter function of the array is more apparent, and the radiance is nearly 0. The experimental results indicate that compared with the convex MMA, the concave MMA has a smaller gain wide angle and working interval; however, its maximum gain value can reach 2.66. This confirms that reflective optical surfaces formed by different microstructures have different reflective properties.

3.2. Analysis of Color Gamut and Chromatic Differences

A Photo Research 705 spectrum radiometer was utilized for the screen-reflected color measurement of the MMAs. A non-contact color measurement was conducted. The 45°/0° color measurement geometrical conditions, recommended by CIE, were adopted to guarantee the illumination uniformity of the screen surface. Two standard D65 light sources were placed at both sides of the measured screen, each forming a 45° angle with the normal of the screen; PR705 was perpendicular to the screen, and the measurement conditions are shown in Figure 5a. D65 and PR-705 were both positioned 2 m from the MMA. The D65 intensity was adjusted to third gear. Light of 700 nm, 546.1 nm, and 435.8 nm wavelengths were projected onto the MMA. A standard white board was first measured to obtain the spectral radiance values. Under the same conditions, the colored light reflected by the MMA was measured.

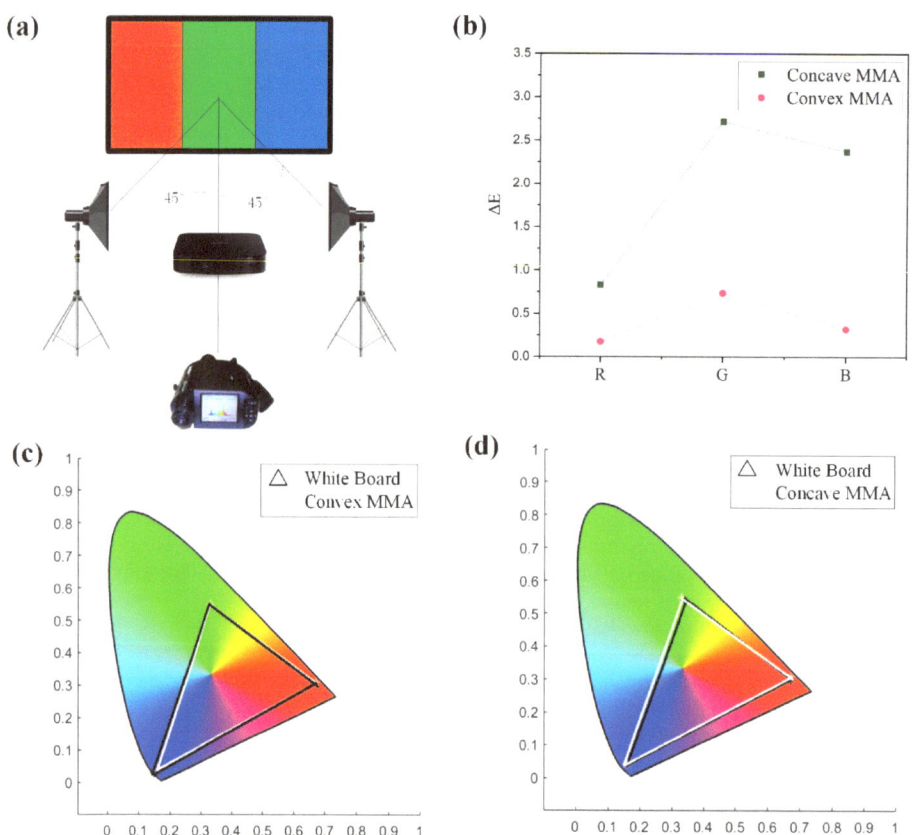

Figure 5. (**a**) is the schematic diagram of testing conditions; (**b**) shows the chromatic differences of the convex/concave MMAs; (**c**) demonstrates the color gamut of the convex MMA; and (**d**) is the color gamut of the concave MMA.

Tristimulus values were employed to calculate the chromatographic coordinates of the colors so as to calculate the color gamut range covered by the colors reflected by the screen (the process of the calculation is shown in the Supporting Information). Figure 5c,d show the color gamut ranges of the convex/concave MMAs, respectively. The white triangular curve is the color gamut range of the standard white board, while the black triangle is that of the MMA. The figures show that the chromaticity coordinates of the colors reflected by the convex MMA are (0.67, 0.31), (0.33, 0.55), and (0.15, 0.03), respectively; the chromaticity coordinates of the colors reflected by the concave MMA are (0.67, 0.30), (0.34, 0.54), and (0.16, 0.04), respectively; and the chromaticity coordinates of the standard white board are (0.67, 0.31), (0.33, 0.55), and (0.16, 0.04), respectively. It can be found that the color gamut range of the convex MMA is nearly the same as that of the white board, while the concave MMA is different in the green and red parts. Generally, their color gamut ranges are consistent, and the color gamut covers 99% of the sRGB. The reason is that the color reproduction is not compromised by the MMA preparation system. This proves that microstructures do not reduce color gamut ranges when modulating the luminance gain, gain ranges, and working ranges.

A chromatic difference calculation was made to verify the color reproduction capacity of the MMAs (the process of the calculation is shown in the Supporting Information), and the results are shown in Figure 5b. The chromatic differences of the convex MMA in the red, green, and blue wavebands are 0.1741, 0.7356, and 0.3196, respectively; the chromatic differences of the concave MMA in the red, green, and blue wavebands are 0.8313, 2.7163, and 2.3720, respectively. The chromatic differences of the convex MMA are less than 0.75 and are extremely small, indicating that the MMA has a strong color reproduction capacity. The chromatic differences of the concave MMA are less than 2.72; the chromatic differences in the green and blue wavebands are larger, but are within acceptable ranges. Generally speaking, the MMAs have different reflective characteristics and, at the same time, acceptable small chromatic differences in reflected colors and a desirable color reproduction capacity.

3.3. Analysis of Image Quality

Besides the luminance and color properties, reflected image quality is an important evaluation parameter for reflective optical systems as well. Human visual perceptions were combined with four image quality evaluation indicators, i.e., edge intensity (EI), average gradient (AG), information entropy (EN), and differential mean opinion score (DMOS), to evaluate the reflective images of MMAs and the Fresnel reflection screen (the meaning of the four quantitative indicators is provided in the Supporting Information).

As shown in Figure 6a, the projected images were changed to a standard square chessboard and a round array test to analyze the imaging properties of diverse geometric structures. A digital camera (Nikon D3200, Nikon, Tokyo, Japan) was used to take photos. The D65 light source and the digital camera were both positioned 2 m from the MMA. The D65 intensity was adjusted to third gear. Figure 6b,c show the results. It is observed that the square and round images have clear and explicit edges, and the definitions and contrasts of the edges are high. Both convex and concave MMAs do not result in distortion as a result of image off-centering. No geometric distortions were found in the convex and concave MMAs, while the imaging effects were consistent with the Fresnel screen imaging effects produced by industrial production. The reason for this is that the processing system in the paper can perfectly control structural shapes and meet usage requirements. As mentioned above, the length and width of each pixel in a large-format MMA are 80 μm and the resolution can reach 317.5/inch. Therefore, they can meet the requirements of existing practical applications.

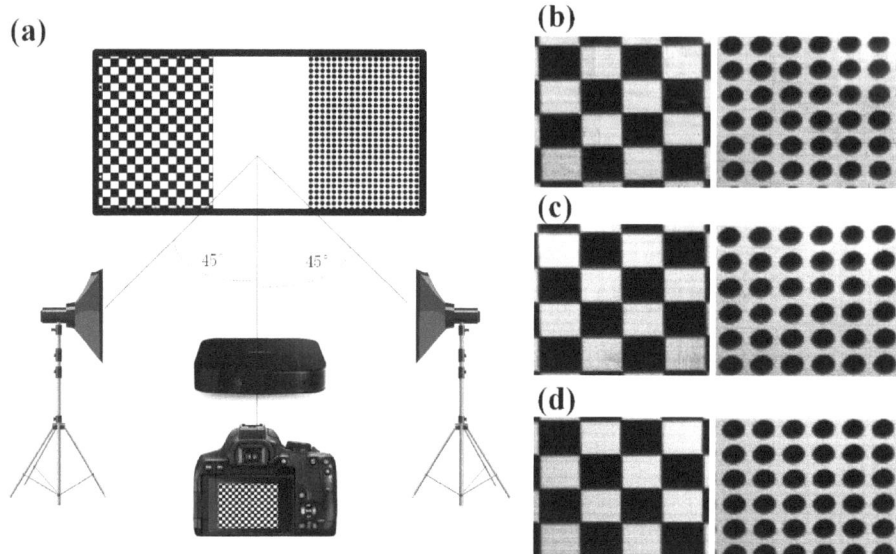

Figure 6. (**a**) is the schematic diagram of testing conditions; (**b**) is about the test results of the convex MMA; (**c**) shows the test results of the concave MMA; and (**d**) demonstrates the test results of the Fresnel optical screen.

An objective evaluation method was employed for quantitative analysis to further evaluate the image quality. Four image quality indicators were employed to evaluate the image quality. The evaluation results are shown in Table 1.

Table 1. Results of quantitative analysis.

Microstructure	Square Image Test Results				Round Image Test Results			
	EI	AG	EN	DMOS	EI	AG	EN	DMOS
Convex	61.42	5.51	6.28	39.50	89.61	8.08	6.95	27.28
Concave	61.56	5.28	6.26	43.39	80.95	7.31	6.95	27.89
Fresnel	61.42	5.50	6.12	46.74	84.04	7.56	6.74	30.06

Square image test: It can be observed that the EI value of the convex MMA is equivalent to that of Fresnel (the width of the unit structure is 120 μm and the height is 10 μm), while that of the concave MMA is slightly larger than that of Fresnel, indicating that MMAs composed of microstructures meet or even slightly exceed Fresnel levels in EI. The AG value of the convex MMA is equivalent to that of Fresnel, while that of the concave MMA is slightly less than that of Fresnel, indicating that MMAs composed of microstructures meet Fresnel levels in AG. The EN value of the convex MMA is the largest, followed by the concave MMA, and that of the Fresnel is the smallest. It can be observed that the DMOS value of Fresnel is the largest, followed by the concave MMA, and that of the convex MMA is the smallest, indicating that the Fresnel-reflected image has the lowest quality and that of the convex MMA has the highest. Round image test: It can be observed that the convex MMA has the highest EI value, the concave MMA has the lowest EI value, and Fresnel is between the two. This situation applies to AG values; however, the concave MMA has the largest EN value, followed by that of the convex MMA. The differences between them are quite small, and the EN value of Fresnel is the lowest. Fresnel DMOS is the largest, followed by the concave MMA, and the convex MMA has the smallest DMOS. Generally, all of the indicators of the convex MMA in the round array test are satisfactory, followed by the concave MMA, while those of Fresnel are the worst. The reason for this may be that the Fresnel structure blocks light to a certain extent and the working range is small, resulting in the worst overall image quality. The luminance gains of the convex MMA are within an interval with smooth changes; the working range is large and the color reproduction capacity is satisfactory, making the overall image quality high. The concave MMA has good color reproduction capacity and the luminance gains are large, but the gain changes are steeper and the working range is smaller than that of the convex MMA. Therefore, its overall image quality is ranked second place. According to the subjective and objective image quality evaluations, the designed concave and convex MMAs have a better imaging effect.

4. Industrial-Level Production and Applications

In this study, an attempt was made to produce the designed convex/concave MMAs at an industrial-level scale, as shown in Figure 7a, and were applied to the projection display segment for comparison with a commercially priced Fresnel optical projection screen. The images were projected to the three screens and photographed at 0°, 15°, 30°, 45°, 60°, 75°, and ~90°, respectively. The results are shown in Figure 7b–h.

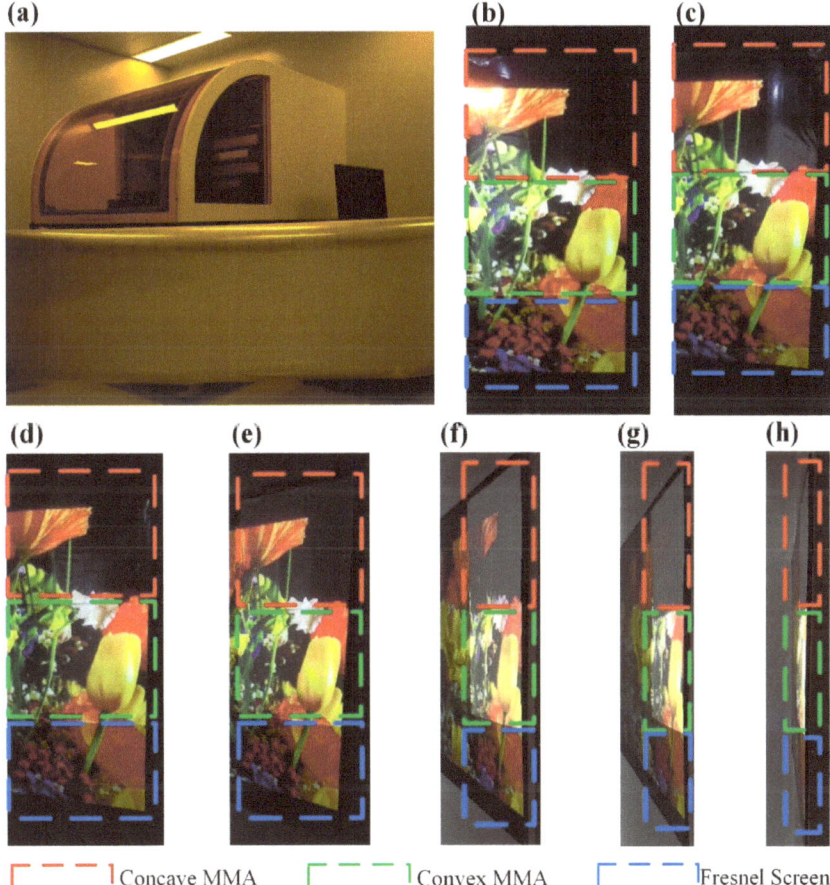

Figure 7. (**a**) is the large-scale MMA prepared for the paper; (**b**–**h**) display the image photographed at 0°, 15°, 30°, 45°, 60°, 75°, and ~90°, respectively.

It can be observed that, at 0°, the concave MMA has the highest image luminance, the convex MMA image luminance is ranked second place, the colors are uniform and exquisite, and the overall effects are desirable. The Fresnel optical screen has the darkest image that is slightly dim; at 15°, the concave MMA has the highest image luminance, but the luminance is lower than that at 0°. The convex MMA image luminance is ranked second place, the colors are uniform and exquisite, and the overall effects are good. The Fresnel optical screen has the darkest image that is slightly dim. At 30°, the concave MMA image luminance continues to decline and the convex MMA layer image luminance is lower as well, but the overall effects are good. The Fresnel optical screen has the darkest image with distortion at some dim parts; at 45°, the concave MMA image luminance continues to decline and becomes dim. At this moment, the convex MMA image has the highest luminance, with overall good image effects. The Fresnel optical screen has the darkest image, with larger image distortion; at 60°, the concave MMA has the darkest image luminance, with poor imaging effects. The convex MMA image remains bright, with good overall effects. The Fresnel optical screen image is the darkest, with larger distortion; at 75°, the concave MMA image luminance continues to decline, and it is difficult to identify the image with poor imaging effects. The convex MMA image has the highest luminance, making it the only bright screen with good overall effects. The Fresnel optical screen image is dark, with larger distortion; at 90°, it is difficult to identify the concave MMA and the Fresnel optical screen, while the convex MMA image has the highest luminance, making it the only sample with signals. Overall, the convex MMA has the best performance when applied to projection displays; concave MMA can be used for retro-reflective marking, where the reflected light is brighter and can be recognized at longer distances.

Table 2 shows a detailed comparison of the MMA with the Fresnel optical screen. Three-dimensional lithography was employed for the convex MMA optical screen, and the machining accuracy and surface roughness were found to be superior to precision machining for existing Fresnel optical screens [24,25]. The machining speed for the convex MMA screen is 233% of that for the latter; moreover, the machining process is simple and the machining freedom is high, as 4096 gradients are available on Axis Z. Existing equipment can form an 800 mm × 800 mm microstructure surface at a time; preparing large-format products has always been a constraint for precision machining. The machining difficulty increases with the increase in the format, while 3D lithography can easily solve the problem of preparing large-format products, not only improving the machining, and R&D efficiency, but also lowering the costs. This shows that the technology system of "3D lithography + UV compression molding transfer" can substantially lower production costs and improve R&D, being of great importance to scientific research and engineering manufacturing.

Table 2. Comparison of various optical screens and machining methods.

Comparative Item	MMA Optical Screen	Fresnel Optical Screen
machining method	3D lithography	precision machining
accuracy	0.3 μm	1.0 μm
surface roughness	8.5 nm	50 nm
machining speed	8929 mm^2/h	3828 mm^2/h
maximum format	800 × 800 mm	determined by the mold
machining freedom	designed at will on Axis Z	determined by the mold
working range	~180°	>30°
gain range	>90°	~30°

5. Conclusions

In summary, the microstructures prepared using 3D lithography are utilized to modulate reflective properties without additional power sources. The luminance gain wide angle of the convex MMA is >90° and the visual angle width is approximate to 180°; with chromatic differences < 0.75 and small geometric distortions, the MMA has reflection characteristics such as wide-range gains and large-interval working ranges. The maximum surface gain of the concave MMA can reach 2.66; with a chromatic difference < 2.72 and small geometric distortions, the MMA has high gain reflection characteristics. This has been verified in the field of laser projection and better results were achieved: we attempted industrial-level production and applied the microstructures to projection display segments with better performance and lower processing costs compared to existing mainstream optical screens. The research in this paper may provide different ideas for the development of innovative reflective optical systems and the promotion of efficient R&D and actual production, providing a highly valuable solution for industrial applications.

6. Experimental Section

Fabrication: We uploaded the design drawings of the MMA into the 3D lithography equipment (Pico Master 100, 4PICO, Sint-Oedenrode, The Netherlands), and prepared a 1 mm thick glass coated with the 12 µm photoresist (AZ4562, Merck, Darmstadt, Germany). The glue application speed was 650 r/min. Next, the glass coated with photoresist was relaxed for 30 min to release the air bubbles in the photoresist. We baked the photoresist at 90 °C for 50 min using a slow heating process. After 10 min cooling, we put the glass coated with the photoresist into the 3D lithography equipment. In the lithography process, the lithography spot was 550 nm, the step was 275 nm, the speed was 100 mm/s, and the exposure intensity was 800 mj/cm^2. Finally, the photoresist was developed for 6 min after the laser exposure. The developer used was AZ400K (AZ400K, Darmstadt, Germany). UV adhesive (Kangdexin Composite Material Group, MAC91A, Beijing, China) was applied to the stencil and irradiated for 35 s using a UV lamp (Kangdexin Composite Material Group, S901, Beijing, China). Roll-to-roll embossing was carried out with self-developed equipment: the mold temperature was 65 °C, the glue filling rate was 0.5 m/s, and the curing time was 35 s.

Characterization: SEM images of the gold-sputtered samples were taken by a field-emission scanning electron microscope (MIRA3, Tescan, Brno, Czech Republic). Surface roughness data were obtained by confocal microscopy (VK-X1000, Keyence, Osaka, Japan). A spectrophotometric radiation brightness meter was also utilized (PR-705, Photo Research, Syracuse, NY, USA).

Supplementary Materials: The following supporting information can be downloaded at: https://www.mdpi.com/article/10.3390/mi15040506/s1, Section S1. The process of Zemax simulation; Section S2. Determine the dimensional parameters of the MMA; Section S3. Calculation method of color gamut; Section S4. Calculation of chromatic differences; Section S5. The meaning of the four quantitative indicators. Figure S1: Simulation results of MMA simulation at different depths: (a–f) are simulation results at depths of 1 µm, 2 µm, 4 µm, 6 µm, 8 µm, and 10 µm, respectively. Figure S2: The change rule of gain with caliber and depth.

Author Contributions: Methodology, Z.X.; Software, Zemax 2019.4, H.D.; Formal analysis, S.C.; Data curation, D.W.; Writing—original draft, H.C.; Writing—review & editing, C.G. All authors have read and agreed to the published version of the manuscript.

Funding: This study was supported by the National Natural Science Foundation of China (62175172, 61575133); Jiangsu Planned Projects for Postdoctoral Research Funds; and the Priority Academic Program Development of Jiangsu Higher Education Institutions.

Data Availability Statement: The original contributions presented in the study are included in the article/supplementary material, further inquiries can be directed to the corresponding authors.

Conflicts of Interest: The authors declare no conflicts of interest.

References

1. Song, Y.; Panas, R.M.; Hopkins, J.B. A review of micromirror arrays. *Precis. Eng.* **2018**, *51*, 729–761. [CrossRef]
2. Darkhanbaatar, N.; Erdenebat, M.-U.; Shin, C.-W.; Kwon, K.-C.; Lee, K.-Y.; Baasantseren, G.; Kim, N. Three-dimensional see-through augmented-reality display system using a holographic micromirror array. *Appl. Opt.* **2021**, *60*, 7545–7551. [CrossRef] [PubMed]

3. Cho, A.R.; Han, A.; Ju, S.; Jeong, H.; Park, J.-H.; Kim, I.; Bu, J.-U.; Ji, C.-H. Electromagnetic biaxial microscanner with mechanical amplification at resonance. *Opt. Express* **2015**, *23*, 16792–16802. [CrossRef] [PubMed]
4. Hopkins, J.B.; Panas, R.M.; Song, Y.; White, C.D. A high-speed large-range tip-tilt-piston micromirror array. *J. Microelectromech. Syst.* **2017**, *26*, 196–205. [CrossRef]
5. Alnakhli, Z.; Lin, R.; Liao, C.-H.; El Labban, A.; Li, X. Reflective metalens with an enhanced off-axis focusing performance. *Opt. Express* **2022**, *30*, 34117–34128. [CrossRef] [PubMed]
6. Panas, R.M.; Hopkins, J.B.; Jackson, J.A.; Uphaus, T.M.; Smith, W.L.; Harvey, C. *Hybrid Additive and Microfabrication of an Advanced Micromirror Array*; American Society of Precision Engineering: Austin, TX, USA, 2015.
7. Hillmer, H.; Al-Qargholi, B.; Khan, M.M.; Worapattrakul, N.; Wilke, H.; Woidt, C.; Tatzel, A. Optical MEMS-based micromirror arrays for active light steering in smart windows. *Jpn. J. Appl. Phys.* **2018**, *57*, 08PA07. [CrossRef]
8. Hopkins, J.B.; Panas, R.M. Array Directed Light-Field Display for Autostereoscopic Viewing. U.S. Patent 9007444 B2, 14 April 2015.
9. Dong, L.; Zhang, Z.; Wang, L.; Weng, Z.; Ouyang, M.; Fu, Y.; Wang, J.; Li, D.; Wang, Z. Fabrication of hierarchical moth-eye structures with durable superhydrophobic property for ultra-broadband visual and mid-infrared applications. *Appl. Opt.* **2019**, *58*, 6706–6712. [CrossRef] [PubMed]
10. Hung, A.C.-L.; Lai, H.Y.-H.; Lin, T.-W.; Fu, S.-G.; Lu, M.S.-C. An electrostatically driven 2D microscanning mirror with capacitive sensing for projection display. *Sens. Actuators A Phys.* **2015**, *222*, 122–129. [CrossRef]
11. Jorissen, L.; Oi, R.; Wakunami, K.; Ichihashi, Y.; Lafruit, G.; Yamamoto, K.; Bekaert, P.; Jackin, B.J. Holographic Micromirror Array with Diffuse Areas for Accurate Calibration of 3D Light-Field Display. *Appl. Sci.* **2020**, *10*, 7188. [CrossRef]
12. Hopkins, J.B.; Panas, R.M. Flexure design for a high-speed large-range tip-tilt-piston micro-mirror array. In Proceedings of the ASPE 29th Annual Meeting, Boston, MA, USA, 9–14 November 2014.
13. Li, Z.; Lin, J.; Liu, Z.; Feng, S.; Liu, Y.; Wang, C.; Liu, Y.; Yang, S. Durable Broadband and Omnidirectional Ultra-antireflective Surfaces. *ACS Appl. Mater. Interfaces* **2018**, *10*, 40180–40188. [CrossRef] [PubMed]
14. Dong, T.; Han, X.; Chen, C.; Li, M.; Fu, Y. Research on a wavefront aberration calculation method for a laser energy gradient attenuator. *Laser Phys. Lett.* **2013**, *10*, 096001. [CrossRef]
15. Zhu, Q.; Fu, Y.; Liu, Z. A bio-inspired model for bidirectional polarisation detection. *Bioinspiration Biomim.* **2018**, *13*, 066002. [CrossRef] [PubMed]
16. Lin, H.; Ouyang, M.; Chen, B.; Zhu, Q.; Wu, J.; Lou, N.; Dong, L.; Wang, Z.; Fu, Y. Design and Fabrication of Moth-Eye Subwavelength Structure with a Waist on Silicon for Broadband and Wide-Angle Anti-Reflection Property. *Coatings* **2018**, *8*, 360. [CrossRef]
17. Wu, J.; Ouyang, M.; Zhao, Y.; Han, Y.; Fu, Y. Mushroom-structured silicon metasurface for broadband superabsorption from UV to NIR. *Opt. Mater.* **2021**, *121*, 111504.1–111504.6. [CrossRef]
18. Beasley, D.B.; Bender, M.; Crosby, J.; McCall, S.; Messer, T.; Saylor, D.A. Advancements in the micromirror array projector technology II. In *Technologies for Synthetic Environments: Hardware-in-the-Loop Testing X*; SPIE—The International Society for Optical Engineering: Bellingham, WA, USA, 2005.
19. Dauderstädt, U.; Dürr, P.; Gehner, A.; Wagner, M.; Schenk, H. Analog Spatial Light Modulators Based on Micromirror Arrays. *Micromachines* **2021**, *12*, 483. [CrossRef] [PubMed]
20. Tang, Y.; Li, J.; Xu, L.; Lee, J.-B.; Xie, H. Review of Electrothermal Micromirrors. *Micromachines* **2022**, *13*, 429. [CrossRef] [PubMed]
21. Lani, S.; Bayat, D.; Despont, M. 2D tilting MEMS micro mirror integrating a piezoresistive sensor position feedback. In *MOEMS and Miniaturized Systems XIV*; Piyawattanametha, W., Park, Y.-H., Eds.; SPIE: Bellingham, WA, USA, 2015; Volume 9375:93750C.
22. Ersumo, N.T.; Yalcin, C.; Antipa, N.; Pégard, N.; Waller, L.; Lopez, D.; Muller, R. A micromirror array with annular partitioning for high-speed random-access axial focusing. *Light Sci. Appl.* **2020**, *9*, 183. [CrossRef] [PubMed]
23. Kappa, J.; Sokoluk, D.; Klingel, S.; Shemelya, C.; Oesterschulze, E.; Rahm, M. Electrically Reconfigurable Micromirror Array for Direct Spatial Light Modulation of Terahertz Waves over a Bandwidth Wider Than 1 THz. *Sci. Rep.* **2019**, *9*, 2597. [CrossRef] [PubMed]
24. Pan, Y.; Zhao, Q.; Guo, B.; Long, Y. High-efficiency machining of silicon carbide Fresnel micro-structure based on improved laser scanning contour ablation method with continuously variable feedrate. *Ceram. Int.* **2021**, *47*, 4062–4075. [CrossRef]
25. Koh, K.M.; Samuel Kim, U. Fresnel Prism on Hess Screen Test. *Case Rep. Ophthalmol. Med.* **2013**, *2013*, 187459. [CrossRef] [PubMed]

Disclaimer/Publisher's Note: The statements, opinions and data contained in all publications are solely those of the individual author(s) and contributor(s) and not of MDPI and/or the editor(s). MDPI and/or the editor(s) disclaim responsibility for any injury to people or property resulting from any ideas, methods, instructions or products referred to in the content.

Article

The Effects of Etchant on via Hole Taper Angle and Selectivity in Selective Laser Etching

Jonghyeok Kim [1,2], Byungjoo Kim [1], Jiyeon Choi [1,2] and Sanghoon Ahn [1,2,*]

[1] Department of Laser & Electron Beam Technologies, Korea Institute of Machinery & Materials, 156 Gajeongbuk-Ro, Yuseong-Gu, Daejeon 34103, Republic of Korea; imj02096@kimm.re.kr (J.K.); byungjookim@kimm.re.kr (B.K.); jchoi@kimm.re.kr (J.C.)

[2] Department of Mechanical Engineering (Robot·Manufacturing Systems), University of Science and Technology, 217 Gajeong-Ro, Yuseong-Gu, Daejeon 34113, Republic of Korea

* Correspondence: shahn@kimm.re.kr

Abstract: This research focuses on the manufacturing of a glass interposer that has gone through glass via (TGV) connection holes. Glass has unique properties that make it suitable for 3D integrated circuit (IC) interposers, which include low permittivity, high transparency, and adjustable thermal expansion coefficient. To date, various studies have suggested numerous techniques to generate holes in glass. In this study, we adopt the selective laser etching (SLE) technique. SLE consists of two processes: local modification via an ultrashort pulsed laser and chemical etching. In our previous study, we found that the process speed can be enhanced by changing the local modification method. For further enhancement in the process speed, in this study, we focus on the chemical etching process. In particular, we try to find a proper etchant for TGV formation. Here, four different etchants (HF, KOH, NaOH, and NH_4F) are compared in order to improve the etching speed. For a quantitative comparison, we adopt the concept of selectivity. The results show that NH_4F has the highest selectivity; therefore, we can tentatively claim that it is a promising candidate etchant for generating TGV. In addition, we also observe a taper angle variation according to the etchant used. The results show that the taper angle of the hole is dependent on the concentration of the etchant as well as the etchant itself. These results may be applicable to various industrial fields that aim to adjust the taper angle of holes.

Keywords: selective laser etching; chemical etching; selectivity; taper angle; ultrashort pulsed laser

Citation: Kim, J.; Kim, B.; Choi, J.; Ahn, S. The Effects of Etchant on via Hole Taper Angle and Selectivity in Selective Laser Etching. *Micromachines* **2024**, *15*, 320. https://doi.org/10.3390/mi15030320

Academic Editor: Lucia Romano

Received: 23 January 2024
Revised: 20 February 2024
Accepted: 23 February 2024
Published: 25 February 2024

Copyright: © 2024 by the authors. Licensee MDPI, Basel, Switzerland. This article is an open access article distributed under the terms and conditions of the Creative Commons Attribution (CC BY) license (https://creativecommons.org/licenses/by/4.0/).

1. Introduction

Recently, in order to enhance IC chip performance, flip chip bonding has been widely adopted. To achieve flip chip bonding, numerous via holes are required. Therefore, a through-silicon via (TSV) has been applied. However, silicon has several disadvantages, such as its relatively high price and electric noise at high radio frequency. On the other hand, glass has unique properties that are suitable for interposer material, namely, low permittivity, high transparency, and adjustable thermal expansion coefficient. Signal noise can be avoided because of its low permittivity, three-dimensional alignment can be easily achieved because of its transparency, and warpage can be prevented because its thermal expansion can be matched with a Si wafer. Therefore, through-glass via (TGV) is becoming an increasingly popular alternative to TSV [1–7].

There are various methods for generating holes in glass [8]: mechanical drilling [9,10], powder blasting, abrasive slurry jet machining (ASJM) [11,12], laser drilling, deep reactive ion etching (DRIE) [13,14], plasma etching, spark-assisted chemical engraving (SACE), vibration-assisted micromachining, laser-induced plasma micromachining (LIPMM) [15], water-assisted micromachining, and selective laser etching [16]. The most important parameter to be considered in mass production is uniformity, although processing time should also be considered. Selective laser etching (SLE) can create uniform holes but currently requires a relatively long processing time.

Selective laser etching (SLE) is a widely used method for creating precise and intricate patterns on various materials. It was first introduced in 2001 by A. Marcinkevičius [17]. This technology involves the use of a laser and etchant to selectively remove material from the substrate, resulting in intricate patterns and shapes [17]. It can also be used to create holes in various materials, including metals, ceramics, silica, and glass [16–18]. It irradiates a laser to generate local modifications on the sample and proceeds with the etching process [17]. The modification area has about a 333 times higher etch rate compared to the non-modification area [16]. This is because the modification area changes in both physical and chemical properties, and the modification area quickly reacts with etchants. The physical and chemical changes include nanograting formation, volume expansion, and refractive index change [19,20]. These enable channel generation inside glass with excellent accuracy [17,21]. SLE is currently used in various fields such as biotechnology, nanotechnology, optics, and IT technology [16,18,21]. Our previous study shows how to enhance the process speed by changing the local modification method. We suggest that adding additional pulse energy after a few hundred picoseconds after the initial pulse can increase the etch rate of the local modification area [22]. In this study, we try to increase the etch rate with the etchant itself. To achieve this, four different etchants are tested, namely, HF solution, NaOH solution, KOH solution, and NH_4F solution. For a quantitative comparison, the selectivity has been adopted. Our results show that NH_4F solution has the highest selectivity and TGV can be formed within 3 h of etching with it. This is three times faster than a previous study that used KOH solution [22].

The glass holes generated via SLE can be also applied to various fields because the taper angle can be adjusted. The taper angle of glass holes is a crucial factor in several industrial applications. For instance, in the semiconductor industry, shower heads are used in the cleaning process, and the taper angle of the shower head holes determines the velocity and direction of the cleaning solution [23]. Similarly, in the field of biotechnology, microneedles with tapered holes are used to deliver drugs or extract fluids from the body. The taper angle of the holes affects the flow rate and penetration depth of the microneedle [24]. The taper angle of the hole also affects the spray pattern. This can be utilized in abrasive processes as well [25,26]. Therefore, research on the taper angle is useful for various industrial applications. In this study, we find that the taper angle varies according to the etchant itself and its concentration. Based on our study, we can tentatively claim that it is possible to increase productivity and adjust a hole's taper angle based on the choice of etchant.

2. Experiments

2.1. Substrate Material

A borosilicate glass (D 263® T eco, SCHOTT, Mainz, Germany) substrate with a thickness of 0.1 mm was used for the TGV process. This glass is a material that uses environmentally friendly cleaning agents instead of arsenic and antimony substances. The company that makes it claims that it has excellent chemical resistance and thermal stability and that it can be used in a wide range of applications that require high-precision parts. In particular, because of its thermal expansion characteristics, it is the best possible candidate material for glass interposers. Therefore, it is a suitable material for use in this study.

2.2. Ultrashort Pulsed Laser

Ultrashort pulsed lasers refers to lasers with pulse durations of one hundred picoseconds or less. An ultrashort pulsed laser has several advantages in glass processing. First, it transfers energy to the glass before thermal diffusion occurs, minimizing residue as a specific heat process and allowing microfabrication without post-processing steps. Second, it has high peak powers due to its short pulse durations, which makes it useful for processing materials with large energy bandgaps, such as glass. The high peak power of the ultrashort pulsed laser induces nonlinear absorption induction in the glass, enabling the internal processing of the glass.

2.3. Local Modification by Ultrashort Pulsed Laser

In this study, Bessel beams were used to increase productivity. Bessel beams have the advantage of a long depth of focus. Therefore, a single pulse of a Bessel beam can produce a full-thickness local modification in glass. Figure 1 is a schematic diagram of the experimental setup for local modification by an ultrashort pulsed laser. A diode-pumped Yb: KGW ultrashort pulse laser (Pharos, PH1-20, photoconversion, center wavelength 1030 nm) is used as the energy source. Our previous study confirmed that double pulses with a 213 ps interval are a suitable local modification process condition for generating TGV [22]. In addition, based on our previous study, a pulse duration of 1 ps is also adopted. The pulse repetition rate is 100 Hz and is synchronized with the motion stage (M-414.2PD, C-863.11, Physik Instrument, Karlsruhe, Germany). The stage movement speed is 10 mm/s. Therefore, the distance between each local modification is 100 µm. A pulse energy of at least 30 µJ is required to produce a local change. The pulse energy on the glass surface is measured and calculated via a power meter (NovaII, Ophir with 30A-BB-18 sensor, Ophir, Jerusalem, Israel). The total pulse energy reach to the sample is 68 µJ (34 µJ + 34 µJ).

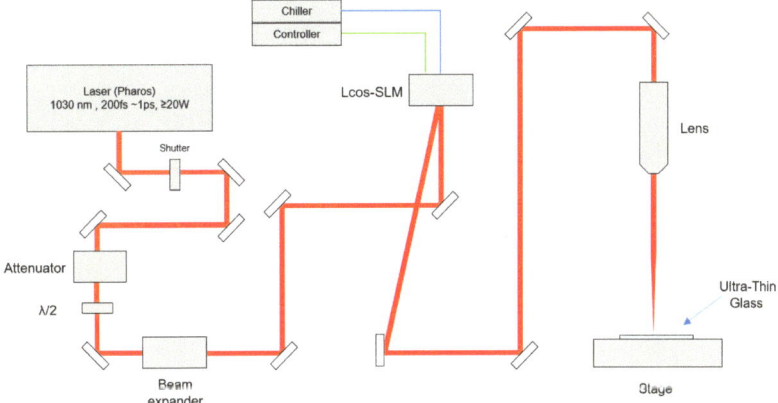

Figure 1. Schematic diagram of the experimental setup for local modification via an ultrashort pulsed laser.

2.4. Bessel Beam Shaping

Beam shaping from a Gaussian beam to a Bessel beam goes through two stages. First, the phase image is applied to the LCOS-SLM modulator (X10468-03, Hamamatsu Photonics, Shizuoka, Japan) to change the beam shape from a Gaussian beam to a donut beam. In this study, phase images are generated with optical engineering programs (VirtualLab Fusion, LightTrans, Jena, Germany). In our experimental setup, an image with 64 phase levels is sufficient to generate a donut beam. Next, the donut-shaped beam is focused with a plano-convex lens (25.4 mm focal length). Eventually, a Bessel beam is formed and measured with a beam profiler (FM100-YAG1064-50x, Metrolux, Berlin, Germany), where the Bessel beam width is 5.8 µm (FWHM) and the beam length is 180 µm.

2.5. Selective Laser Etching

In this study, we use the SLE technique to generate TGV. Normally, KOH solution is used to etch glass [27]. However, the purpose of this study is to enhance the SLE process speed by modifying the etching process. Thus, various solutions were tested, namely, HF solution, NaOH solution, KOH solution, and NH_4F solution.

According to the Arrhenius equation, elevating the temperature of the etchant increases the frequency of collisions and accelerates chemical reactions. Therefore, the etching temperature was set to 10 °C below the boiling point of each etchant. For safety reasons, there was a 10 °C buffer.

After the local modification process with an ultrashort laser, the modified glass sample was immersed in a Teflon jar, which was filled with the etchant. Then, the jar was placed in an oil bath (WHB-6, Dai Han Scientific, Seoul, Republic of Korea) which was filled with a heat transfer fluid (Therminol D12®, Kingsport, TN, USA) for a certain amount of hours. As mentioned above, the etching process was performed at a suitable temperature for each etchant. After the etching process was finished, we cleaned the glass with DI water and IPA. Then, we placed the glass sample on the polyester wiper and removed the deionized (DI) water and isopropyl alcohol (IPA), as well as the remaining residue, with compressed nitrogen gas. This etching system uses a safety valve that keeps the internal pressure constant in the Teflon jar, as previously patented [28] (Figure 2).

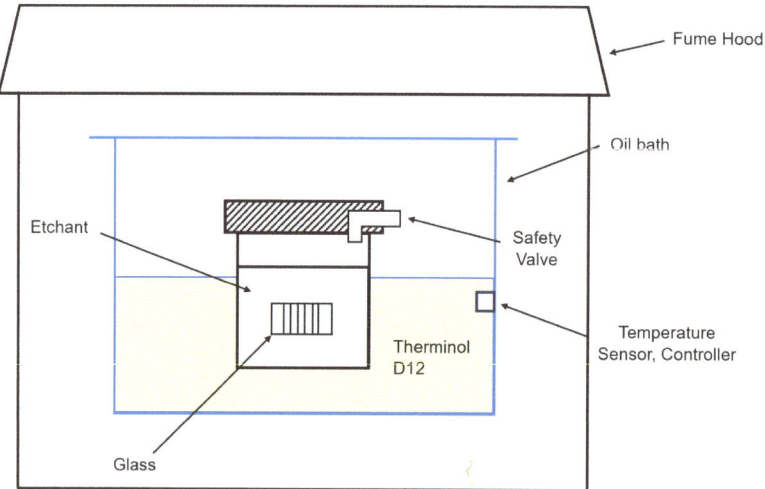

Figure 2. Schematic diagram of the etching process [28].

3. Results and Discussion

3.1. Etchants and Selectivity

The goal of this study was to enhance TGV productivity. In this study, the SLE process was adopted. We generated a local modification inside the glass via an ultrashort pulsed laser and applied chemical etching. As mentioned above, four different etchants (HF, NaOH, KOH, and NH_4F) were tested in order to increase the etching speed.

For a quantitative comparison between each etchant, the concept of selectivity was adopted. Selectivity was calculated based on the etch rate of the modification area and non-modification area. Since each etchant has different characteristics, we decided to compare the fastest etching conditions. The calculated selectivity results are presented in Table 1 and Figure 3.

$$Selectivity = \frac{Modification\ area\ etch\ rate\ +\ Non\ Modification\ area\ etch\ rate}{Non\ Modification\ area\ etch\ rate}$$

Table 1. Etching conditions and etch rate and selectivity for each etchant.

Etchant	Modification Etch Rate (μm/h)	Non-Modification Etch Rate (μm/h)	Selectivity
HF	191.0	143.9	2.3
NaOH	5.0	1.7	3.9
KOH	12.5	4.1	4.1
NH_4F	33.3	10.5	4.2

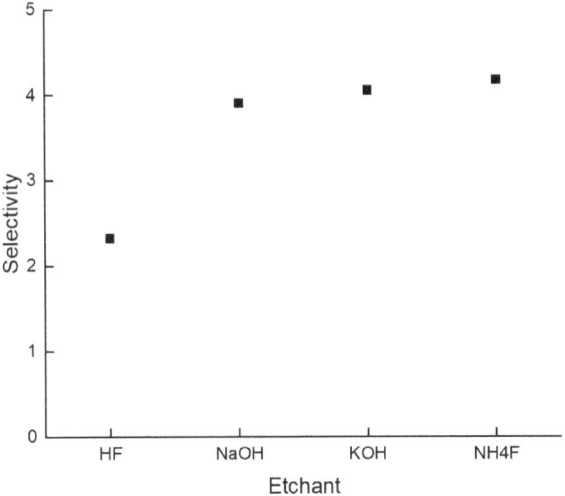

Figure 3. Selectivity of HF, NaOH, KOH, and NH_4F solutions.

For HF, etching was carried out at room temperature for 15 min with a 10 M solution. The etch rate of the modification area was measured as 191.0 μm/h, and the etch rate of the non-modification area was 143.9 μm/h. As a result, the selectivity of the HF 10 M solution was calculated as 2.3. Figure 4a shows the etching result for HF. Since the etch rate of HF was too fast and the selectivity was relatively low, a via hole could not be formed.

For NaOH, etching was carried out at 90 °C for 20 h with a 3 M solution. The etch rate of the modification area was measured as 5.0 μm/h, and the etch rate of the non-modification area was 1.7 μm/h. As a result, the selectivity of the NaOH 3 M solution was calculated as 3.9. Figure 4b shows the etching result for NaOH. After 20 h of etching, hourglass-shaped via holes were generated.

For KOH, etching was carried out at 110 °C for 9 h with an 8 M solution. The etch rate of the modification area was measured as 12.5 μm/h, and the etch rate of the non-modification area was 4.1 μm/h. As a result, the selectivity of the KOH 8 M solution was calculated as 4.1. Figure 4c shows the etching result for KOH. After 9 h of etching, similar to NaOH cases, hourglass-shaped via holes were generated. However, the taper angle was different. We discuss this issue in Section 3.2.

For NH_4F, etching was carried out at 85 °C for 3 h with an 8 M solution. The etch rate of the modification area was measured as 33.3 μm/h, and the etch rate of the non-modification area was 10.5 μm/h. As a result, the selectivity of the NH_4F 8 M solution was calculated as 4.2. Figure 4d shows the etching result for NH_4F. In the case of NH_4F etching, it only takes 3 h to generate a via hole. Even though the etch rate of NH_4F is the second fastest of the four etchants, it has the highest selectivity value. Thus, we can conclude that selectivity is a reasonable parameter for determining etching efficiency. We can also tentatively claim that the NH_4F 8 M solution is the most efficient etchant for generating TGV of the four etchants.

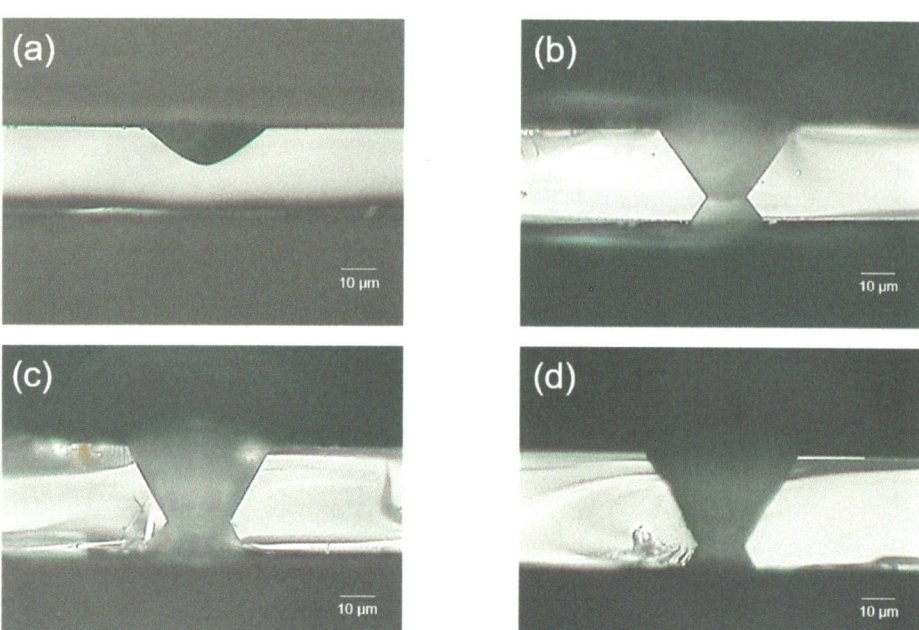

Figure 4. Via hole formed by each etchant. (**a**–**d**) Optical microscope images. (**a**) HF 10 M solution; (**b**) NaOH 3 M solution; (**c**) KOH 8 M solution; (**d**) NH$_4$F 8 M solution.

3.2. Etchants and Hole Taper Angle

As shown in Figure 4, taper angles are different for each etchant. Thus, we tried to establish the feasibility of changing the taper angle with the etchant and its concentration. Based on this study, we conclude that the taper angle can be determined by the molar concentration of each etchant (HF, NaOH, KOH, and NH$_4$F).

For the HF solution, 1 M (3.5 mL (49% HF solution) + 96.5 mL (deionized (DI) water)), 3 M (10.6 mL (49% HF solution) + 89.4 mL (deionized (DI) water)), 5 M (17.75 mL (49% HF solution) + 82.25 mL (deionized (DI) water)), and 10 M (35.5 mL (49% HF solution) + 64.5 mL (deionized (DI) water)) solutions were tested. Since the HF solution is a strong acid, the etching process was performed at room temperature for safety reasons. Moreover, the etch rate of the HF solution was too fast; therefore, the glass was dipped for only 10 min in order to observe the taper angle. The measured angles were 50° for 1 M solution, 53° for the 3 M solution, 53° for the 5 M solution (Figure 5a), and 41° for the 10 M solution. In addition, when the via hole was formed, a taper angle of 34° was measured. In this case, samples were etched in a 10 M HF solution for 15 min at room temperature. It was the lowest taper angle among all cases. The etching results are summarized in Figure 6.

In the case of the NH$_4$F solution, 1 M (8.42 mL (40% NH$_4$F solution) + 91.58 mL (deionized (DI) water)), 3 M (25.25 mL (40% NH$_4$F solution) + 74.75 mL (deionized (DI) water)), 5 M (42.09 mL (40% NH$_4$F solution) + 57.91 mL (deionized (DI) water)) (Figure 5b), 6 M (50.5 mL (40% NH$_4$F solution) + 49.5 mL (deionized (DI) water)), 7 M (58.9 mL (40% NH$_4$F solution + 41.1 mL (deionized (DI) water)), and 8 M (67.3 mL (40% NH$_4$F solution) + 32.7 mL (deionized (DI) water)) solutions were tested. For a 40% NH$_4$F solution, the boiling point is 109 °C, so for safety reasons, etching was performed at 85 °C for 2 h. The measured angles were 47° for the 1 M solution, 55° for the 3 M solution, 53° for the 5 M solution, 53° for the 6 M solution, 53° for the 7 M solution, and 50° for the 8 M solution (Figure 6). When a via hole was formed, a taper angle of 60° was measured. In this case, samples were etched in an 8 M solution for 3 h at 85 °C. We expected a higher etch rate to have a higher taper angle. However, it is independent. The most efficient etching solution had the second lowest taper angle (47°~60°).

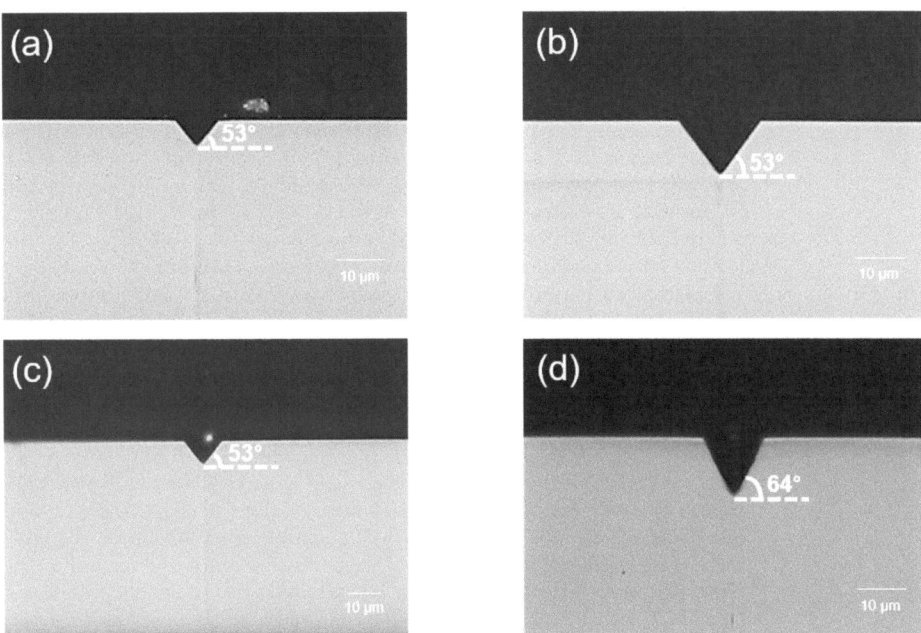

Figure 5. Via hole formed via each etchant. (**a**–**d**) Optical microscope images. (**a**) HF 5 M solution; (**b**) NH$_4$F 5 M solution; (**c**) NaOH 5 M solution; (**d**) KOH 5 M solution.

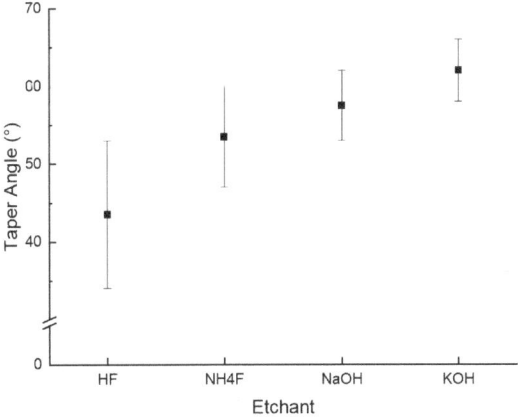

Figure 6. Taper angle range of HF, NaOH, KOH, and NH$_4$F solutions.

In the case of the NaOH solution, 1 M (13 mL (20% NaOH solution) + 87 mL (deionized (DI) water)), 2 M (26 mL (20% NaOH solution) + 74 mL (deionized (DI) water)), 3 M (40 mL (20% NaOH solution) + 60 mL (deionized (DI) water)), 4 M (53 mL (20% NaOH solution) + 47 mL (deionized (DI) water)), and 5 M (100 mL (20% NaOH solution) solutions were tested. A 20% NaOH solution has a boiling point of 110 °C; thus, for safety reasons, etching was performed at 100 °C for 3 h. The measured angles were 58° for the 1 M solution, 57° for the 2 M solution, 62° for the 3 M solution (Figure 5c), 55° for the 4 M solution, and

53° for the 5 M solution. When a via hole was formed, a taper angle of 55° was measured (Figure 6). In this case, samples were etched in a 3 M solution for 20 h.

In the case of the KOH solution, 1 M (8.5 mL (45% KOH solution) + 91.5 mL (deionized (DI) water)), 2 M (17 mL (45%KOH solution) + 83 mL (deionized (DI) water)), 3 M (25 mL (45% KOH solution) + 75 mL (deionized (DI) water)), 4 M (34 mL (45% KOH solution) + 66 mL (deionized (DI) water)), 5 M (42 mL (45% KOH solution) + 58 mL (deionized (DI) water)), 8 M (68 mL (45% KOH solution) + 32 mL (deionized (DI) water)), and 10 M (85 mL (45% KOH solution) + 15 mL (deionized (DI) water)) were tested. For a 45% KOH solution, the boiling point is 132 °C, so for safety reasons, the 1–5 M solution was etched at 100 °C, the 8 M solution at 110 °C, and the 10 M solution at 115 °C for 5 h. The measured angles were 58° for the 1 M solution, 66° for the 2 M solution, 60° for the 3 M solution, 63° for the 4 M solution, 64° for the 5 M solution, 63° for the 8 M solution, and 58° for the 10 M solution (Figures 5d and 6). When a via hole was formed, a taper angle of 63° was measured. In this case, samples were etched in an 8 M solution for 9 h at 110 °C. The results indicated that KOH has the highest taper angle.

In Figure 6, it can be observed that HF has the lowest taper angle and KOH has the highest taper angle. In our experimental setup, a sample was etched by HF at room temperature, NH_4F at 85 °C, NaOH at 100 °C, and KOH at 100–115 °C. Thus, temperature might also have an effect on determining the taper angle other than etchant. Therefore, we performed etching by using the KOH solution with different temperatures. At 100 °C, a taper angle of 58–64° was measured, while a 58° taper angle was measured at 115 °C. Thus, we can conclude that temperature does not determine the taper angle. In addition, the results indicate that the concentration of the etchant determines the taper angle. However, there is no trend or linear relationship between the concentration and the taper angle (Figure 7). As shown in Figure 7, there is a certain concentration that generates the highest taper angle, and it varies with each etchant. However, in most cases, the highest taper angle is generated at the 3–5 M concentration of each etchant, except for KOH.

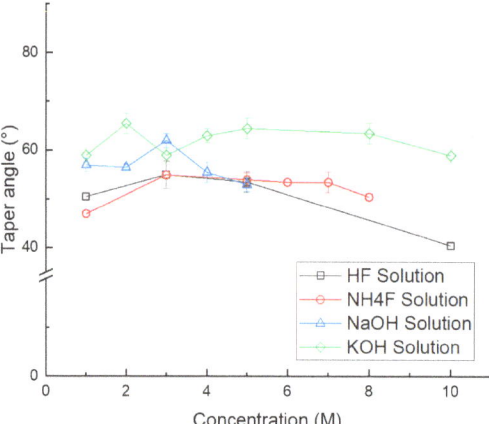

Figure 7. Taper angle based on concentration of HF, NaOH, KOH, and NH_4F solutions.

4. Conclusions

This study presents the etching results according to four different etchants and etching conditions. The four etchants are HF, NH_4F, NaOH, and KOH solutions, and the etching conditions are the concentration of each etchant and the etching temperature. We conducted this study to find the optimal process condition to enhance the productivity of TGV in glass interposers. The results can be summarized as follows:

(1) The results show that the most efficient etchant is NH_4F, and the TGV could be generated within 3 h through etching with the 8 M NH_4F solution at 85 °C. As

mentioned above, we could generate TGV three times faster than demonstrated in previous studies. In addition, we found that selectivity is the most trustworthy parameter for representing etching efficiency.

(2) The results also reveal that the taper angle of a blind hole is affected by the etchant. The etchant itself determines the taper angle. HF, NH$_4$F, NaOH, and KOH solutions generated 41°–53°, 47°–60°, 53°–62°, and 58°–66° taper angles, respectively. This study might be helpful for those who want to generate holes with certain angles. However, we still need to understand the principle underlying the phenomena identified in this study.

Author Contributions: J.K.: writing—original draft, data curation—local modification and chemical etching; B.K.: data curation—donut beam generation with SLM; J.C.: data analysis—SLE; S.A.: conceptual design of study and article revision. All authors have read and agreed to the published version of the manuscript.

Funding: This work was supported by the TA CR (CZ)-KIAT (KR) bilateral R&D project funded by the Ministry of Trade, Industry and Energy (MOTIE, Korea) (Project Number: P066500002) and the Basic Research Fund of Korea Institute of Machinery and Materials (Project Number: NK248F).

Data Availability Statement: The data underlying the results presented in this paper are not publicly available at this time but may be obtained from the authors upon reasonable request.

Acknowledgments: Sanggil Ryu and Sangwon Shim participated in the bilateral project and helped the authors to understand the industrial needs of TGV.

Conflicts of Interest: The authors declare no conflict of interest.

References

1. Töpper, M.; Ndip, I.; Erxleben, R.; Brusberg, L.; Nissen, N.; Schröder, H.; Yamamoto, H.; Todt, G.; Reichl, H. 3-D Thin Film Interposer Based on TGV (Through Glass Vias): An Alternative to Si-Interposer. In Proceedings of the 2010 Proceedings 60th Electronic Components and Technology Conference (ECTC), Las Vegas, NV, USA, 1–4 June 2010; IEEE: Piscataway, NJ, USA, 2010; pp. 66–73.
2. Keech, J.; Chaparala, S.; Shorey, A.; Piech, G.; Pollard, S. Fabrication of 3D-IC Interposers. In Proceedings of the 2013 IEEE 63rd Electronic Components and Technology Conference, Las Vegas, NV, USA, 28–31 May 2013; pp. 1829–1833.
3. Delmdahl, R.; Paetzel, R. Laser Drilling of High-Density Through Glass Vias (TGVs) for 2.5D and 3D Packaging. *J. Microelectron. Packag. Soc.* **2014**, *21*, 53 57. [CrossRef]
4. Lueck, M.; Huffman, A.; Shorey, A. Through glass vias (TGV) and aspects of reliability. In Proceedings of the 2015 IEEE 65th Electronic Components and Technology Conference (ECTC), San Diego, CA, USA, 26–29 May 2015; pp. 672–677.
5. Shorey, A.B.; Lu, R. Progress and Application of through Glass via (TGV) Technology. In Proceedings of the 2016 Pan Pacific Microelectronics Symposium (Pan Pacific), Big Island, HI, USA, 25–28 January 2016; pp. 1–6.
6. Watanabe, A.O.; Ali, M.; Zhang, R.; Ravichandran, S.; Kakutani, T.; Raj, P.M.; Tummala, R.R.; Swaminathan, M. Glass-Based IC-Embedded Antenna-Integrated Packages for 28-GHz High-Speed Data Communications. In Proceedings of the 2020 IEEE 70th Electronic Components and Technology Conference (ECTC), Orlando, FL, USA, 3–30 June 2020; pp. 89–94.
7. Sato, Y.; Imajyo, N.; Ishikawa, K.; Tummala, R.; Hori, M. Laser-Drilling Formation of through-Glass-via (TGV) on Polymer-Laminated Glass. *J. Mater. Sci. Mater. Electron.* **2019**, *30*, 10183–10190. [CrossRef]
8. Hof, L.A.; Ziki, J.A. Micro-Hole Drilling on Glass Substrates—A Review. *Micromachines* **2017**, *8*, 53. [CrossRef]
9. Chen, S.T.; Jiang, Z.H.; Wu, Y.Y.; Yang, H.Y. Development of a Grindingdrilling Technique for Holing Optical Grade Glass. *Int. J. Mach. Tools Manuf.* **2011**, *51*, 95–103. [CrossRef]
10. Park, B.J.; Choi, Y.J.; Chu, C.N. Prevention of exit crack in micro drilling of soda-lime glass. *CIRP Ann.* **2002**, *51*, 347–350. [CrossRef]
11. Kowsari, K.; Nouraei, H.; James, D.F.; Spelt, J.K.; Papini, M. Abrasive Slurry Jet Micro-Machining of Holes in Brittle and Ductile Materials. *J. Mater. Process. Technol.* **2014**, *214*, 1909–1920. [CrossRef]
12. Nouraei, H.; Kowsari, K.; Spelt, J.K.; Papini, M. Surface Evolution Models for Abrasive Slurry Jet Micro-Machining of Channels and Holes in Glass. *Wear* **2014**, *309*, 65–73. [CrossRef]
13. Li, X.; Abe, T.; Esashi, M. Deep Reactive Ion Etching of Pyrex Glass Using SF6 Plasma. *Sens. Actuators A Phys.* **2001**, *87*, 139–145. [CrossRef]
14. Kolari, K. Deep Plasma Etching of Glass with a Silicon Shadow Mask. *Sens. Actuators A Phys.* **2008**, *141*, 677–684. [CrossRef]
15. Malhotra, R.; Saxena, I.; Ehmann, K.; Cao, J. Laser-Induced Plasma Micro-Machining (LIPMM) for Enhanced Productivity and Flexibility in Laser-Based Micro-Machining Processes. *CIRP Ann.* **2013**, *62*, 211–214. [CrossRef]

16. Kim, S.; Kim, J.; Joung, Y.H.; Ahn, S.; Choi, J.; Koo, C. Optimization of Selective Laser-Induced Etching (SLE) for Fabrication of 3D Glass Microfluidic Device with Multi-Layer Micro Channels. *Micro Nano Syst. Lett.* **2019**, *7*, 15. [CrossRef]
17. Marcinkevicius, A.; Juodkazis, S.; Watanabe, M.; Miwa, M.; Matsuo, S.; Misawa, H.; Nishii, J. Microfabrication in Silica. *Opt. Lett.* **2001**, *26*, 277–279.
18. Hörstmann-Jungemann, M.; Gottmann, J.; Wortmann, D. Nano- and Microstructuring of SiO_2 and Sapphire with Fs-Laser Induced Selective Etching. *J. Laser Micro/Nanoeng.* **2009**, *4*, 135–140. [CrossRef]
19. Hermans, M.; Gottmann, J.; Riedel, F. Selective, Laser-Induced Etching of Fused Silica at High Scan-Speeds Using KOH. *J. Laser Micro/Nanoeng.* **2014**, *9*, 126–131. [CrossRef]
20. Hnatovsky, C.; Taylor, R.S.; Simova, E.; Rajeev, P.P.; Rayner, D.M.; Bhardwaj, V.R.; Corkum, P.B. Fabrication of Microchannels in Glass Using Focused Femtosecond Laser Radiation and Selective Chemical Etching. *Appl. Phys. A* **2006**, *84*, 47–61. [CrossRef]
21. Gottmann, J.; Hermans, M.; Repiev, N.; Ortmann, J. Selective Laser-Induced Etching of 3D Precision Quartz Glass Components for Microfluidic Applications—Up-Scaling of Complexity and Speed. *Micromachines* **2017**, *8*, 110. [CrossRef]
22. Kim, J.; Kim, S.; Kim, B.; Choi, J.; Ahn, S. Study of Through Glass Via (TGV) Using Bessel Beam, Ultrashort Two-Pulses of Laser and Selective Chemical Etching. *Micromachines* **2023**, *14*, 1766. [CrossRef] [PubMed]
23. Turboexpo, A.; Germany, M. The Effects of Injection Angle and Hole Exit Shape on Turbine Nozzle Pressure Side Film Cooling. In Proceedings of the ASME Turbo Expo 2000, Munich, Germany, 8–11 May 2020; p. 11.
24. Ji, J.; Tay, F.E.H.; Miao, J.; Iliescu, C. Microfabricated Microneedle with Porous Tip for Drug Delivery. *J. Micromechanics Microengineering* **2006**, *16*, 958–964. [CrossRef]
25. Matsumura, T.; Muramatsu, T.; Fueki, S. Abrasive Water Jet Machining of Glass with Stagnation Effect. *CIRP Ann.* **2011**, *60*, 355–358. [CrossRef]
26. Alberdi, A.; Suárez, A.; Artaza, T.; Escobar-Palafox, G.A.; Ridgway, K. Composite Cutting with Abrasive Water Jet. *Procedia Eng.* **2013**, *63*, 421–429. [CrossRef]
27. Ross, C.A.; MacLachlan, D.G.; Choudhury, D.; Thomson, R.R. Optimisation of Ultrafast Laser Assisted Etching in Fused Silica. *Opt. Express* **2018**, *26*, 24343. [CrossRef] [PubMed]
28. Ahn, S.; Kim, S.I.; Choi, J.; Park, C. Etching Apparatus and Interposer Manufacturing System Including the Same. Patent KR10-2423292, 2022.

Disclaimer/Publisher's Note: The statements, opinions and data contained in all publications are solely those of the individual author(s) and contributor(s) and not of MDPI and/or the editor(s). MDPI and/or the editor(s) disclaim responsibility for any injury to people or property resulting from any ideas, methods, instructions or products referred to in the content.

Review

Review on Abrasive Machining Technology of SiC Ceramic Composites

Huiyun Zhang, Zhigang Zhao, Jiaojiao Li, Linzheng Ye and Yao Liu *

School of Mechanical Engineering, North University of China, Taiyuan 030051, China; szys2057324@126.com (H.Z.); zhao105820@163.com (Z.Z.); lijiaojiao1004@gmail.com (J.L.); yelinzheng@nuc.edu.cn (L.Y.)
* Correspondence: liuyao@nuc.edu.cn

Abstract: Ceramic matrix composites have the advantages of low density, high specific strength, high specific die, high-temperature resistance, wear resistance, chemical corrosion resistance, etc., which are widely used in aerospace, energy, transportation, and other fields. CMCs have become an important choice for engine components and other high-temperature component manufacturing. However, ceramic matrix composite is a kind of multi-phase structure, anisotropy, high hardness material, due to the brittleness of the ceramic matrix, the weak bonding force between fiber and matrix, and the anisotropy of composite material. Burr, delamination, tearing, chips, and other surface damage tend to generate in the machining, resulting in surface quality and strength decline. This paper reviewed the latest abrasive machining technology for SiC ceramic composites. The characteristics and research directions of the main abrasive machining technology, including grinding, laser-assisted grinding, ultrasonic-assisted grinding, and abrasive waterjet machining, are introduced first. Then, the commonly used numerical simulation research for modeling and simulating the machining of ceramic matrix composites is briefly summarized. Finally, the processing difficulties and research hotspots of ceramic matrix composites are summarized.

Keywords: ceramic matrix composites; material removal mechanism; numerical simulation; abrasive machining

Citation: Zhang, H.; Zhao, Z.; Li, J.; Ye, L.; Liu, Y. Review on Abrasive Machining Technology of SiC Ceramic Composites. *Micromachines* **2024**, *15*, 106. https://doi.org/10.3390/mi15010106

Academic Editor: Ha Duong Ngo

Received: 11 December 2023
Revised: 22 December 2023
Accepted: 24 December 2023
Published: 7 January 2024

Copyright: © 2024 by the authors. Licensee MDPI, Basel, Switzerland. This article is an open access article distributed under the terms and conditions of the Creative Commons Attribution (CC BY) license (https://creativecommons.org/licenses/by/4.0/).

1. Introduction

Ceramic matrix composites (CMCs), mainly including SiC_f/SiC, SiC_p/SiC, C_p/SiC, and C_f/SiC, have the advantages of low density, high specific strength, high elastic modulus, high temperature resistance, wear resistance, and chemical resistance [1]. As the result of fiber reinforcement, the CMC materials have improved fracture toughness and are widely used in aerospace, energy, transportation, and other fields to resist the thermal shock in high temperature environments. Additionally, CMCs are lightweight (about 1/3 of traditional metal materials) and have high combustion temperature, thus offering significant advantages in terms of lightweight design, fuel efficiency, and the gas emissions of aviation engines, and making them the preferred materials for high-temperature components of aviation engines.

As illustrated in Figure 1a, the use of CMC in blades, guides, hoods, nozzles, and combustion chambers in high-temperature sections of an aeroengine, increases engine thrust by 25%, saves fuel consumption by 10%, and decreases nitrogen oxide emissions. As shown in Figure 1b, the CMCs are used to manufacture non-structural parts such as thermal barrier coating and thermal conductive sheets of the engine, and key structural components, such as engine blades, guides, nozzles, and combustion chambers, to reduce the weight of the whole engine and to improve its durability and life. In Figure 1c, in Japan's 100 kW automotive ceramic gas turbine (CGT) project, CMCs were used to manufacture five components, including turbine rotor, rear plate, port bushing, expansion refiner, and internal scroll support [2]. It has demonstrated the feasibility of the application of CMCs in operational engines and their superiority in terms of thermal shock resistance and particle impact resistance capabilities.

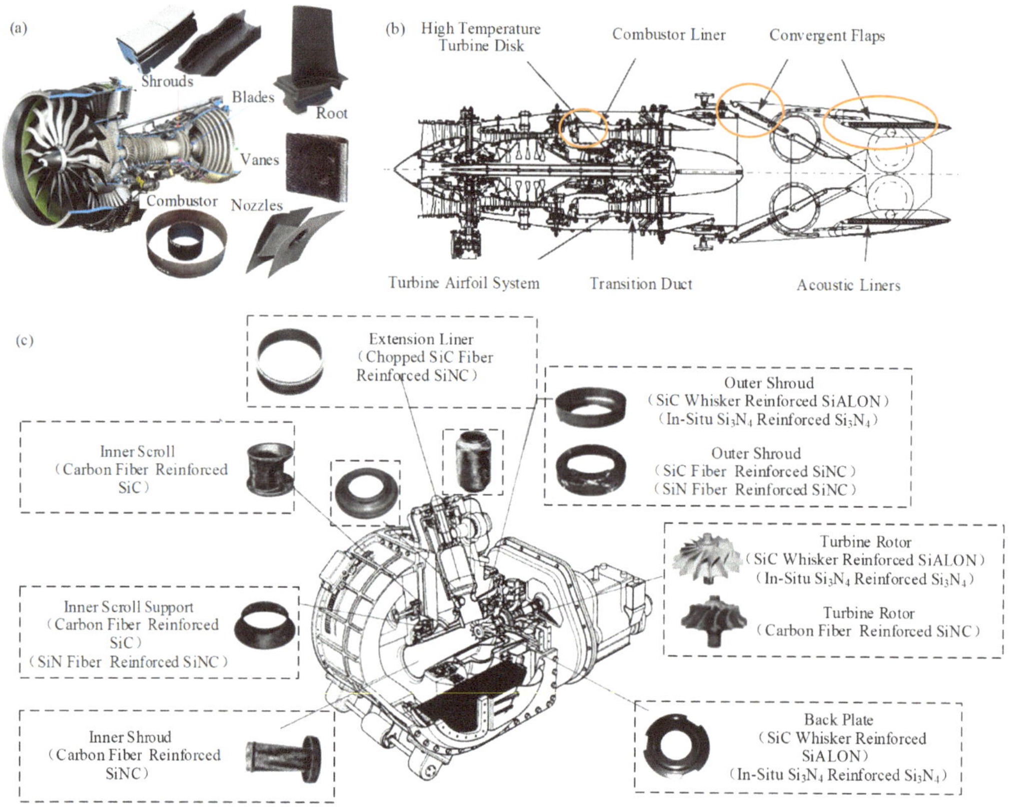

Figure 1. The application of CMCs: (**a**) high-temperature parts of aero-engines; (**b**) hot zone of the jet engine; (**c**) Main CMCs components of CGT.

CMCs are very difficult to machine due to the complex woven structure, anisotropic properties, and high hardness. CMC components are usually fabricated in a rough mold and are then machined to the design dimension, shape accuracy, and surface condition. Burr, delamination, crack, collapse, tear, and other surface damage easily occur, resulting in a machining quality decline, which creates a challenge for CMC machining. Simultaneously, the material removal mechanism of CMCs is very complicated, which seriously limits research progress. Fiber in CMC not only enhances the mechanical properties of the matrix, but also changes the material removal mechanism. In the grinding process, the damage usually starts from the matrix. The different material properties of the matrix and fiber show different removal modes, and the interactions transferred through the interface affect each other. Therefore, exploring the material removal mechanism is an effective way to realize lower damage and greater efficient machining of the CMC materials.

In this paper, the recent abrasive machining experimental study of CMC abrasive machining technologies, including the conventional grinding (CG), laser-assisted grinding (LAG), abrasive waterjet machining (AWJM), and ultrasonic vibration-assisted grinding (UVAG), were reviewed to reveal the material removal mechanism. Then, the CMC simulation studies, including FEM, SPH, DEM, and a hybrid of two or three methods, were introduced. Finally, the challenging aspects and pivotal points of the abrasive machining mechanism are summarized.

2. Experimental Study of CMC Abrasive Machining

To meet the requirements of complex shapes and dimensional tolerances, CMC materials need to be machined. To improve the machining surface integrity and manufacturing efficiency of CMC, research was conducted, both domestically and abroad. Abrasive machining is the most suitable method for the precision machining of CMC materials [3]. This section introduces the latest achievement in conventional grinding (CG), laser-assisted grinding (LAG), abrasive waterjet machining (AWJ), and the ultrasonic vibration-assisted grinding (UVAG) of CMCs.

2.1. Conventional Grinding

CG is the most common method for the precision machining of CMC [4]. The ground surface characteristics and material removal mechanism of CMCs have been analyzed. In the CMC grinding process, the orientation of fibers exhibits a crucial influence on the ground surface quality [5]. Qu et al. [6] studied the effects of carbon fiber orientation and grinding parameters on grinding force and surface quality during the grinding of C_f/SiC, as shown in Figure 2a. The results indicated that grinding depth has significant effects on the surface quality. Specifically, the surface quality decreases and the grinding forces increase with increasing grinding depth. In addition, greater grinding surface quality is observed at $\beta = 90°$. The poorer ground surfaces are obtained at $\alpha = 0°$. Cao et al. [7] studied the grinding process of woven CMC. The results showed that the best surface quality is in the 90° direction with the fiber. Hu et al. [8] concluded that the local orientation of fibers would also affect the surface quality of processed materials. Yin et al. [9] used the single grain scribing on SiC_f/SiC with different wheel speeds to reveal the material removal mechanism and fiber breakage behaviors. The results showed that the surface material densification and smearing could be suppressed by increasing grinding speed. When grinding along the fiber longitudinal direction, fibers experience plowing and would be covered by the smearing of the matrix. Increasing wheel speed enhances the brittle fracture and breakage of the fibers. In high-speed grinding, fibers present brittle fractures and the matrix is torn off. When grinding transverses to the fiber longitudinal direction, increasing wheel speed leads to the complete removal of the fibers and a few cutting-off fiber end residuals on the groove bottom surface, which improves the surface finish. Zhang et al. [5] compared the surface morphology and grinding mechanism of C_f/SiC composites in transverse, normal, and longitudinal directions (in Figure 2b), and found that the fibers were most easily removed during transverse grinding, and the ground surface quality was the best in this direction. During normal direction grinding, fibers and matrix are mainly removed in the form of small fragments, and the ground surface quality is worse than that in the transverse direction. In longitudinal grinding, the composite material has the worst ground surface quality. Liu et al. [10] proposed a three-dimensional surface profile characterization through experiments, which can describe the surface quality of composite materials accurately.

Liu et al. [11] investigated the effect of fiber angle (FA) on grinding force, surface morphology, and roughness through surface grinding experiments. As shown in Figure 2c, the fracture characteristics of the grinding surface include matrix cracking, fiber fracture, and interfacial debonding, revealing that the main removal mechanism of 2D C_f/SiC. Fiber fracture was more severe at 30° and 45° FA than at 0° FA. Those differences can be attributed to the fact that the damage is not synchronized during the grinding process due to the different mechanical properties of the matrix and fibers. Figure 2d shows the surface morphology of 0° and 90° fiber bundles at different grinding angles. During the grinding process, the scraping force shows a periodic variation of 0° and 90° fiber weave structure when the scraping angle is 0°. However, when the scribing angle is 45°, the force is relatively more stable and the surface quality is better.

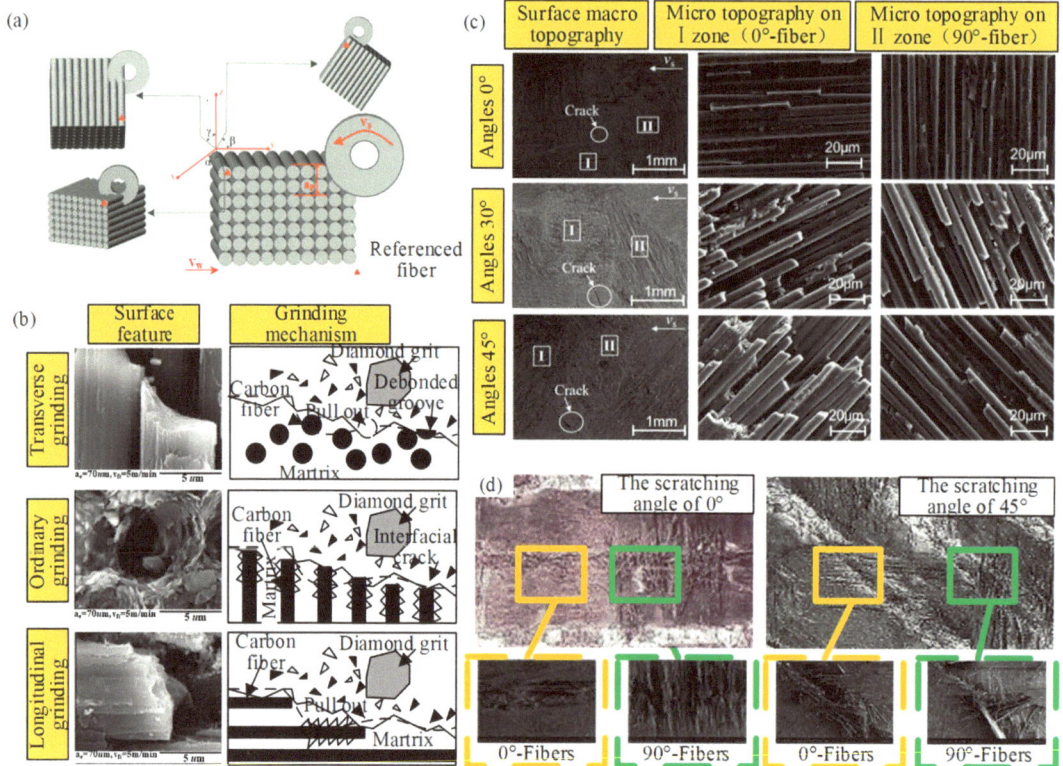

Figure 2. (**a**) Schematic diagram of the grinding direction and datum plane studied by Qu et al. [6]; (**b**) surface morphology and material removal mechanism of C_f/SiC in transverse, normal, and longitudinal direction by Zhang et al. [5]; (**c**) SEM of the ground surface in different fiber orientations; (**d**) surface morphology of 0° and 90° fiber bundles at different scribing directions by Liu et al. [10].

In the grinding of CMCs, grinding depth, feed speed, friction force, lubricant, and tool parameters have important effects on surface quality and fracture characteristics. Wang et al. [12] established a force model for the surface grinding of unidirectional C_f/SiC composites, and concluded that the grinding force is inversely proportional to the wheel speed and proportional to the grinding depth and feed speed. Zhang et al. [13] established a model between grinding force and depth in fiber-reinforced composites. With the increase of grinding depth, the grinding force exhibits an almost linear relationship along the direction of the reinforced fiber, which affects the ground surface quality. Zhang et al. [5] performed the experimental research on the grinding mechanism of woven CMC and revealed that both the grinding force and the surface roughness increased with the feed speed and the grinding depth, and decreased with the wheel speed. Tawakoli et al. [14] investigated the CMC grinding forces and surface quality during conventional and intermittent grinding by using the normal and fan-shaped wheel. The results found that the conventional grinding wheel could obtain better surface roughness and the fan-shaped grinding wheel could significantly reduce the grinding force and grinding temperature. These results revealed that intermittent grinding could reduce scratching, the plowing phenomenon, and specific energy, which can provide a high surface quality.

The CMC grinding damage modes can be analyzed through indentation fracture mechanics [15]. Yang et al. [16] carried out orthogonal grinding of C_f/SiC to observe the ground surface topography and found that the main material removal mode was brittle fracture, including matrix cracking, fiber break, fiber wear, and interface bonding. Pineau et al. [17] built a machining force prediction model based on the experimental results of the shear strength and predicted the crack occurrence of woven CMC based on the virtual test results. Lamon et al. [18] proposed the main damage mode of two-dimensional woven C_f/SiC based on the micromechanical method of the mechanical behavior of brittle matrix composites.

Due to the high hardness and low thermal conductivity of C_f/SiC composites, the grinding wheel absorbs and stores a lot of heat during dry grinding, resulting in the reduction of grain number, sharpness, and life. In addition, due to the passivation of the grinding wheel, the surface quality of the machined surface decreases [19]. Hu et al. [20] studied the cutting force, surface integrity, and machining defects in the milling process of 2D C_f/SiC composites with the commercialized PCD tool. The results showed that fiber fracture, matrix damage, and fiber-matrix debonding are the main types of material failure modes. Xu et al. [21] studied the dry cutting temperature of C_f/SiC with a straight-edge PCD tool. The study showed that the highest temperature reached around 732 °C, with a significant increase in cutting force, which makes the surface quality poor. Qu et al. [22] investigated the effect of minimum quantity lubrication (MQL) on the grinding performance of unidirectional carbon fiber-reinforced ceramic matrix composites, as shown in Figure 3a, which includes abrasive, lubricant, and material. The surface morphology was observed using scanning electron microscopy (in Figure 3b). The main fracture patterns of carbon fiber are smooth separation, fracture, detachment, and pulling out.

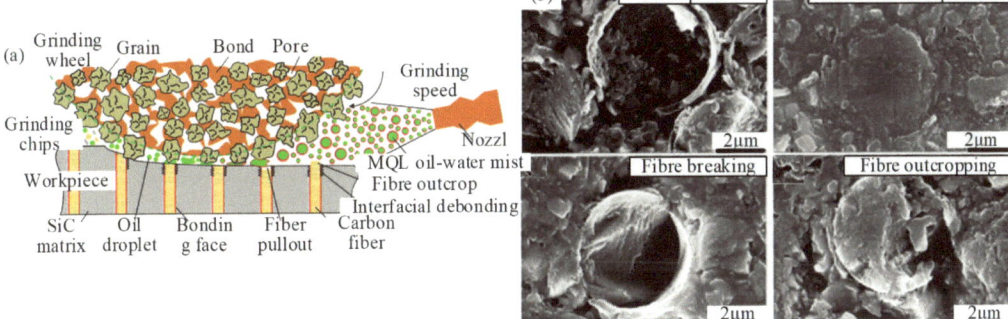

Figure 3. (**a**) Schematic diagram of the ground surface under the MQL system by Qu et al. [22]; (**b**) SEM photo of typical fracture of carbon fiber under MQL.

CMC grinding has shown serious tool wear, low machining efficiency, and poor machining quality. In particular, defects such as poor dimensional consistency, surface cracks, fragmentation, delamination, fiber shedding, fiber pulling out, and chipping seriously affect the processing quality and surface accuracy of the material [23]. Therefore, in the grinding process, the appropriate grinding depth and cutting angle should be carefully selected.

2.2. Laser-Assisted Grinding (LAG)

Due to the high brittleness and anisotropy of the CMC, LAG acts as a feasible process with which to improve the machinability of materials, which induces the strength loss of materials through high temperatures. LAG uses the thermal effect of the laser to soften the material in the area to be ground, which can reduce the cutting force and improve the machinability of the material. Compared to traditional processing, LAG can reduce the force, surface roughness, and production costs.

Zhang et al. [24] designed a single-factor experiment to understand the micro-structure evolution and ablation behaviors of the microgrooves processed by a picosecond laser. This study provided a theoretical basis and practice guidance for LAG of the SiC_f/SiC composites. Figure 4a shows the amorphous SiO_2 smoke, with extremely fine particles. The dust attached to the treated surface caused the occurrence of pileup, leading to a slight increase in Ra value along the machined microgrooves. When the laser parameters (pulse energy, repetition frequency, scanning times, and scanning speed) were adjusted to increase the microgroove depth, the SiC vapored in the subsurface layer was recrystallized during rapid cooling, as presented in Figure 4b. Compared to the ultrahigh hardness SiC fiber and SiC matrix, the SiO_2 smoke dust or recrystallized SiC were powdery due to the ablation, which were easy to remove and were expected to greatly improve the surface quality. An et al. [25] analyzed the products of continuous laser ablating CMC, and the factors affecting the depth of ablation layer were analyzed. As shown in Figure 4c, according to the morphology and types of ablation products, the laser ablation of SiC_f/SiC can be divided into the coagulated layer (Layer 1), the re-crystallized SiC layer (Layer 2), the heat-affected layer (Layer 3), and the non-affected layer (Layer 4). In Layer 1, gaseous SiC and oxygen in the air undergo an active oxidation reaction at high temperatures to form smoky amorphous SiO_2 condensation deposited on the surface of the ablated layer. Layer 2 is mainly composed of re-crystallized SiC particles with a micro-size scale. A small amount of SiO_2 soot is attached to the particles.

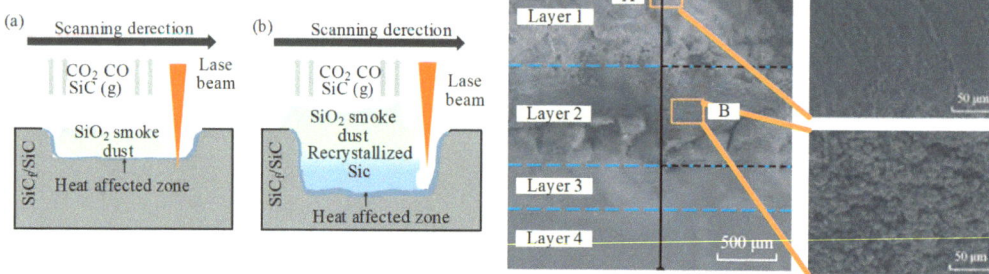

Figure 4. (**a**,**b**) Schematic diagram of laser-induced ablation under LAG; (**c**) the products of continuous laser ablating CMCs under LAG by An et al. [25].

As shown in Figure 5, Zhou et al. [26] proposed a laser-induced ablation-assisted grinding (LIAAG), which is based on the chemical properties of materials. This method utilizes lasers to ablate workpieces before grinding, aiming for high efficiency, minimal damage, and reduced abrasive wear, and found that C_f/SiC composites were chemically transformed into relatively loose and homogeneous ablation products (SiO_2 and recrystallized SiC) at high laser ablation temperatures. In Figure 5c, surface morphologies displayed the microfracture and crushing of carbon fibers and SiC matrixes, and the grinding-induced damages, such as macro fracture, fiber pulling out, and interface debonding. In Figure 5f, the abrasive belt was primarily worn in micro-adhesion and micro-abrasion, rather than cleavage fracture and fall-off in traditional grinding. The surface integrity was improved greatly, and the abrasive wear was reduced significantly, which provided a vital high-performance processing method for CMC components.

Li et al. [27] introduced the laser-assisted precision grinding technology to improve the processing quality of 3D woven C_f/SiC composites. The laser process parameters were adjusted to control the depth of the thermally induced damage layer and to reduce the hard brittleness of the material. In the study, experiments were carried out to investigate the effect of laser parameters on material damage and the effect of LAG processes. As shown in Figure 6a, due to the material anisotropy, the depth and width of the thermal-damage slot are 0° fiber > 90° fiber > normal fiber and 90° fiber > 0° fiber > normal fiber, respectively. In addition, laser irradiation causes complex reactions, such as sublimation, decomposition, and oxidation, on the surface of the material and generates SiO_2 and Si as well as recrystallized SiC, resulting in the formation of a porous SiC layer (thermal metamorphic layer) on the subsurface, as shown in Figure 6b. In the LAG, the normal grinding force, the tangential grinding force, the specific grinding energy, and surface roughness are reduced by a maximum of 35.6%, 43.6%, 43.58%, and 24.22%, respectively, compared to the conventional grinding (CG) process. As shown in Figure 6c–e, the material removal of 3D C_f/SiC composite material is dominated by brittle removal, which is mainly dominated by fiber breakage, pulling out, layer breakage, matrix cracking, and interface debonding. However, a trend toward ductile domain removal has begun to emerge.

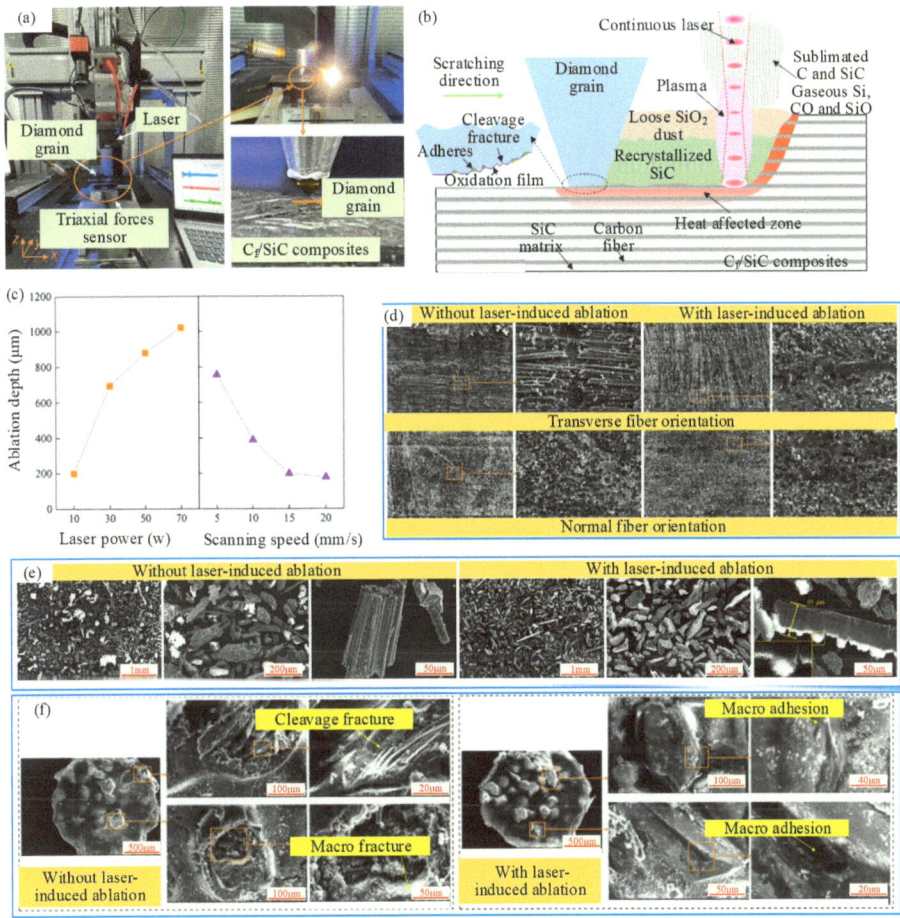

Figure 5. Laser-induced ablation-assisted grinding (LIAAG) by Zhou et al. [26]: (**a**) Laser-induced ablation-assisted grain scratching testing machine; (**b**) material removal mechanism of C_f/SiC composites during LIAAG; (**c**) temperature field and ablation depth during laser ablation; (**d**) SEM images of C_f/SiC composites surface after grinding; (**e**) SEM images of grinding chips; (**f**) wear morphology of diamond abrasive grains.

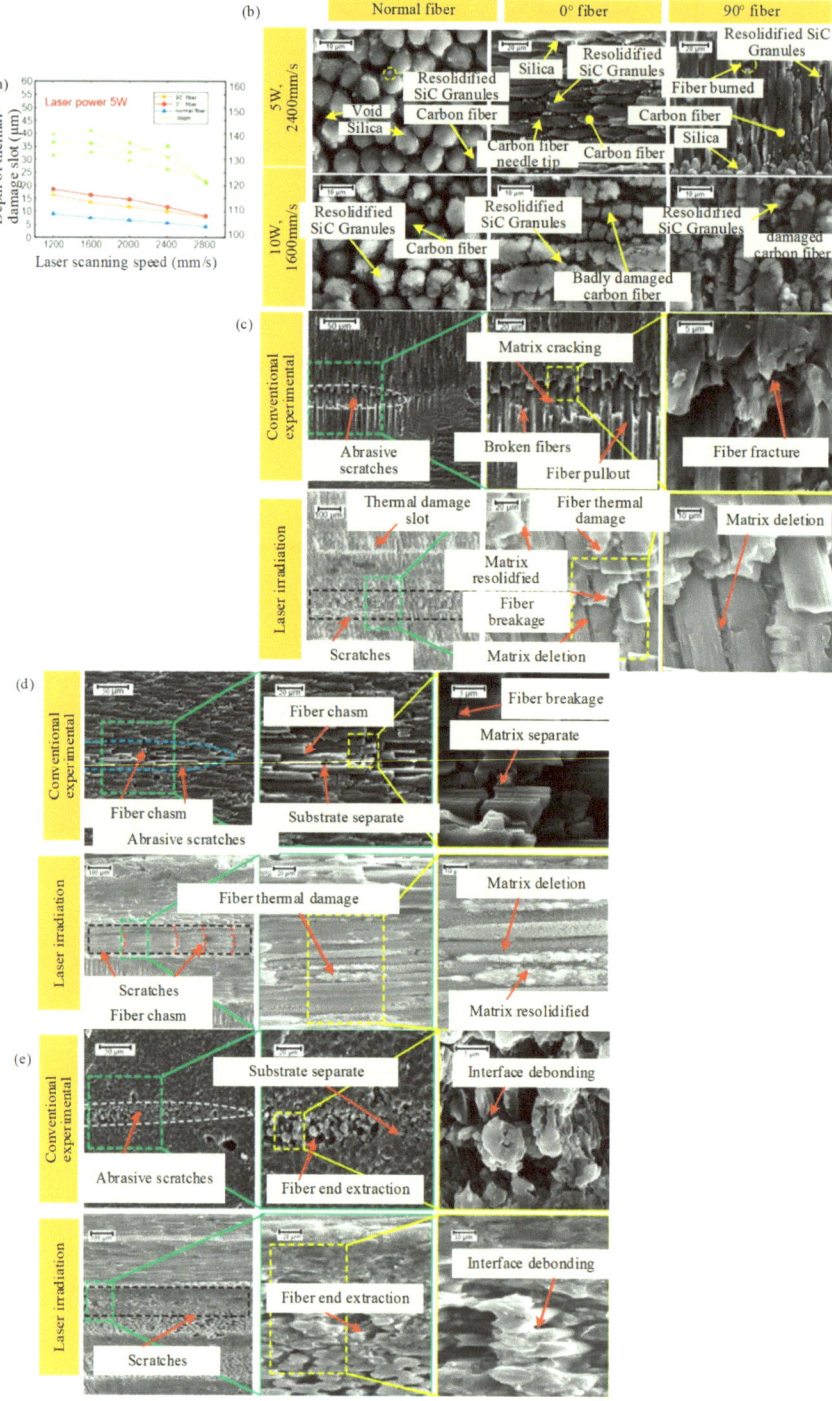

Figure 6. Laser-assisted grinding to improve the processing quality of 3D C_f/SiC composites: (**a**) Relationship between laser power, scanning speed and thermal-damage slot size; (**b**) micro-structure

of fiber surface after laser irradiation; (**c**) microscopic profile of 0° fiber area scratches; (**d**) microscopic profile of 90° fiber area scratches; (**e**) microscopic profile of normal fiber area scratches.

Kong et al. [28] investigated the cutting performance of CG and LAG of SiC$_f$/SiC using electroplated diamond grinding heads. A three-dimensional transient heat transfer model based on a Gaussian heat source was developed to investigate the distribution of temperature fields on both the surface and subsurface of SiC$_f$/SiC subjected to laser irradiation. As seen in Figure 7, after laser irradiation, the surface and subsurface temperatures of the workpiece reach > 1000 °C, which is sufficient for the oxidation reaction and softening of the material. In Figure 7c,d, the effects of laser heating temperature on the workpiece surface on the grinding forces were analyzed. The axial and feed grinding forces were more than 40% lower under LAG than CG, due to the removal mechanism of the SiC matrix changing from brittle to ductile and the oxidation reactions occurring in the SiC$_f$/SiC composites. The material removal mechanism was analyzed by observing the morphology of machined surfaces, as shown in Figure 7e–g, which showed that ductile removal from the SiC matrix occurs during LAG. In terms of abrasive wear, the mean height of exposed abrasive grains from the machined surface was reduced by 1.02 μm and 12.52 μm in LAG and CG, respectively. Under CG, the abrasive grains mainly exhibit cleavage fractures; however, under LAG, micro-abrasion is the main wear form.

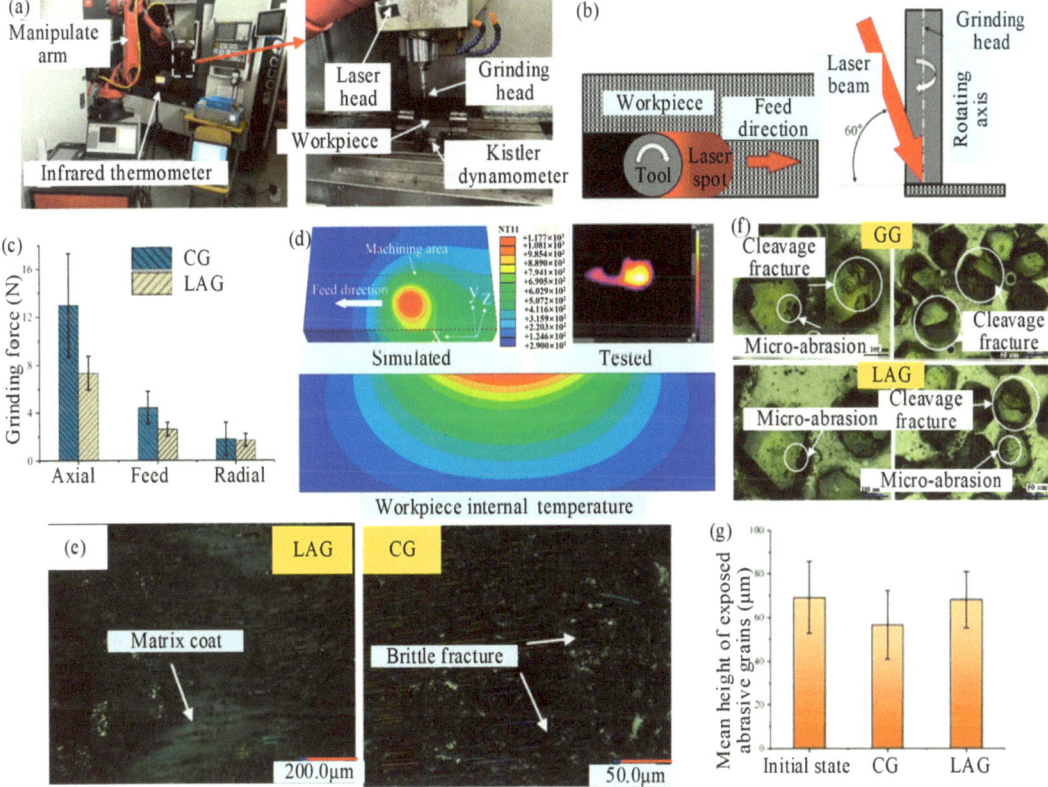

Figure 7. The cutting performance of CG and LAG of SiC$_f$/SiC by Kong et al. [28]: (**a**) LAG system; (**b**) schematic diagram of LAG; (**c**) comparison of CG and LAG forces; (**d**) laser temperature field; (**e**) surface morphology; (**f**) grinding head abrasive grains wear forms; (**g**) mean heights of exposed abrasive grains.

LAG can reduce the CMC surface temperature gradient and hardness, which improves the grinding conditions and achieves a smooth ground surface. After laser preheating, the surface temperature of the processed material increase, and the hardness of the material is reduced. To avoid rapid temperature reduction, it is necessary to minimize the time difference between the laser and grinding.

It can be seen from the above research that the grinding force of the CMC is significantly reduced under laser heating. The tool wear and machining defects are also reduced. However, due to CMCs with high melting points and high hardness, the temperature required for material softening, melting, and even gasification is very high—higher than 1000 °C. Therefore, the laser heat-affected zone is large, in which the physical and mechanical properties of the CMC are changed. The material in the heat-affected zone must be removed to obtain the required surface. Moreover, interface cracks and surface oxidation caused by the laser thermal effect made LAG difficult to apply to the shape machining of typical components of complex curved surfaces [29].

2.3. Abrasive Waterjet Machining (AWJM)

AWJM is a new technology that has been developed rapidly in the past 30 years. It uses high-pressure and high-speed water jets to impact the workpiece to achieve cutting, perforating, and surface material removal [30]. For CMCs, AWJM technology has unique advantages. It belongs to the category of non-contact processing and avoids the problem of tool wear. In addition, the processing temperature is low, which can greatly reduce the thermal effects zones. The study by Hashish et al. [31] shows that the AWJM can successfully cut CMCs. However, there are still several problems that exist, such as serious nozzle wear, and micro and macro defects.

Hashish et al. [32] cut the SiC_f/SiC with AWJM at different impact velocities, and found that the size of the edge breakage decreased with the impact velocity decrease. The taper decreases with the increase of abrasive hardness and injection pressure, and increases with the cutting speed increase. In addition, there is a parallel relationship between corrugation and taper. The influences of abrasive hardness, injection pressure, and cutting speed on corrugation are similar to those on taper. In addition, by comparing the influence of #80 and #120 abrasive grain on the edge breakage, it was found that the small grain size can reduce the edge breakage defect.

Zhang et al. [33] explored the accurate control of hole shape for AWJM hole-making of C_f/SiC based on experimental and mathematical analysis methods, as shown in Figure 8. The results reveal that $D_{difference}$ is influenced by the standoff distance, followed by the traverse speed. However, influence of the pressure and the abrasive flow rate is rare. The traverse speed, pressure, and abrasive flow rate affect the $D_{difference}$ by changing the total energy of the jet. The standoff distance mainly affects the $D_{difference}$ by changing the effective impact area, which is fundamentally different from other process parameters. In Figure 8b, when the jet is cutting circular trajectories, there is a deflection effect, which causes D_{jet} to be larger than $D_{trajectory}$ and affects hole size accuracy. The deflection effect increases with the increase of traverse speed and standoff distance and decreases with the increase of waterjet pressure and $D_{trajectory}$. Therefore, when holes are cut at high traverse speed, low pressure, and a tiny trajectory diameter, the deflection effect must be considered. However, this study does not relate to surface damage in composite materials processing.

Ramulu et al. [34] researched the cutting force and surface micro-structure of composite materials, and material tensile behavior, under AWJM machining and conventional milling processing. The results showed that AWJM machining exhibits superior stability in the machining process and yields better surface quality. The cutting force is much lower than that of CG. Ren et al. [35] summarized the advantages and mechanism of AWJM machining ceramics and found that surface roughness could be controlled by controlling the relationship between the fibers' axial direction and the AWJM direction. When the AWJM direction is parallel to the axis of the fibers, at the top of the drilling hole, the material processing surface is relatively flat, and the fracture surface of the fiber and the matrix is consistent under the action of the AWJM. Inside the borehole, prominent broken fibers were seen, and the SiC matrix between the fibers was removed and carried away by the AWJM. At the exit of the borehole, the fibers pull out, break, and strip from the matrix. When the direction is perpendicular to the axial direction of the fibers, the drilling hole is relatively flat from the top to bottom. The matrix around the fiber is not spalling, and the fiber pulls out. However, the fiber and the matrix threshing phenomenon is reduced.

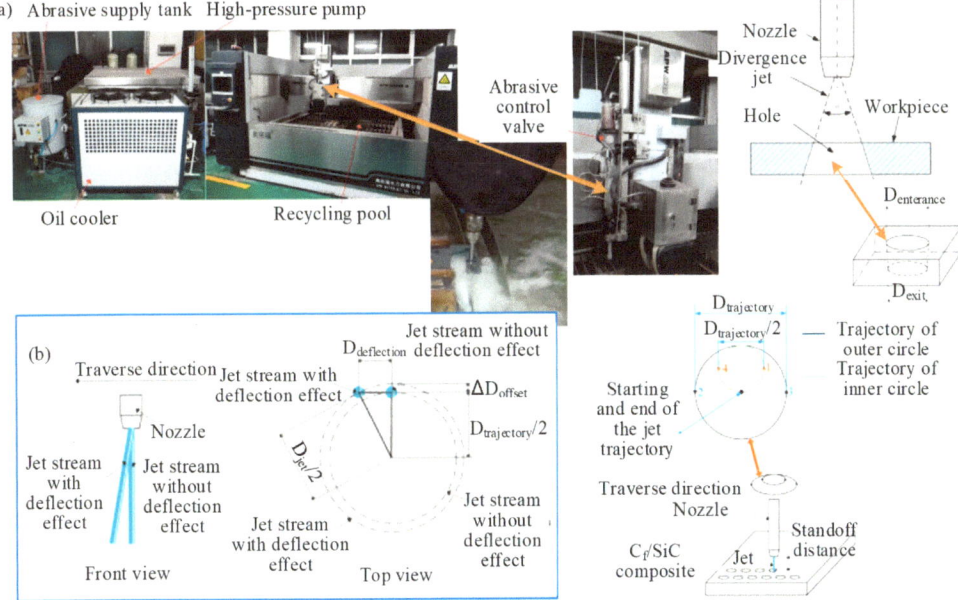

Figure 8. AWJM hole-making of C_f/SiC by Zhang et al. [33]: (**a**) Experimental setup and processing diagram of AWJM; (**b**) schematic diagram of deflection effect.

Generally, AWJM has many advantages, such as the material not being affected by heat during processing, which can effectively improve the quality of in-hole processing. However, when processing deeper holes, the surface of the workpiece is prone to burr and chipping. The entrance and exit dimensional error may be large and the processing quality is relatively low [36].

2.4. Ultrasonic Vibration-Assisted Grinding (UVAG)

Ultrasonic processing technology employs an ultrasonic generator to convert electrical energy into ultrasonic waves that oscillate at a specific frequency and vibrate through an amplification tool (variable amplitude bar). The ultrasonic wave was generated in UVAG. The suspended particles in the working fluid impact the surface of the workpiece and remove excess material. UVAG can machine the insulation material and complicated three-dimensional structures, regardless of material hardness. To solve the problems of poor machining quality and serious tool wear in CMC material processing, researchers both domestically and aboard have conducted comparative experiments between UVAG and CG to observe and test its technological parameters.

Kang et al. [37] experimented with the UVAG of SiC_f/SiC material on the end face of the diamond grinding wheel. The experimental setup and operational principles are shown in Figure 9, aimed at examining the removal mechanism, grinding force, surface morphology, and surface roughness. The results show that the appropriate ultrasonic amplitude can effectively reduce the grinding force, induce the fracture of SiC fibers, and improve the surface finish. However, when the amplitude is too large, the surface impact is large, and the surface quality declines. Ding et al. [38] studied the surface/subsurface breakage formation mechanism and machining quality of C/SiC composite conducted by UVAG and CG tests. The results showed that main breakage types of different angle fibers in ground surface were lamellar brittle fracture. Compared to CG, these breakages were reduced by UVAG which can reduce grinding force. Moreover, the ground surface roughness obtained by UVAG was lower than CG and the maximum reduction was 12%. Zhang et al. [39] investigated the material removal and breakage mechanism in UVAG of two-dimensional woven carbon-fiber-reinforced silicon carbide matrix composites (2D-C_f/SiC). The results show that the predominant material removal mode in UVAG is brittle fracture. The forms of material breakage are matrix cracking, fiber fracture, fiber pull-out, interfacial debonding, and interfacial fracture. Compared with CG, the normal force, tangential force, and surface roughness in UVAG decreased by approximately 20%, 18%, and 9%, respectively. Bertsche et al. [40] studied the material removal rate, cutting force, tool wear, and surface roughness of rotary ultrasonic machining of CMC. Compared to conventional cutting processes, UVAG effectively reduced cutting force and tool wear, and significantly improved surface roughness. Huang et al. [41] compared six commonly used tools for micro-hole drilling of SiCf/SiC CMC assisted by ultrasonic vibration: carbide drill, PCD drill, electroplated diamond abrasive tool, grinding drill, coated grinding drill, and PDC tool. As shown in Figure 10, the influence of different kinds of tools on cutting forces, machining accuracy, and tool wear was investigated. The results showed that the machining accuracy of the PDC tool is best, followed by the electroplated diamond abrasive tool. The machining accuracy of the grinding drill and coated grinding drill is poor with a bit of difference; PDC and coated grinding drill have less tool wear, while the grinding drill has slightly more severe tool wear. However, the tool wear of the electroplated diamond abrasive tool is severe. Wang et al. [42] conducted rotary ultrasonic machining of C_f/SiC composites to analyze the micro-structural characteristics of the hole surfaces under various fiber directions, ultrasonic amplitudes, and spindle speeds. The results show that the fiber direction and spindle speed have significant effects on the surface morphology of the material. The introduction of ultrasonic vibration improves the surface quality of C_f/SiC composites, and higher ultrasonic amplitude and lower spindle speed are helpful for improving the surface quality of the hole.

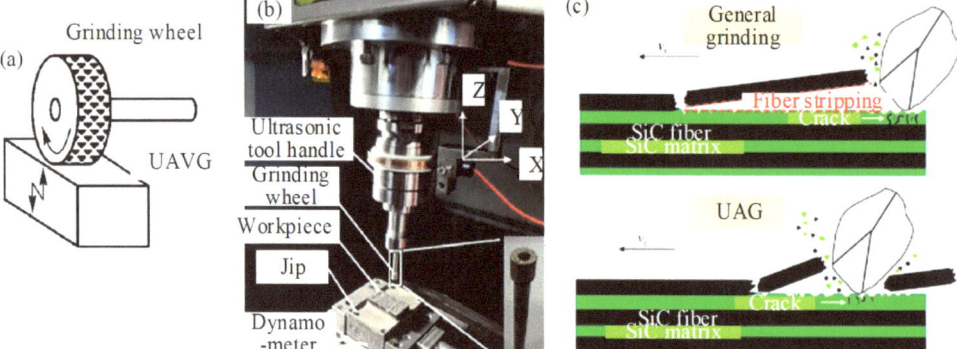

Figure 9. The experiment on UVAG of SiC$_f$/SiC material by Kang et al. [37]: (**a**) a type of UVAG; (**b**) UVAG test platform; (**c**) schematic diagram of grinding removal process of SiC$_f$/SiC composites.

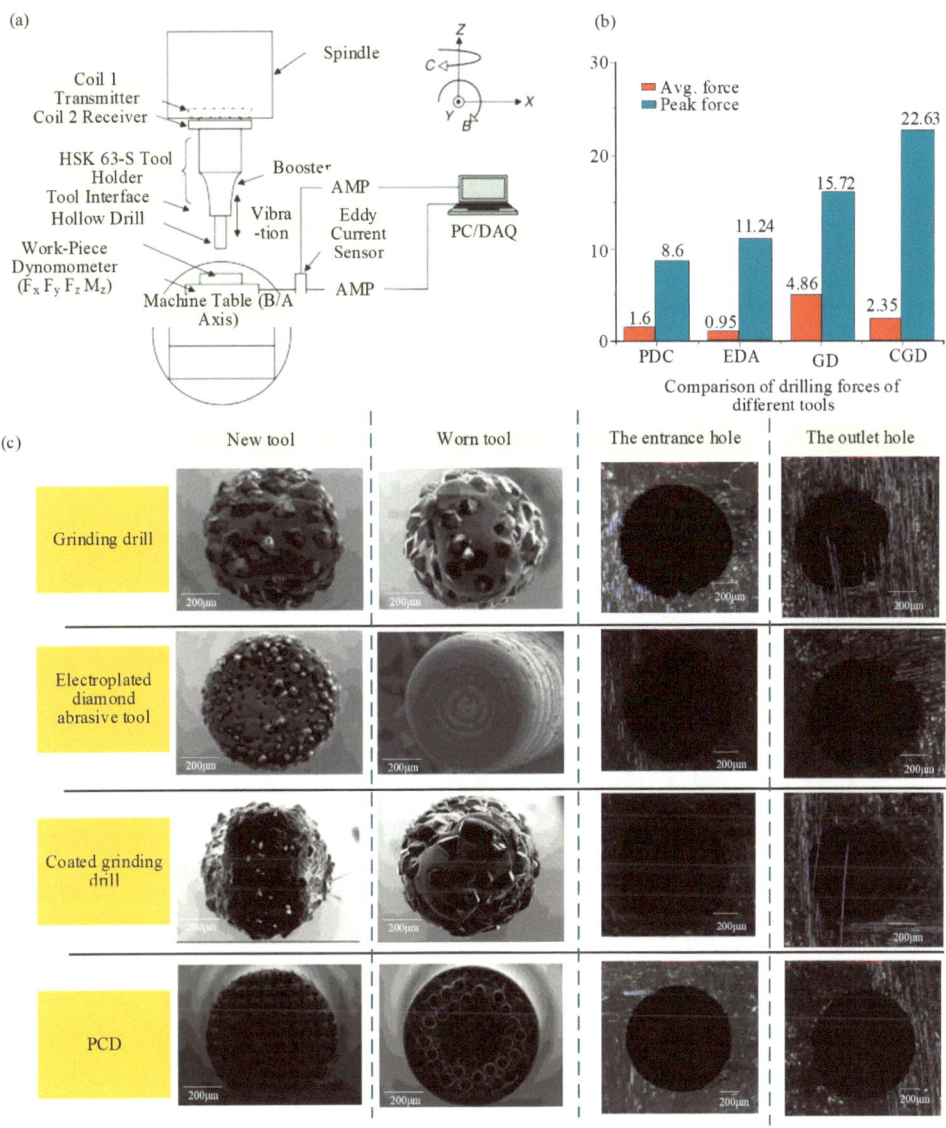

Figure 10. Different tools for micro-hole drilling of SiCf/SiC CMC assisted by ultrasonic vibration by Bertsche et al. [40] and Huang et al. [43]: (**a**) Experiment setup; (**b**) Comparison of drilling forces of different tools; (**c**) Hole morphology and wear condition of Different tools.

Under the same conditions, UVAG can reduce the cutting force and cutting temperature, which can effectively reduce the chip, burr, crack, fiber stripping, and processing damage of CMCs, and improve the processing precision and quality. Therefore, UVAG technology has been widely used in the precision machining of CMCs. And it is mainly suitable for the processing of small holes or micro-structures. However, as the interaction between tool and material in UVAG becomes more complex, the realization of the judgment and control of material removal mode is difficult. Moreover, the aggravated wear of the tool induced by high-frequency micro-vibration and the very low efficiency of large-diameter ceramic holes processing are the main drawbacks of UVAG [23,43].

2.5. Summary for Experiment Study

In summary, as shown in Table 1, compared to CG, non-CG technologies have obvious advantages. However, there are also certain limitations. For example, LAG requires strict control of technical parameters such as optimal power; UVAG is limited by ultrasonic critical speed, and the processing efficiency is low. AWJM in the processing of deep holes appear as a large dimensional error and surface burr, and the chip phenomenon.

Table 1. Comparison of different CMC grinding technologies.

Method	CG	LAG	UVAG	AWJM
Advantage	Wide application range, simple process, and high processing efficiency.	Reduce material hardness and improve processing performance.	Good processing quality, small surface loss.	Super-hard abrasive high-speed impact workpiece surface to achieve removal processing, no thermal effect.
Shortcoming	Rely on diamond tools, which are expensive and prone to serious wear; and It is difficult to process parts with complex shapes and high dimensional accuracy.	The heat-affected zone material must be removed to obtain the desired surface; High temperature will reduce the cutting performance of the tool, in particular lead to diamond graphitization.	The processing efficiency is low and the processing range is limited.	Large impact force, easy to break the edge and damage the surface of the workpiece.

Compared with the single special processing technology, the composite processing technology can show more excellent results. At present, several composite machining technologies have been developed, such as WJM and laser composite machining technology [44], ultrasonic and EDM composite machining technology [45], laser and ellipsoid ultrasonic vibration composite machining technology [46], ultrasonic vibration and electrolytic linear dressing (ELID) grinding composite machining technology [47], and other new composite machining technologies. These technologies have their advantages. There is still a lack of a process method that can simultaneously take into account the surface quality, efficiency, and cost of CMC processing. Therefore, the theoretical research of the new multi-energy field composite machining method with high efficiency and quality and its applications are still the research hotspots in the field of aerospace high-tech manufacturing.

3. Simulation Study of CMC Abrasive Machining

The microscopic results of CMC materials' processing can be obtained directly through experimental testing, such as the machinability, tool wear, forces, and surface quality. And with the aid of microscopes, subsurface damages induced by machining can also be revealed. Therefore, extensive experimental works on CMC machining have been conducted in the past decades to investigate and understand the process of machining composites. However, experimental investigations cannot explore the instant occurrence of material removal, tool-matrix-fiber phase interaction, and damage during grinding. This has significantly limited the depth of understanding and the level of process design. In addition, micro-structural examinations and experimental investigations are expensive and time-consuming. The advances in numerical analysis enabled comprehensive studies of composite grinding through the mechanic's model. The numerical techniques used to study the CMC grinding process mainly include the finite element method (FEM), smoothed particle hydrodynamics (SPH), molecular dynamics (MD) analysis, and discrete element method (DEM) analysis.

3.1. Finite Element Method (FEM)

The finite element method (FEM) studies the real physical systems' (geometry and load states) interaction through numerical analysis. The dynamic explicit algorithm is used in FEM due to the extreme deformation and complicated interactions among the matrix, fiber, interface, and tools in CMC machining. Grinding is performed continuously by tiny, abrasive cutting edges. Experimental observation and analysis of the grinding process is very difficult due to the large number of grains, irregular geometry, high grinding speed, and small and inconsistent grinding depth. The use of FEM can save a lot of time when determining experimental processing parameters. The FEM can obtain the ground surface morphology, force, temperature distribution, chip formation, and coupling relationship of parameters.

Zhang et al. [48] studied the mechanical properties of C_f/SiC composites and the influence of interface phase on mechanical properties by using a cohesion model and the Oliver–Pharr method. The interface strength and thickness influence on the load-displacement curve during nanoindentation was analyzed. It was found that, due to the presence of the interface, the in-situ mechanical properties of carbon fiber materials were changed. The hardness and elastic modulus of carbon fiber materials near the interface were increased. The maximum load, material hardness, and elastic modulus were positively correlated with the interface strength and interface thickness. The simulation provides a theoretical basis and research method for the study of grinding parameters and material removal mechanisms of C_f/SiC composites.

Liao et al. [49] simulated the single grain scribing process, as shown in Figure 11. The chip formation and temperature field distribution of negative rake angle were simulated by FEM. The results show that the chip formation can be explained by cutting theory when the negative front angle is $-15°\sim-40°$. However, it is more appropriate to apply the hardness indentation principle proposed by Shaw [50] to describe the large negative rake angle. The shear angle decreases and the shear strain increases with the increase of the negative rake angle. Meanwhile, in large negative rake angle grinding, the highest temperature appears at the contact point between the abrasive tip and the workpiece, which is higher than the small negative rake angle grinding temperature. With the increase of the negative rake angle of abrasive particles, the energy consumed in chip formation increases. Wu et al. [51], through FEM, concluded that, under wet grinding conditions, grinding fluid can reduce the wheel sliding force (tangential force), which can reduce the grinding zone temperature. Long et al. [52] simulated and analyzed the temperature field in the process of high-speed grinding of engineering ceramics by using the FEM. The result showed that there was a high-temperature gradient in the shallow surface, which would generate a large thermal tensile stress and lead to machining defects such as thermal cracks.

Fang et al. [53] established a multi-interface stress transfer simulation of CMC materials by FEM, through a unit-cell model. The shear stress and optimized interface stress propagation were studied. As shown in Figure 12, at the middle position plane of the interface phase, the shear stress rapidly increases to the maximum value near the forced spot. Then, a decrease along the fiber direction occurs and tends to zero about $2R_f$ away from the forced spot. The above trends of shear stress are the same under different interfacial thicknesses, with the increasing thickness of the interface phase. The maximum value of shear stress gradually decreases. The position reaching the maximum value is gradually far away from the forced spot. In the radial direction of fiber, the shear stress exhibits a rapid increase near the interface between the fiber and the interface phase, reaching its maximum value at the interface. For the PyC interface phase, the shear stress decreases along the radial direction. After crossing the interfacial phase and the interface, the shear stress continues to decrease in the matrix along the radial direction. Liu et al. [54] established a scratching simulation model of SiC_f/SiC based on the JH-2 model in ABAQUS (in Figure 13). The magnitude of the scratching force under different scratching speeds, depths, and different diamond types was analyzed, as well as the influence of fiber orientation. Li et al. [55] established a two-dimensional simulation model of a single diamond grain scribing C_f/SiC in ABAQUS, and the influence of wheel speed and grinding depth on grinding force and workpiece surface morphology was analyzed. Ellahi et al. [56] analyzed the cutting performance of C_f/SiC with a PCD tool by the FEM. In Figure 14, the fracture modes in parallel and vertical to the fiber direction and the corresponding force distribution near the cutting edge were analyzed. At cutting angle $\theta = 0°$, with respect to the fiber direction, and the machining direction parallel to the fiber direction, the material is removed mainly due to the debonding of fiber and matrix. Initially, the matrix fracture and crushing of carbon fibers occur near the cutting edge, and cracks propagate along the fiber direction. In the uncut chip area, the bonding of fibers and matrix becomes weak as the fiber axial strength is far stronger than the bonding strength of fibers and matrix. Stresses mostly occur at the tooltip and propagate along the fiber direction. At a 90° fiber angle, the rake face of the tool is perpendicular to the fiber direction and carbon fibers are subjected to tensile stresses. Initially, cracks occur ahead of the tooltip and move in the feed direction. The shear strength of fibers is lower than axial tensile strength. Therefore, when fibers are subjected to the tensile stress, fibers begin to elongate and eventually break as the force surpasses the fibers' shear strength. The friction between the cutting tool and workpiece is maximum at that point and the material has been removed mainly due to the shearing of fiber bundles. Stresses are mostly distributed at the rake face of the cutting tool and the maximum stress is concentrated around the tool edge. This research provides an efficient way to analyze the cutting performance of C_f/SiC composite.

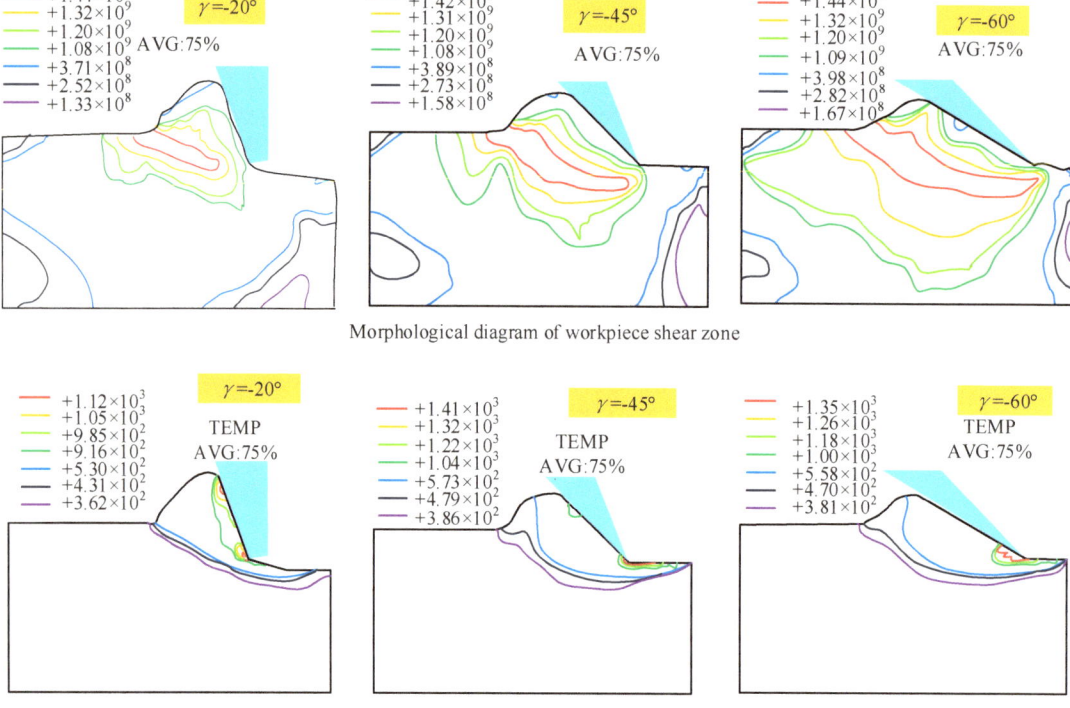

Figure 11. The morphology and temperature distribution of the workpiece shearing zone during abrasive grinding.

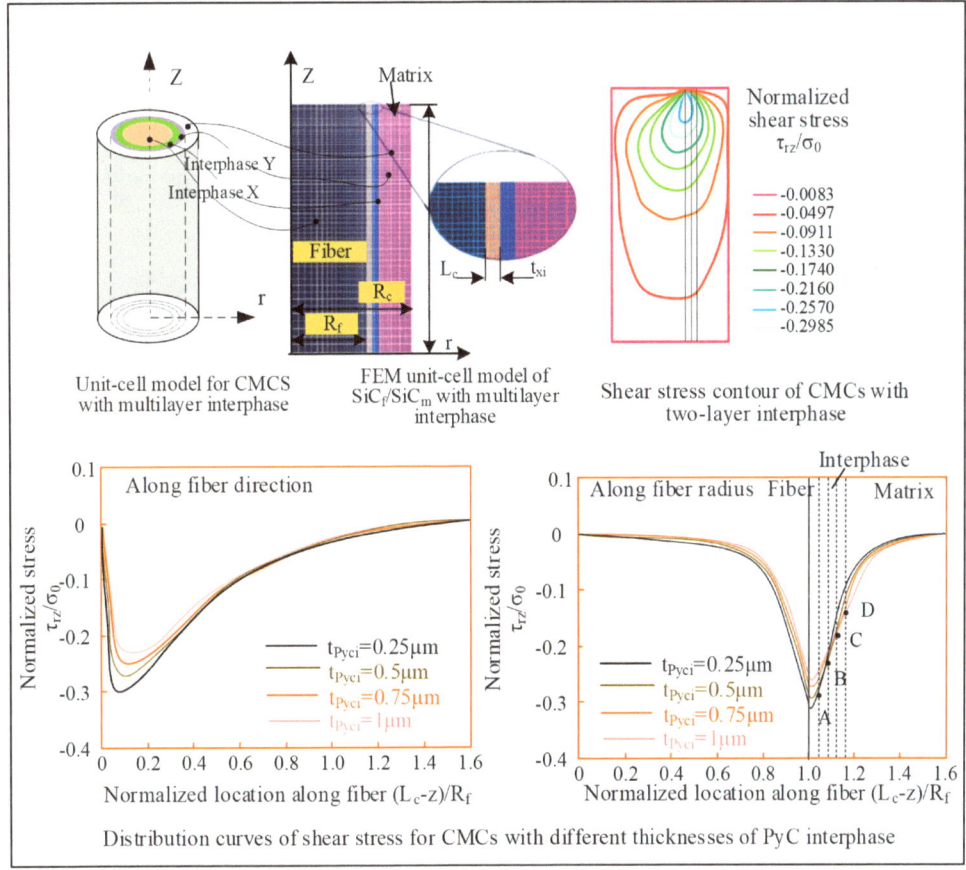

Figure 12. Multi-interface stress transfer simulation of CMC materials.

However, all the above studies only established two-dimensional simulation models. CMC materials have a complex structure composed of matrix, fiber, and interface layer. The deformations are complicated during processing. The effect of weaving structure on material properties is very important. Huang et al. [57] established a 3D woven C_f/SiC composite model and conducted a comparative study on single-grain scratching simulation with and without ultrasonic assistance. This research provides a reference for the establishment of geometric and constitutive models for carbon fiber, SiC ceramic matrix, and interface layer.

3.2. Smoothed Particle Hydrodynamics (SPH)

SPH is a meshless Lagrange method developed in 1977 [58]. It uses a group of particles to describe a continuous fluid (or solid). Each particle carries various physical quantities, including mass and velocity. By solving the dynamic equation of particles and tracking the trajectory of each particle, the mechanical behavior of the system can be obtained.

Unlike FEM, SPH can present precision and stable solutions to problems involving deformable boundaries, especially for large deformation and crack propagation, by approximating the governing equations of particles. SPH is more favorable for a cutting simulation owing to the particle-based algorithm, eliminating the element's deformation between particles [59].

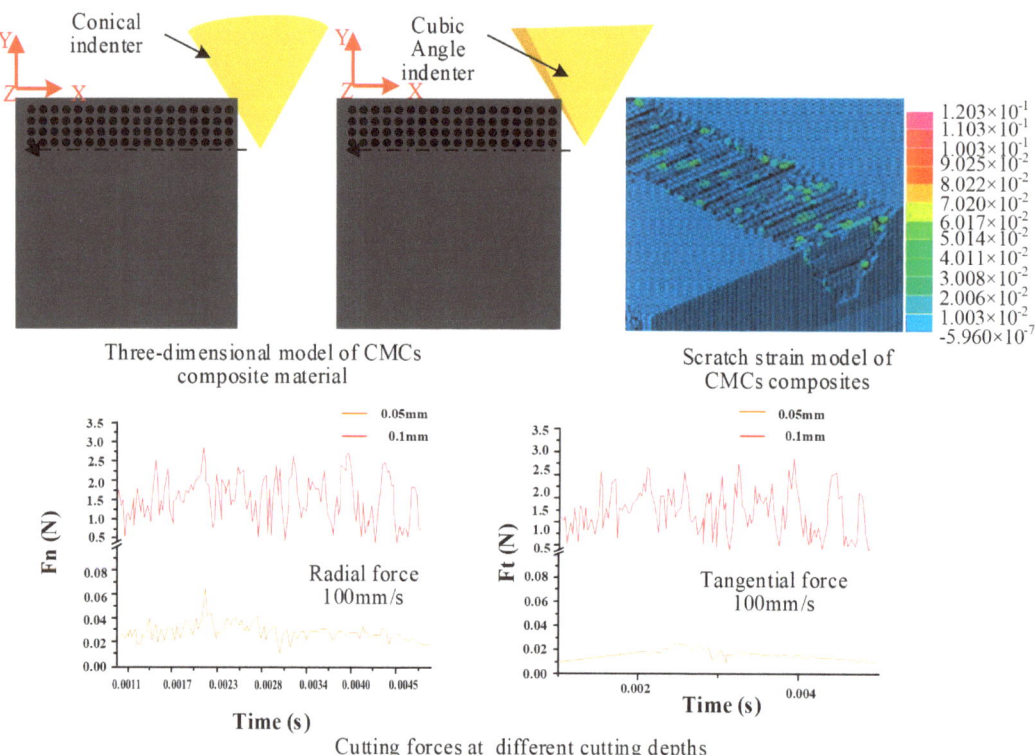

Figure 13. Scratching simulation of SiC$_f$/SiC based on JH-2 model.

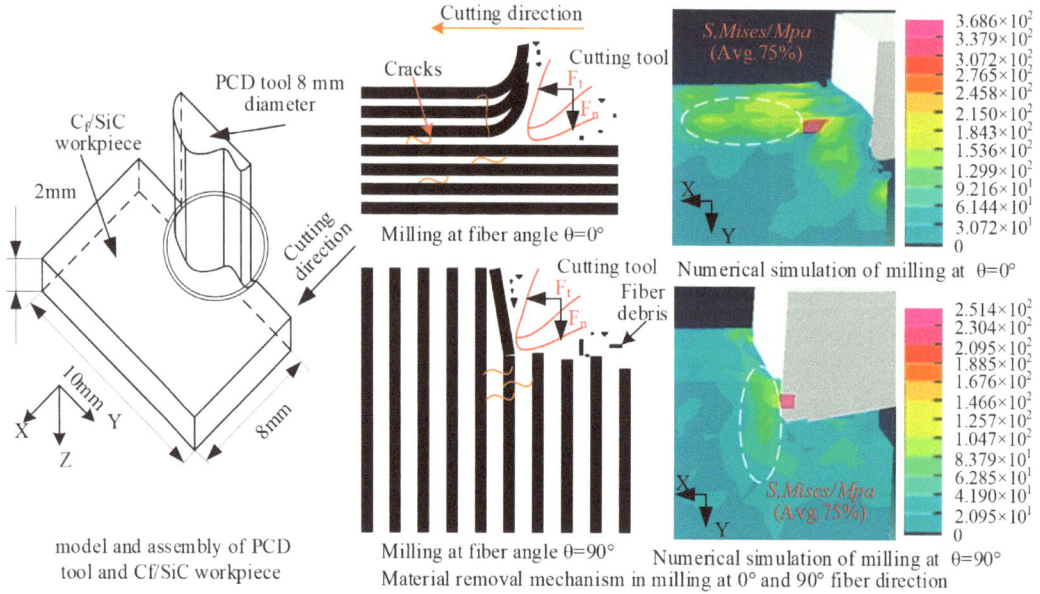

Figure 14. C$_f$/SiC cutting simulation with PCD tool.

Liu et al. [60] built a three-dimensional single abrasive grain scratch model to analyze the SiC grinding mechanism, including the material removal process, speed effect, ground surface roughness, and scratching force by using SPH (in Figure 15a). The simulation results showed that the material removal process went through the pure ductile mode, brittle assisted ductile mode, and brittle mode with the increase of the cutting depth. The critical cutting depth of ductile–brittle transition was approximately 0.35 μm based on the variations in ground surface crack conditions, surface roughness, and maximum scratching force. And increasing the scratching speed promoted the transformation of deep and narrow longitudinal cracks into shallow and wide transverse cracks on the surface, which improved the surface quality. Duan et al. [61] constructed a three-dimensional FEM and SPH model of a single diamond scratching. As shown in Figure 15b, the 3D boundary was solved by coupling multiple SPH particles into an iron-solid cell consisting of multiple nodes. And in Figure 15c, Liang et al. [62] simulated the single grain scribing process with SiC ceramics by using SPH-coupled FEM. These studies provide effective methods with which to study grinding processes and ground surface quality for SiC and other brittle material.

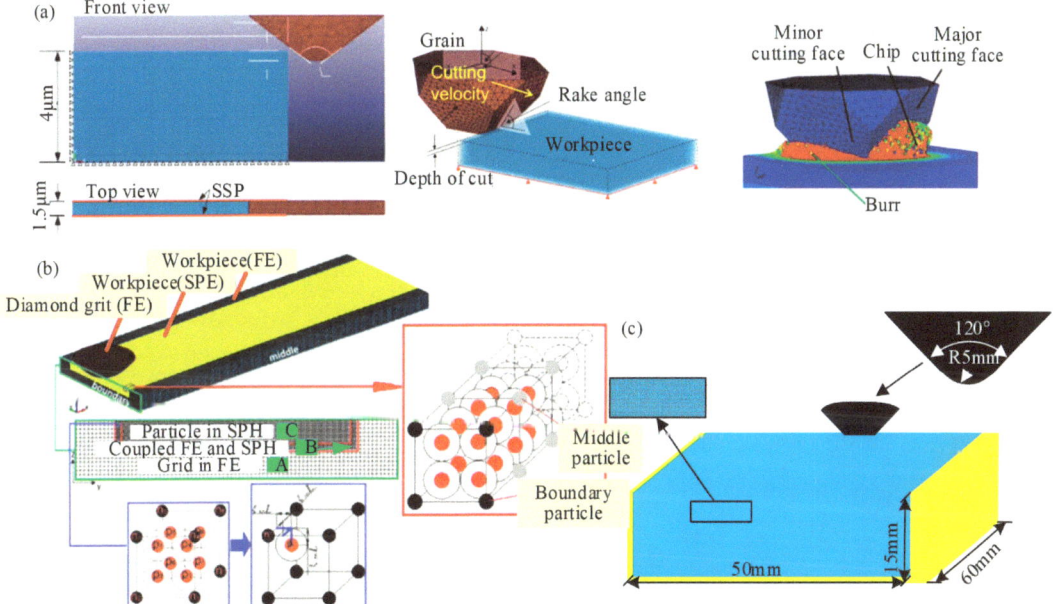

Figure 15. SPH model of single abrasive grain grinding silicon carbide ceramics. (a) Schematic for scratching model by Liu et al. [60]; (b) The 3D simulation model for single diamond scratching by Duan et al. [61]; (c) SPH grinding model before scratching by Liang et al. [62].

Zheng et al. [63] developed an accurate and efficient modeling platform for simulating mechanical properties of particle-reinforced matrix composites. As shown in Figure 16, a master–slave method is adopted within SPH formulation for imposing the essential boundary conditions and other linear displacement constraints. It was proven that the optimized master–slave (MS) method can provide better accuracy for the displacement constraints. The optimization with modified boundary interpolation technique (i.e., MS2) is computationally as efficient as the PF method. This study provided a better choice for implementing essential boundary conditions and other displacement constraints to overcome the potential problems encountered by the penalty function (PF) method. Takabi et al. [64] instructed a method to build the SPH model of both ductile and brittle materials with damage criteria representing crack initiation and post-failure behavior for machining analysis. The effects of damage definition on the chip morphology and cutting forces demonstrate that an appropriate damage criterion must be taken into account for SPH cutting simulations, despite the natural separation of particles, and regardless of the ductility of the material. Although SPH is feasible in machining simulations, however, it underestimates forces, instabilities, and numerical issues with boundary particles, which can lead to inaccurate prediction. Shi et al. [65] simulated the grinding fracture process of the fiber and matrix of C_f/SiC composites through the JH-2 model established by SPH. The microcracks initiation and propagation were analyzed and it was found that the radial crack was deflected due to the existence of carbon fibers. Zhou et al. [66] explored the grinding removal mechanism of 2.5D C_f/SiC composites in different grinding directions based on experiments and SPH simulations, and found that the main removal mechanism is brittle fracture. The main damage modes are matrix cracking, interface debonding, fibers fracture, and fibers pulling out. And because of the existence of fibers, when the crack propagates towards the interface between fiber and matrix, the crack deflects. Grinding 2.5D C/SiC composites, the fibers fracture and fibers pulling out are obvious when the fibers' direction is perpendicular to the feeding direction and the ground surface, and the ground surface roughness is the smallest in this direction. But in the other two directions, fiber fracture is the main factor, accompanied by a small amount of fibers pulling out.

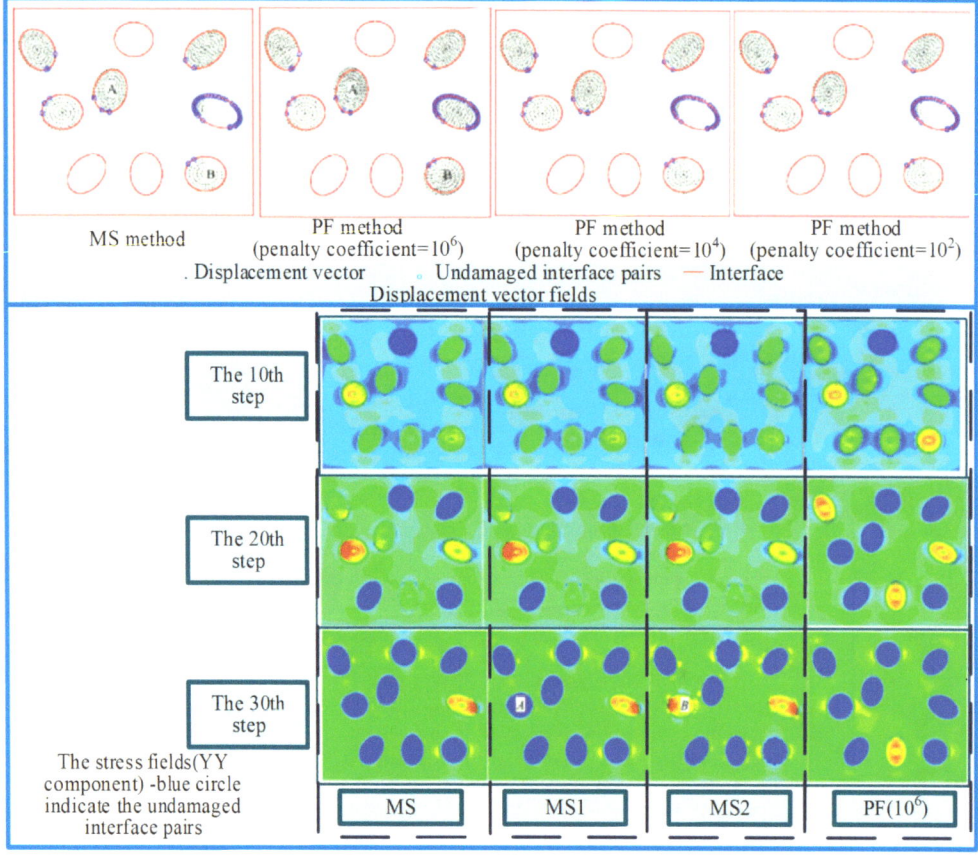

Figure 16. Modelling for particle-reinforced matrix by Zheng et al. [57].

Despite its natural superiority in addressing the large deformation problems, SPH has low computational efficiency due to the kernel approximation compared to the conventional FEM.

3.3. Molecular Dynamics (MD)

In the 1990s, a large number of researchers began to study SiC ductile grinding. Due to the high experimental cost and long processing time in the ductile domain, many researchers began to consider the use of simulation methods, such as MD simulation, to explain the ductile removal mechanism of brittle materials.

MD analysis is a powerful tool for studying complex microscopic systems. This technique can obtain the motion trajectories of microscopic particles. MD analysis, as a theoretical research method, is very vital in studying nanoscale grinding process and has been successfully used to study the microscopic mechanism of tool wear, surface quality, and subsurface damage.

Based on the Tersoff [67] potential function, Li et al. [68] studied the effect of amorphous carbon interfacial layer thickness on the fracture mechanical behavior of SiC_{NF}/SiC composites through MD calculations, and found that increasing the thickness of the interfacial layer would reduce the stress concentration coefficient of the fibers, increase the fracture energy, and transform the brittle fracture mode of the cracked penetrating fibers into the fiber pulling out failure mode, which can enhance the effect of reinforcing toughness, as shown in Figure 17.

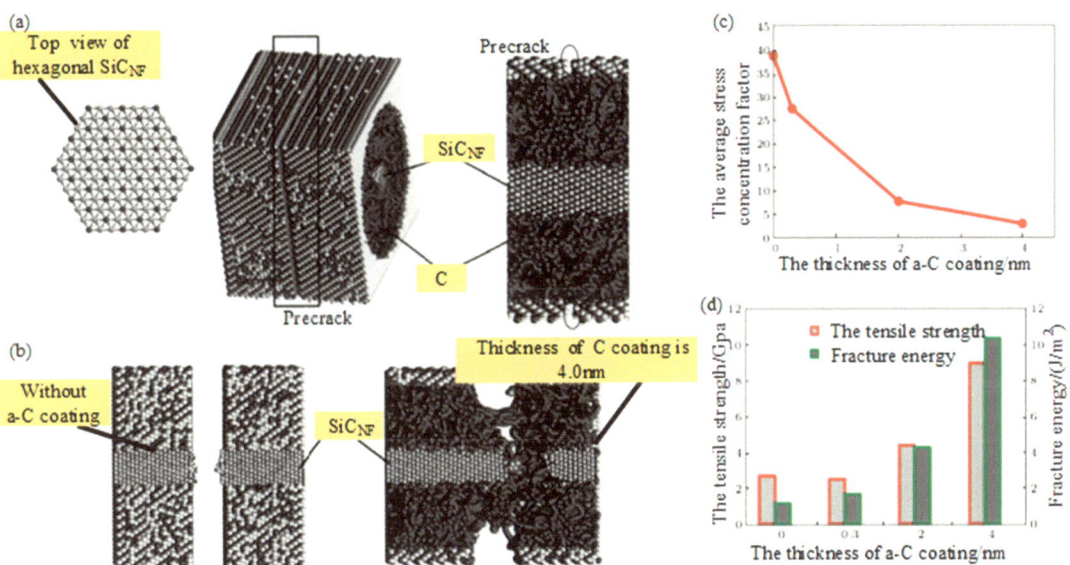

Figure 17. The effect of amorphous carbon interfacial layer thickness on the fracture mechanical behavior of SiC_{NF}/SiC composites through MD calculations studied by Li et al. [68]: (a) Atomic configuration of SiC_{NF}/SiC nanocomposite with predefect in matrix; (b) atomic configuration of fracture surface in SiC_{NF}/SiC nanocomposite with C coating in different thickness; (c) the average stress concentration factor of SiC_{NF} vs. the thickness of C coating; (d) the tensile strength and fracture energy of SiC_{NF}/SiC nanocomposite as a function of the thickness of C coating.

Miao et al. [69] investigated the tensile mechanical properties of vertically aligned CNTs/SiC nanocomposites (VSNs) through an in-situ transmission electron microscopy (TEM) tensile test and MD simulation. As shown in Figure 18a, molecular models of VSNs used the Teroff empirical bond-order potential to describe interactions of SiC and interface between carbon nanotubes (CNTs) and the SiC matrix. And the interatomic forces in CNTs were described by the adaptive intermolecular reactive empirical bond order potential. During the tensile process, the periodic boundary condition was implemented in all directions and an NVT ensemble with a Nose–Hoover thermostat was employed to maintain a constant temperature of 300 K. Figure 18b,c shows the molecular models and results of tensile, pulling out, and peel-off of CNTs. The tensile mechanical performances of VSNs along the ∥ direction are higher than those along the ⊥ direction. Meanwhile, the fracture modes of VSNs are different along these two directions. The CNTs can bridge at the crack surfaces and further bear the stress transfer along ∥ direction. However, the weak interaction between CNTs and SiC matrix plays a key role in the tensile failure along the ⊥ direction.

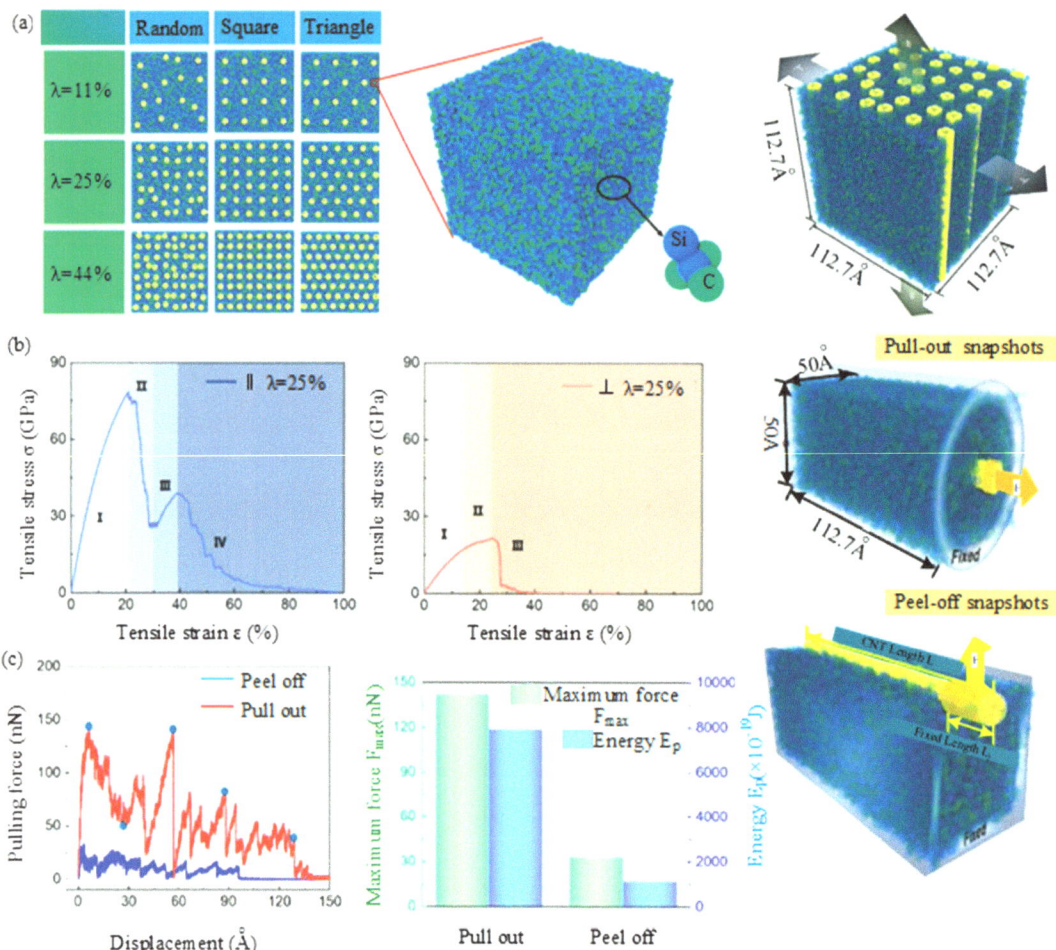

Figure 18. The tensile mechanical properties of vertically aligned CNTs/SiC nanocomposites (VSNs) through MD simulation by Miao et al. [69]: (**a**) Molecular models of VSNs with different carbon nanotubes (CNTs) distribution types and contents and tensile models of VSNs along ∥ and ⊥ directions; (**b**) tensile stress–strain curves and tensile snapshots with stress distribution along ∥ and ⊥ directions; (**c**) molecular models for pull-out and peel-off of CNTs from SiC matrix.

At present, the MD simulation is mainly applied to the cutting depth at the submicro and nano levels. And due to the maximum computing power being only a few cubic microns, the simulation scale is greatly limited. Meanwhile, the MD simulation is suitable for studying the movement of less than 10^{-9} s. So, the cutting speed used in the scratch experiment is only a few hundred microns per second, which cannot meet the speed requirements (several hundred meters per second) of the simulation experiment. In addition, in the MD simulation, the material must have a high deformation rate to ensure the accuracy of the calculation time, which is rarely used in the non-equilibrium state simulation of polymer processing models.

In summary, at the micro-scale, MD simulation methods are widely used in polymer structure design and composite material design. However, in the non-equilibrium state simulation for polymer processing, the calculation accuracy and force field model need to be improved.

3.4. Discrete Element Method (DEM)

During the machining process of the CMC, the interaction between wheel and workpiece gives rise to defects such as machining surface roughness and surface/sub-surface cracks. Experimental measurements or theoretical analysis of machining processes are difficult. The simulations like FEM and SPH are computationally expensive. Therefore, to deeply understand the grinding mechanism of the CMC, DEM is used to study the generation and propagation of micro-cracks during machining.

DEM was originally developed by Cundall and Stracke [70] in 1979 for the analysis of rock mechanics problems, and has been implemented in many other fields, such as DEM simulating particles of rocks, clayey soils, and ceramics.

Jiang et al. [71] simulated the grinding process of ceramics to understand the grinding mechanism, through building a two-dimensional particle flow program as a simulation platform. The process parameters used in the DEM model of the horizontal spindle plunge grinding process are shown in Figure 19a. The relationship among the grinding force and ground surface crack number with the grinding time is shown in Figure 19b, and the number of cracks and the average grinding force increase steadily with the increase of grinding time, and the grinding force of F_x increases faster than that of F_y due to the grinding chips filling the clearance of the grinding wheel. In addition, when the grinding force changes dramatically, the crack increases rapidly, resulting in the fracture of the ceramic parts. As shown by the red dot in Figure 19c, the grinding wheel moving forward, the crack growth, and the damage appear on the machining surface. Some micro-cracks can be found along the grinding track. The microcracks extend along the contact surface of the particles and form macroscopic cracks on the machined surface, resulting in the removal of the front-end material. It can be seen that the microcracks do not form obvious macroscopic intermediate cracks. The grains disintegrate into dispersed particles. In the grinding zone, most of the grains are crushed into particles, which are then quickly thrown away from the surface of the workpiece by the tangential force. Qiang et al. [72] established the DEM model of silicon carbide ceramics and the ultra-precision cutting model of the single point diamond. A dynamic simulation was carried out. The influence of the residual stress distribution on the direction of the workpiece depth under different cutting conditions, such as tool rake angle, cutting speed, and cutting depth, were analyzed. This study proved that a residual stress analysis is feasible with the DEM.

Li et al. [73], respectively, used DEM and the bonded particle model to establish and calibrate the discrete element models of the SiC ceramic matrix and carbon fiber. The displacement softening contact model was used to characterize the bilinear constitutive relationship of interlayer and fiber/matrix interface element damage. The production and expansion of the matrix crack and the dis-adhesion of the interface can be visually demonstrated. This study proved that the displacement-softened contact model can be used to study the elastoplastic of the interface of composite materials, which is feasible for studying C_f/SiC composites with DEM.

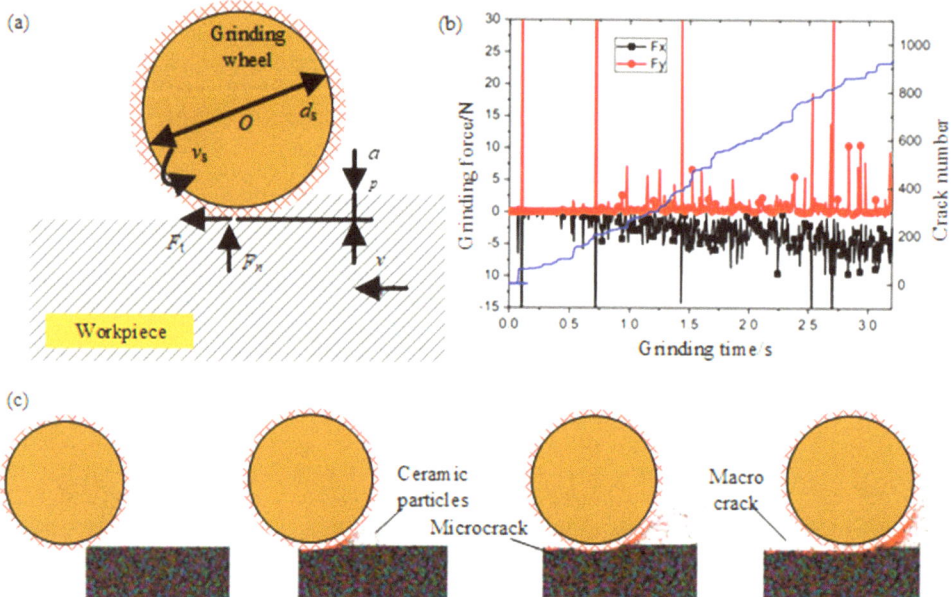

Figure 19. (**a**) DEM model for SiC ceramics; (**b**) simulation of grinding process; (**c**) relationship between grinding force and crack number by Jiang et al. [71].

DEM speed is fast and the storage space required is small, which is especially suitable for solving large displacement and nonlinear problems. In the DEM application of composite machining, the main issue lies in the fact that the mathematical formulation of this method requires a particular refinement. However, compared with the FEM, DEM is more advantageous in simulating the chipping of brittle matrix materials. On the contrary, stress and strain distribution estimations are normally less accurate.

3.5. Summary of Simulation Study

To sum up, all those numerical methods have been used to model machining processes. Typically, the FEM is used to simulate the micro-scale machining. The MD is used for the simulations of nanoscale machining processes. The length scale of the model dimensions in the DEM, which can be macroscale or micro-scale, depends on particle sizes and computational capacity. As discussed above, the DEM and MD simulations usually need to handle the interactions between massive numbers of particles and, hence, the computational cost can be extremely high when the total number of particles increases to a certain level. To maintain an efficient calculation, the total particle number should be appropriately selected. The model dimension should reasonably reflect the physical problem considered. A comparison between the FEM, SPH, DEM, and MD is shown in Table 2.

In addition, multiscale modeling combining atomistic simulation with continuum simulation to capture material deformation at different length scales has also been studied. The hybrid FEM-MD method is one of these methodologies. As shown in Figure 20, Wang et al. [74] proposed a multiscale method combining MD and FEM simulation to describe the separation behavior of the SiC/PyC interface and predict the stress–strain response in SiC_f/SiC composites; it provides a way to constitute relation prediction.

Table 2. A comparison between the FEM, SPH, DEM, and MD.

Method	FEM	SPH	DEM	MD
Basic theory	Continuum mechanics	Meshless Lagrange method, Kernel Function and Describing a continuous fluid (or solid) with a swarm of interacting particles	Newton's law of motion and the relationship between force and relative displacement between neighbor particles	Newton's law of motion and the potential function
Timestep	~μs	~μs	~μs	~fs
Length scale	Macroscale to mesoscale	Macroscale to mesoscale	Macroscale to mesoscale	Nanoscale
Common usage	Continuum materials and composites	Continuum materials and composites	Granular and discontinuous materials, composites	Nanomaterials
Limitation	Cannot well represent the discreteness, fracture, and damage processes in materials	It requires a relative high computation and a lot of time to calibrate the parameter	Requires a relatively high computation and a lot of time to calibrate the parameter	Requires a huge amount of computation and limits to nanometric sizes

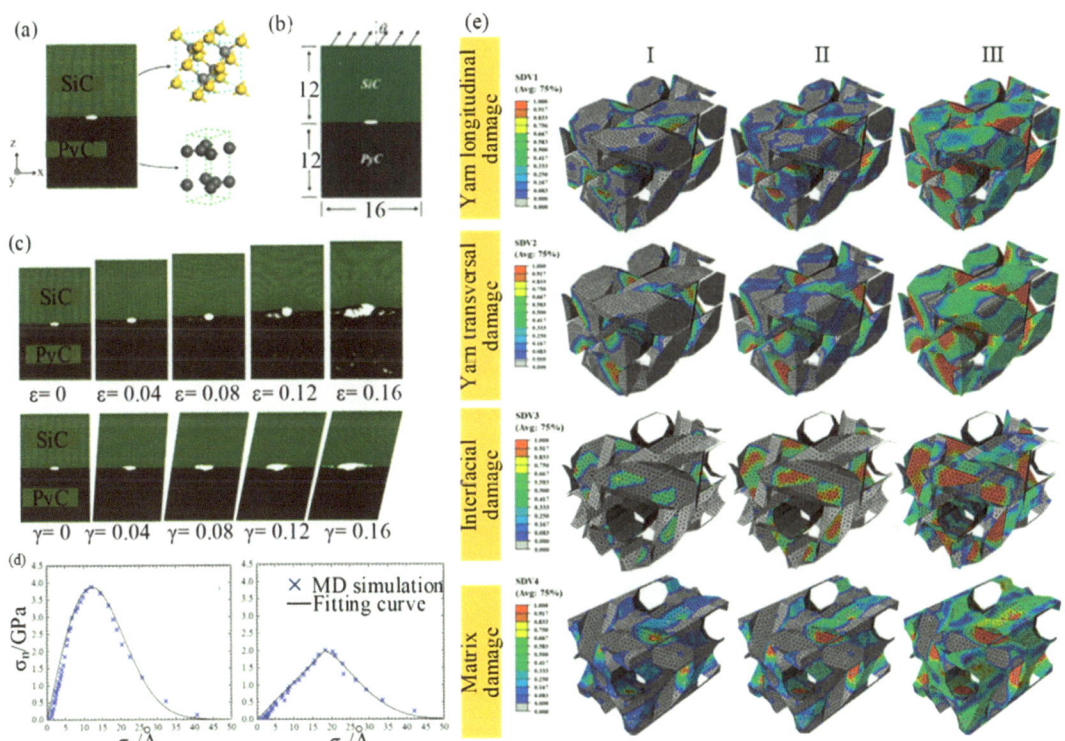

Figure 20. The hybrid of FEM-MD method by Wang et al. [74]: (**a**) MD simulation model of SiC/PyC interface; (**b**) load on MD simulation model; (**c**) damage evolution process of interface crack under pure tensile and shear load; (**d**) on the basis of MD simulation, the traction-separation laws (TSLs) for normal and tangential in interface; (**e**) damage evolution of 3D woven composite FEM model (I: ε_Z = 0.03%; II: ε_Z = 0.23%; III: ε_Z = 0.75%).

4. Conclusions

In this paper, the research on CMC grinding technology is reviewed. Firstly, the research status of CG and non-CG technology is introduced, and their advantages and disadvantages are compared. Summarized in Table 1, those technologies have their own advantages, but there is still a lack of a process method that can simultaneously take into account the quality, efficiency, and cost of CMCs processing. Therefore, the theoretical research and application of the new multi-energy field composite machining method with high efficiency and high quality are still a research hotspot in the field of aerospace high-tech artificial manufacturing. Then, this paper briefly introduces four kinds of simulation methods from basic theory, research status, and application scope. The different characteristics, such as computing power and model scale, give the four simulation methods their own application scope and shortcomings. Similar to machining technology, it is a trend to apply the multi-scale simulation method to the study of CMCs machining.

In addition to the studies on techniques and simulation methods, the particularity structure and physical properties of CMCs materials also deserve our attention The removal mechanism of CMC materials is complex and changeable due to their structural characteristics. Different physical properties, reinforcement, matrix materials, and the two-phase interface have diverse machinability, which change the removal mechanism. The matrix and reinforcing fibers are not necessarily removed synchronously during the grinding process. On the other hand, the anisotropy and filamentous toughening phase structure would cause random interface debonding and cracking events during a machining process. In addition, in the case of the particularity structure, the variation of fiber orientations relative to cutting directions, random waviness of individual fibers, and uneven fiber dispersion in the matrix bring about significant variations of material removal mechanisms during machining. Meanwhile, in terms of tool wear and machining rate, the high abrasion of the reinforcements to a cutting tool leads to excessive tool wear, which, in turn, brings about unsteady machining and significant subsurface damage such as delamination, reinforcement fracture, and burr formation, as well as varying surface integrity.

Therefore, the reinforced fibers not only improve the mechanical properties of the CMC materials, but also change the removal mechanism. Using new technology to improve processing quality, understanding the influence of fibers' orientation and distribution on the material removal mechanism and removal mode, and establishing an appropriate mechanical processing model are important research directions for the future. The difficulties this research must face and the directions it may go in are given in Table 3.

Table 3. Research difficulties and direction for grinding technology for SiC CMC materials.

Main Difficulty	General Problem	Research Directions
Diverse machinability	The great difference in physical properties between SiC matrix and C fiber, and the matrix and reinforcing fiber are not necessarily removed synchronously.	The effect of reinforcing fibers on the removal mechanism and mode of the material.
The anisotropy and filamentous toughening phase structure	The reinforcement-matrix interphase material with random interface debonding and cracking events during a machining process.	The mechanical model of the two-phase interface and the transformation of the removal mode at the bonded surfaces during grinding.
Fiber orientation	The variation of fiber orientations relative to cutting directions, random waviness of individual fibers, and uneven fiber dispersion in the matrix bring about significant variations of material removal mechanisms.	Grinding mechanism and surface quality on different fiber orientations.
Tool wear	The high abrasion of the reinforcements to a cutting tool leads to excessive tool wear.	Application of non-CG and research on new grinding mechanism; research on grinding quality under different grinding tools.

Author Contributions: H.Z. analyzed the Machining Technology of the ceramic composites and wrote the manuscript, Z.Z., J.L., L.Y. and Y.L. summarized and reviewed the manuscript. All authors participated in preparing the paper and the results discussions. All authors have read and agreed to the published version of the manuscript.

Funding: This study was supported by the Opening Foundation of Shanxi Provincial Key Laboratory for Advanced Manufacturing Technology (No. XJZZ202312, XJZZ202308) and the Natural Science Foundation of China (No. 51905498).

Conflicts of Interest: The authors declare that they have no known competing financial interests or personal relationships that could have appeared to influence the work reported in this paper.

References

1. Naslain, R. Design, Preparation and Properties of Non-Oxide CMCs for Application in Engines and Nuclear Reactor: An Overview. *Compos. Sci. Technol.* **2004**, *64*, 155–170. [CrossRef]
2. Itoh, T.; Kimura, H. Status of the Automotive Ceramic Gas Turbine Development Program. *ASME J. Eng. Gas Turbines Power* **1993**, *115*, 42–50. [CrossRef]
3. Wang, C.; Li, K.; Chen, H.; Ma, X.M. Research Progress of Processing Technology for Fiber Reinforced Ceramic Matrix Composites. *Aeronaut. Manuf. Technol.* **2016**, *59*, 55–60. [CrossRef]
4. Azarhoushang, B.; Tawakoli, T. Development of a novel ultrasonic unit for grinding of ceramic matrix composites. *Int. J. Adv. Manuf. Technol.* **2011**, *57*, 945–955. [CrossRef]
5. Zhang, L.; Ren, C.; Ji, C.; Wang, Z.; Chen, G. Effect of fiber orientations on surface grinding process of unidirectional C/SiC composites. *Appl. Surf. Sci.* **2016**, *366*, 424–431. [CrossRef]
6. Qu, S.; Gong, Y.; Yang, Y.; Wen, X.; Yin, G. Grinding characteristics and removal mechanisms of unidirectional carbon fiber reinforced silicon carbide ceramic matrix composites. *Ceram. Int.* **2018**, *45*, 3059–3071. [CrossRef]
7. Cao, X.; Lin, B.; Zhang, X. Investigations on grinding process of woven ceramic matrix composite based on reinforced fiber orientations. *Compos. Part B Eng.* **2015**, *71*, 84–192. [CrossRef]
8. Hu, N.S.; Zhang, L.C. A study on the grindability of multidirectional carbon fibre-reinforced plastics. *J. Mater. Process. Technol.* **2003**, *140*, 152–156. [CrossRef]
9. Yin, J.; Xu, J.; Ding, W.; Su, H. Effects of grinding speed on the material removal mechanism in single grain grinding of SiCf/SiC ceramic matrix composite. *Ceram. Int.* **2021**, *47*, 12795–12802. [CrossRef]
10. Liu, Q.; Huang, G.; Fang, C.; Cui, C.; Xu, X. Experimental investigations on grinding characteristics and removal mechanisms of 2D-Cf/C-SiC composites based on reinforced fiber orientations. *Ceram. Int.* **2017**, *43*, 15266–15274. [CrossRef]
11. Liu, Q.; Huang, G.; Xu, X.; Fang, C.; Cui, C. Influence of grinding fiber angles on grinding of the 2D-Cf/C–SiC composites. *Ceram. Int.* **2018**, *44*, 12774–12782. [CrossRef]
12. Wang, T.; Wang, S.; Qiao, W.L.; Zhang, L.F.; Zhen, T.T. Research on Grinding Force Model of Plane Grinding for Unidirectional C/SiC Composites. *China Mech. Eng.* **2019**, *30*, 2017–2021 Available online: https://kns.cnki.net/kcms/detail/42.1294.TH.20190903.1707.020.html (accessed on 27 April 2018).
13. Zhang, L.C.; Zhang, H.J.; Wang, X.M. A force prediction model for cutting unidirectional fiber-reinforced plastics. *Mach. Sci. Technol.* **2001**, *5*, 293–305. [CrossRef]
14. Tawakoli, T.; Azarhoushang, B. Intermittent grinding of ceramic matrix composites (CMCs) utilizing a developed segmented wheel. *Int. J. Mach. Tools Manuf.* **2011**, *51*, 112–119. [CrossRef]
15. Ma, L.; Yu, A.; Gu, L.; Wang, H.; Chen, J. Mechanism of compound fracture and removal in grinding process for low-expansion glass ceramics. *Int. J. Adv. Manuf. Technol.* **2017**, *91*, 2303–2313. [CrossRef]
16. Qu, S.; Gong, Y.D.; Yang, Y.Y.; Cai, M.; Sun, Y. Surface topography and roughness of silicon carbide ceramic matrix composites. *Ceram. Int.* **2018**, *44*, 14742–14753. [CrossRef]
17. Pineau, P.; Couégnat, G.; Lamon, J. Virtual testing applied to transverse multiple cracking of tows in woven ceramic composites. *Mech. Res. Commun.* **2011**, *38*, 579–585. [CrossRef]
18. Lamon, J. A Micromechanics-Based Approach to the Mechanical Behavior of Brittle-Matrix Composites. *Compos. Sci. Technol.* **2001**, *61*, 2259–2272. [CrossRef]
19. Zhang, Y.; Li, C.; Jia, D.; Zhang, D.; Zhang, X. Experimental evaluation of the lubrication performance of MoS2/CNT nanofluid for minimal quantity lubrication in Ni-based alloy grinding. *Int. J. Mach. Tools Manuf.* **2015**, *99*, 19–33. [CrossRef]
20. Hu, M.; Ming, W.W.; An, Q.L.; Chen, M. Experimental study on milling performance of 2D C/SiC composites using polycrystalline diamond tools. *Ceram. Int.* **2019**, *45*, 10581–10588. [CrossRef]
21. Xu, L.; Zhao, G.L.; Zhang, J.Q.; Wang, K.; Wang, X.Y.; Hao, X.Q. Feasibility study on cryogenic milling of carbon fiber reinforced silicon carbide composites. *Trans. Nanjing Univ. Aeronaut. Astronaut.* **2020**, *37*, 1005–1120. [CrossRef]
22. Qu, S.; Gong, Y.; Yang, Y.; Sun, Y.; Wen, X.; Qi, Y. Investigating Minimum Quantity Lubrication in Unidirectional Cf/SiC composite grinding. *Ceram. Int.* **2020**, *46*, 3582–3591. [CrossRef]
23. An, Q.; Chen, J.; Ming, W.; Chen, M. Machining of SiC ceramic matrix composites: A review. *Chinese. J. Aeronaut.* **2021**, *34*, 540–567. [CrossRef]

24. Zhang, C.; Li, H.; Chen, J.; An, Q.L.; Chen, M. Microstructure evolution and ablation mechanism of SiCf/SiC ceramic matrix composite microgrooves processed by pico-second laser. *Ceram. Int.* **2023**, *49*, 37356–37365. [CrossRef]
25. An, Q.L.; Li, H.; Chen, J.; Ming, W.W.; Chen, M. Multi-energy field assisted high-efficiency and low damage milling process of ceramic matrix composites. *Aeronaut. Manuf. Technol.* **2023**, *66*, 40–51. [CrossRef]
26. Zhou, K.; Xu, J.Y.; Xiao, G.J.; Huang, Y. A novel low-damage and low-abrasive wear processing method of Cf/SiC ceramic matrix composites: Laser-induced ablation-assisted grinding. *J. Mater. Process. Technol.* **2022**, *302*, 117503. [CrossRef]
27. Li, W.; Long, C.J.; Ma, W.Q.; Ke, F.F.; Feng, W. Key technologies for laser-assisted precision grinding of 3D C/C-SiC composites. *J. Eur. Ceram. Soc.* **2023**, *43*, 4322–4335. [CrossRef]
28. Kong, X.J.; Liu, S.W.; Hou, N.; Zhao, M.; Liu, N.; Wang, M.H. Cutting performance and tool wear in laser-assisted grinding of SiCf/SiC ceramic matrix composites. *Mater. Res. Express* **2022**, *9*, 125601. [CrossRef]
29. Dong, Z.G.; Wang, Z.W.; Ran, Y.C. Advances in Ultrasonic Vibration-Assisted Milling of Carbon Fiber Reinforced. *J. Mech. Eng.* **2023**, *59*, 1–31. Available online: https://link.cnki.net/urlid/11.2187.TH.20231129.1504.034 (accessed on 1 December 2023).
30. Chen, G.M. The Feature and Application of NC High Pressure Water Jet Cutting Technology. *Mach. Tool Hydraul.* **2007**, *8*, 44–68.
31. Hashish, M. Turning with abrasive water jets—A first investigation. *J. Eng. Ind.* **1987**, *9*, 281–290. [CrossRef]
32. Hashish, M.; Kotchon, A.; Ramulu, M. Status of AWJ machining of CMCS and hard materials. In Proceedings of the INTERTECH 2015, Indianapolis, India, 19–20 May 2015.
33. Zhang, Y.F.; Liu, D.; Zhang, W.J.; Zhu, H.T.; Huang, C.Z. Hole characteristics and surface damage formation mechanisms of Cf/SiC composites machined by abrasive waterjet. *Ceram. Int.* **2022**, *48*, 5488–5498. [CrossRef]
34. Ramulu, M.; Jenkins, M.G.; Guo, Z. Abrasive water jet machining mechanisms in continuous-fiber ceramic composites. *J. Compos. Technol. Res.* **2001**, *23*, 82–91. [CrossRef]
35. Ren, Y.M. Research on abrasive water jet machining of structural ceramics and ceramic matrix composites. *Intern. Combust. Engine Parts* **2018**, *15*, 116–118. [CrossRef]
36. Wang, X.B.; Li, L.-L.; Zhao, B.; Song, C.-S. Research progress on processing technology and surface and subsurface damage mechanism of ceramic matrix composites. *Surf. Technol.* **2021**, *50*, 17–34.
37. Kang, R.K.; Zhao, F.; Bao, Y.; Sun, H.Q.; Dong, Z.G. Ultrasonic assisted grinding of SiCf/SiC composites. *Diam. Abras. Eng.* **2019**, *39*, 85–91. [CrossRef]
38. Ding, K.; Fu, Y.C.; Su, H.H.; Cui, F.F.; Li, Q.L.; Lei, W.N.; Xu, H.X. Study on surface/subsurface breakage in ultrasonic assisted grinding of C/SiC composites. *Int. J. Adv. Manuf. Technol.* **2017**, *91*, 3095–3105. [CrossRef]
39. Zhang, M.; Pang, Z.; Jia, Y.; Shan, C. Understanding the machining characteristic of plain weave ceramic matrix composite in ultrasonic-assisted grinding. *Ceram. Int.* **2022**, *48*, 5557–5573. [CrossRef]
40. Bertsche, E.; Ehmann, K.; Malukhin, K. Ultrasonic slot machining of a silicon carbide matrix composite. *Int. J. Adv. Manuf. Technol.* **2013**, *66*, 1119–1134. [CrossRef]
41. Huang, B.; Wang, W.; Jiang, R.; Xiong, Y.; Liu, C. Experimental study on ultrasonic vibration–assisted drilling micro-hole of SiCf/SiC ceramic matrix composites. *Int. J. Adv. Manuf. Technol.* **2022**, *120*, 8031–8044. [CrossRef]
42. Wang, J.J.; Zhang, J.F.; Feng, P.F. Effects of tool vibration on fiber fracture in rotary ultrasonic machining of C/SiC ceramic matrix composites. *Compos. Part B Eng.* **2017**, *129*, 233–242. [CrossRef]
43. Zhai, Z.Y.; Mei, X.S.; Wang, W.J.; Cui, J.L. Research advancement on laser etching technology of silicon carbide ceramic matrix composite. *Chin. J. Lasers* **2020**, *47*, 24–34. [CrossRef]
44. Lu, X.Z.; Jiang, K.Y. Research and application development of compound energy field processing Laser microjet. *Sci. China Phys. Mech. Astron.* **2020**, *50*, 30–43. Available online: https://kns.cnki.net/kcms/detail/detail.aspx?FileName=JGXK202003003&DbName=CJFQ2020 (accessed on 27 May 2019). [CrossRef]
45. Che, J.T.; Zhou, T.F.; Zhu, X.J.; Kong, W.J.; Wang, Z.B.; Xie, X.D. Experimental study on horizontal ultrasonic electrical discharge machining. *J. Mater. Process. Tech.* **2016**, *231*, 312–318. [CrossRef]
46. Ishida, T.; Noma, K.; Kakinuma, Y.; Aoyama, T.; Hamada, S.; Ogawa, H.; Higaino, T. Helical Milling of Carbon Fiber Reinforced Plastics Using Ultrasonic Vibration and Liquid Nitrogen. *Procedia CIRP* **2014**, *24*, 13–18. [CrossRef]
47. Zhao, B.; Chen, F.; Jia, X.F.; Zhao, C.Y.; Wang, X.B. Surface quality prediction model of nano-composite ceramics in ultrasonic vibration-assisted ELID mirror grinding. *J. Mech. Sci. Tech.* **2017**, *31*, 1877–1884. [CrossRef]
48. Zhang, H.B.; Ren, C.Z.; Zhang, L.F.; Li, Y.C. Finite Element Simulation of Nanoindentation of C/SiC Composites. *J. Mater. Sci. Eng.* **2016**, *34*, 49–53+74. [CrossRef]
49. Liao, X.T.; Liu, D.F.; Tang, J.Y. Finite element analysis of grinding chip formation of single abrasive grain with negative rake angle. *Mod. Manuf. Eng.* **2009**, *4*, 36–41. [CrossRef]
50. Pang, N. *Ultrasonic Grinding Composite Machining of New Ceramic Materials*; Northeastern University: Shenyang, China, 1997.
51. Wu, Y.H.; Wang, H.; Li, S.H.; Sun, J.; Wang, H. Grinding Thermal Characteristics and Surface Forming Mechanism of Silicon Nitride Ceramics. *Surf. Technol.* **2019**, *48*, 360–368. [CrossRef]
52. Long, H.; Guo, L.; Wang, C. Study on improving the accuracy of finite element simulation of ceramic high-speed grinding temperature. *Manuf. Technol. Mach. Tool* **2022**, *12*, 153–158.
53. Fang, G.W.; Gao, X.G.; Song, Y.D. Finite element simulation of stress transfer through the multilayer interphase in ceramic matrix composites. *Acta Mater. Compos. Sin.* **2018**, *35*, 3415–3422. [CrossRef]

54. Liu, S.H. Research on Removal Mechanism of SiCf/SiC Ceramic Matrix Composite. Master's Thesis, Hunan University, Changsha, China, 2021. [CrossRef]
55. Li, J.D.; Ren, C.Z.; Lv, Z.; Zhang, L.F. Finite Element Simulation of Single Diamond Abrasive Surface Grinding C/SiC. *J. Mater. Sci. Eng.* **2014**, *32*, 686–689+715. [CrossRef]
56. Zohaib, E.; Zhao, G.L.; Long, Z.W.; Jamil, M.; He, N. Milling Performance of Cf/SiC Composites: Three-Dimensional Finite Element Analysis and Experimental Validation. *Trans. Nanjing Univ. Aeronaut. Astronaut.* **2023**, *40*, 253–263. [CrossRef]
57. Huang, X.C.; Su, H.H.; He, J.Y.; Xu, P.F. Three Dimensional Finite Element Simulation Modification of Single AbrasiveParticle Ultrasonic Vibration Assisted Scratching C/SiC Composites. *Mach. Build. Autom.* **2022**, *51*, 98–100+112. [CrossRef]
58. Liu, R.J.; Huang, H.; Zhao, C.X.; Wang, H.H.; Li, G.Z. Study on micropore preparation of ceramic matrix composites by ultra-short pulse laser processing. *Electromachining Mould* **2019**, *2*, 55–58. Available online: https://kns.cnki.net/kcms/detail/detail.aspx?FileName=DJGU201902015&DbName=CJFQ2019 (accessed on 9 January 2019).
59. Feng, Z.P.; Xiao, Q. Research progress of ultrasonic machining technology. *Surf. Technol.* **2020**, *49*, 161–172. [CrossRef]
60. Liu, Y.; Li, B.Z.; Wu, C.J.; Kong, L.F.; Zheng, Y.H. Smoothed particle hydrodynamics simulation and experimental analysis of SiC ceramic grinding mechanism. *Ceram. Int.* **2018**, *44*, 12194–12203. [CrossRef]
61. Duan, N.; Yu, Y.Q.; Wang, W.S.; Xu, X.P. SPH and FE coupled 3D simulation of monocrystal SiC scratching by single diamond grit. *Int. J. Refract. Met. Hard Mater.* **2017**, *64*, 279–293. [CrossRef]
62. Liang, Z.Q.; Mi, Z.Y.; Wang, X.B.; Wang, X.B.; Zhou, T.F.; Wu, Y.B.; Zhao, W.X. Numerical Investigations on the Grinding Forces in Ultrasonic Assisted Grinding of SiC Ceramics by Using SPH Method. *Adv. Mater. Res.* **2014**, *1017*, 735–740. [CrossRef]
63. Zheng, Z.J.; Kulasegaram, S.; Chen, F.; Chen, Y.Q. An efficient SPH methodology for modelling mechanical characteristics of particulate composites. *Def. Technol.* **2021**, *17*, 135–146. [CrossRef]
64. Takabi, B.; Tajdari, M.; Tai, B.L. Numerical study of smoothed particle hydrodynamics method in orthogonal cutting simulations effects of damage criteria and particle density. *J. Manuf. Process.* **2017**, *30*, 523–531. [CrossRef]
65. Shi, Z.L.; Xiang, D.H.; Feng, D.R.; Wu, B.F.; Zhang, Z.M.; Gao, G.F.; Zhao, B. Finite Element and Experimental Analysis of Ultrasonic Vibration Milling of High-Volume Fraction SiCp/Al Composites. *Int. J. Precis. Eng. Manuf.* **2021**, *22*, 1777–1789. [CrossRef]
66. Zhou, Y.G.; Tian, C.C.; Li, H.Y.; Ma, L.J.; Li, M.; Yin, G.Q. Study on removal mechanism and surface quality of grinding carbon fiber toughened ceramic matrix composite. *J. Braz. Soc. Mech. Sci. Eng.* **2022**, *44*, 476. [CrossRef]
67. Tersoff, J. Modeling solid-state chemistry: Interatomic potentials for multicomponent systems. *Phys. Rev. B* **1990**, *41*, 3248. [CrossRef]
68. Li, L.L.; Xia, Z.H.; Yang, Y.Q.; Han, M. Molecular dynamics study on tensile behavior of SiC nanofiber/C/SiC nanocomposites. *Acta Phys. Sin.* **2015**, *64*, 117101. [CrossRef]
69. Miao, L.L.; Yang, L.W.; Wang, C.Y.; Zhao, G.X.; Li, J.J.; Zhao, Y.S.; Sui, C.; He, X.D.; Xu, Z.H.; Wang, C. Anisotropic tensile mechanics of vertically aligned carbon nanotube reinforced silicon carbide ceramic nanocomposites. *Carbon* **2022**, *199*, 241–248. [CrossRef]
70. Cundall, P.A.; Strack, O.D.L. A discrete numerical model for granular assemblies. *Géotechnique* **1979**, *29*, 47–65. [CrossRef]
71. Jiang, S.O.; Li, T.T.; Tan, Y.Q. A DEM Methodology for Simulating the Grinding Process of SiC Ceramics. *Proc. Eng.* **2015**, *102*, 1803–1810. [CrossRef]
72. Qiang, S.Q.; Tan, Y.Q.; Yang, D.M. Discrete element method simulation of residual stresses in SiC single-point diamond ultra-precision machining. *J. Chin. Ceram. Soc.* **2010**, *38*, 918–923+930. [CrossRef]
73. Li, L.T.; Tan, Y.Q.; Qiang, S.Q. Study on Interfaces Properties of C/SiC Ceramic Matrix Composites Using Discrete Element Method. *Mater. Rep.* **2012**, *26*, 148–156. Available online: https://kns.cnki.net/kcms/detail/detail.aspx?FileName=CLDB201222038&DbName=CJFQ2012 (accessed on 25 November 2012).
74. Wang, R.Q.; Han, J.B.; Mao, J.X.; Hu, D.Y.; Liu, X.; Guo, X.J. A molecular dynamics based cohesive zone model for interface failure under monotonic tension of 3D four direction SiCf/SiC composites. *Compos. Struct.* **2021**, *274*, 114397. [CrossRef]

Disclaimer/Publisher's Note: The statements, opinions and data contained in all publications are solely those of the individual author(s) and contributor(s) and not of MDPI and/or the editor(s). MDPI and/or the editor(s) disclaim responsibility for any injury to people or property resulting from any ideas, methods, instructions or products referred to in the content.

Article

An Improved 3D OPC Method for the Fabrication of High-Fidelity Micro Fresnel Lenses

Fei Peng [1,2], Chao Sun [1,2], Hui Wan [1,2] and Chengqun Gui [1,2,*]

[1] Institute of Technological Sciences, Wuhan University, Wuhan 430072, China; 2020106520010@whu.edu.cn (F.P.); 2018300003075@whu.edu.cn (C.S.); wanhui_hb@whu.edu.cn (H.W.)
[2] Hubei Key Laboratory of Electronic Manufacturing and Packaging Integration, Wuhan University, Wuhan 430072, China
* Correspondence: cheng.gui@whu.edu.cn

Abstract: Based on three-dimensional optical proximity correction (3D OPC), recent advancements in 3D lithography have enabled the high-fidelity customization of 3D micro-optical elements. However, the micron-to-millimeter-scale structures represented by the Fresnel lens design bring more stringent requirements for 3D OPC, which poses significant challenges to the accuracy of models and the efficiency of algorithms. Thus, a lithographic model based on optical imaging and photochemical reaction curves is developed in this paper, and a subdomain division method with a statistics principle is proposed to improve the efficiency and accuracy of 3D OPC. Both the simulation and the experimental results show the superiority of the proposed 3D OPC method in the fabrication of Fresnel lenses. The computation memory requirements of the 3D OPC are reduced to below 1%, and the profile error of the fabricated Fresnel lens is reduced 79.98%. Applying the Fresnel lenses to an imaging system, the average peak signal to noise ratio (PSNR) of the image is increased by 18.92%, and the average contrast of the image is enhanced by 36%. We believe that the proposed 3D OPC method can be extended to the fabrication of vision-correcting ophthalmological lenses.

Keywords: 3D lithography; 3D OPC; Fresnel lens; PSNR

Citation: Peng, F.; Sun, C.; Wan, H.; Gui, C. An Improved 3D OPC Method for the Fabrication of High-Fidelity Micro Fresnel Lenses. *Micromachines* **2023**, *14*, 2220. https://doi.org/10.3390/mi14122220

Academic Editors: Yao Liu and Jinjie Zhou

Received: 9 November 2023
Revised: 3 December 2023
Accepted: 5 December 2023
Published: 9 December 2023

Copyright: © 2023 by the authors. Licensee MDPI, Basel, Switzerland. This article is an open access article distributed under the terms and conditions of the Creative Commons Attribution (CC BY) license (https://creativecommons.org/licenses/by/4.0/).

1. Introduction

Three-dimensional lithography based on laser direct writing, or mask-less lithography, refers to the process of the selective exposure of a thick photoresist from the features of the computer-aided design (CAD) model. This technique can effectively modulate the light intensity on the surface and inside the photoresist, in order to obtain a three-dimensional photoresist pattern after development. The characteristics of a smaller volume, lower manufacturing costs, and fast manufacturing efficiency make it widely used in two-dimensional (2D) micro-manufacturing applications, such as microfluidics [1], metamaterials [2], and reticles for X-ray lithography [3,4]. Recently, the flexibility of LDW-based free form grayscale mask design has attracted widespread attention and rapid development in the potential of three-dimensional (3D) microfabrication. Examples include customized micro-lens arrays (MLA) for wave-front sensing [5,6], virtual reality/augmented reality displays [7,8], super hydrophobic lenses for the humid outdoor environment [9], bionic compound eye lenses [10,11] for zoom imaging, Fresnel lenses for visual correction [12,13], Fresnel zone plates for lensless imaging [14,15], and achromatic lenses based on the Fresnel design [16–19]. All the integrated optical devices described above require collaborative optimization in design and manufacturing. Therefore, models based on different mechanisms are proposed and used to predict the microstructural morphology—for example, time- and space-based molecular diffusion models [20], 3D models based on resin cross-linking [21], kinetics-based dill exposure models [22–24], and 3D models based on optical proximity effects [25,26]. It is worth noting that the first two methods mentioned above, starting from more microscopic photochemical reactions and spatio-temporal effects, are more scientific and rigorous, and they have achieved significant results in multi-photon lithography. In grayscale 3D lithography, the optical proximity correction (OPC) 3D model is commonly used to compensate for the morphology of 3D structures. Although this model weakens the influence of time and space in predicting the morphology, the engineering approximation between the exposure energy and structural morphology effectively compensates for lithographic morphology errors. However, when simulating millimeter-scale structures, the existing 3D OPC method requires a huge amount of computing power and terabyte memory storage. Thus, it is necessary to develop a 3D OPC method that requires fewer computational resources, while retaining the high geometric fidelity of microstructures.

Figure 1a illustrates the schematic diagram of a 3D lithography system with 405 nm illumination. First, the CAD mask carrying the microstructure layout is uniformly sampled, and the exposure scheme is generated. Second, the optical signal is transferred to the photoresist-covered substrate using an electrical modulation system, and the platform scans and steps at a constant speed. Finally, the photoresist patterns are formed after a series of developing processes. Taking the red dotted line of the CAD mask as an example, the grayscale 3D lithography mechanism based on LDW is shown in Figure 1b. The selective exposure on the surface of the photoresist causes a change in the light intensity in z-direction $I(z)$, and the photochemical reaction results in the response of the photoactive compound concentration (PAC) in z-direction $M(z)$. As a result, a 3D photoresist image $I_p(z)$ is obtained after the development process. However, the photoresist image is distorted due to the optical proximity effect (OPE) and the photochemical effect. Figure 1c shows the compensation effect of 3D OPC, which improves the fidelity of the photoresist pattern by pre-distorting the CAD mask pattern and inserting assistant features around the structure. Recently, the model-based OPC has been proposed to improve the lithographic fidelity [27,28]; the regularization term with exposed dosage [29] and interior-point optimization with a barrier function [30] have been developed to accelerate optimization; and the neural network-based compensation method [31] has been researched for the prediction of CAD masks. All these methods provide superior theories for later generations and become essential to compensate for the undesired distortions of lithography, but the computational and storage efficiency is still unacceptable. For example, in an $N \times N$ OPC problem, the derivative calculation requires $N^2 \times N^2$ matrix storage. The OPC of projection lithography also has computational efficiency problems, and a series of algorithms have been developed. The development of these algorithms can serve as guidelines for the investigation of 3D OPC. Examples of such algorithms are the conjugate gradient methods [32,33], the augmented Lagrangian methods [34], the compressive sensing methods [35,36], the semi-implicit methods [37,38], and the model-driven neural network methods [39,40]. Inspired by these algorithms, Peng et al. simplified the derivatives as a matrix form and proposed the 3D OPC method based on 3D lithography [41]. Jidling et al. focused on the memory efficiency and a constrained gradient search method (L-BFGS-B) with pattern segmentation was proposed [42]. At the same time, Freymann et al. optimized the resin cross-linked 3D model based on the Downhill Simplex Algorithm (DSA) [21], which also reduced the memory requirements and computational resources. However, these algorithms still require substantial memory and computational resources, especially when optimizing at the millimeter scale. In addition, the existing 3D OPC models do not consider the interplay between optical and photochemical reaction processes, leading to inaccurate model results. Therefore, it is necessary to develop an algorithm that accurately models and optimizes such complex designs and can effectively utilize computing and memory resources.

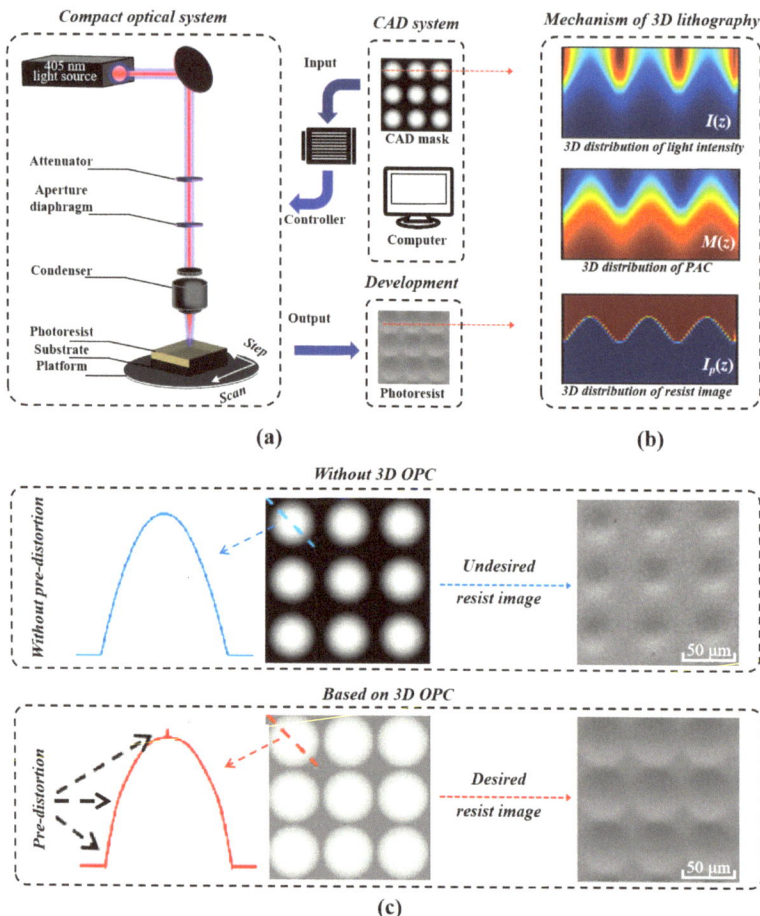

Figure 1. (a) Schematic of 3D lithography system, in which a 405 nm light source is used and the layout on the CAD mask is transferred to the photoresist by a scanning exposure and development process. (b) The 3D distribution of light intensity and photoactive compound concentration (PAC) in the photoresist is changed by selective exposure, and the 3D photoresist pattern is formed after development. (c) The CAD mask compensation method is based on three-dimensional optical proximity correction (3D OPC), where the pixelated mask is numerically solved by convex optimization to improve the print fidelity.

With the development of optical devices towards flatness and customization, the Fresnel design has garnered attention from researchers, because the Fresnel design enables the creation of lightweight devices while maintaining a significantly larger field of view (FOV) [14–17]. Moreover, the advancement of back-end computational models and algorithms has greatly improved the imaging quality of Fresnel designs, leading to their widespread application [18,19]. Considering the potential of the Fresnel structure's applications and its geometrical profile properties, this paper focuses on the establishment of numerical models for 3D OPC and the optimization of central symmetric structures based on the Fresnel design. In particular, the imaging model can be described as the convolution operation of the point spread function (PSF) and CAD mask [43–45]. The photochemical reaction of a thick resist can be described by the Dill model [22–24], which considers the change in the refractive index of the photoresist during exposure, the energy absorption in the photoresist, and the concentration distribution of photosensitivity. Thus, the nonlinear relationship between aerial images and printed photoresist images can be calibrated. To deal with the memory requirements, a subdomain division method with statistics is proposed, which combines the principle of statistics and perceives the overall region through subdomains. Subsequently, total variation (TV) is used during optimization to ensure the continuity of the subdomains. After optimization, the subdomain-based CAD mask is searched, and the optimal global CAD mask is perceived. Finally, the Fresnel lenses are transferred onto PDMS, and the transferred concave Fresnel lenses are applied in a vision correction system. The fabrication of the Fresnel lenses and the experimental results of the visual correction systems show the superiority of the proposed 3D OPC method.

2. Computational Lithography Model with Subdomain Division

In the numerical model of 3D lithography, the intensity of the laser beam on the focal plane (surface of photoresist) is analytically expressed as an ideal Gaussian distribution; the thickness of the photoresist is assumed to be uniform; the reflection from the substrate is ignored in the thick film; and a series of optical effects, such as scattering and cavity formation, is also neglected. Thus, the lithographic process from CAD mask patterns to photoresist can be simplified into two numerical models: the optical scanning (aerial image I_a formation) based on a programmable logic controller, and the resist effect (photoresist image I_p formation) described by the Dill exposure model.

As illustrated in Figure 2, the 3D OPC framework with a central symmetric centimeter target can be divided into three steps. Figure 2a depicts the parameter calibration of lithography, which mainly includes the light spot distribution and the static nonlinear response (depth curve) of the photoresist approximated by the Dill exposure model. In Figure 2b, a 3D OPC model with total variation (TV) is established, which penalizes discontinuities in all directions; through this TV term, the centrally symmetric exposure dose distribution is optimized. Subsequently, the target structure is segmented as M_{sub}^{target} to ensure the efficiency of optimization, and the optimal distribution of the sub-CAD mask is synthesized through 3D OPC. The scheme of subdomain segmentation and the global recovery of the mask is shown in Figure 2c.

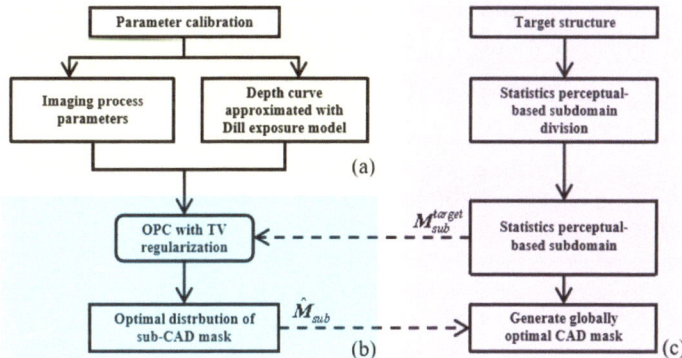

Figure 2. (a) Parameter calibration based on imaging model and Dill exposure approximation model. (b) The 3D OPC model with total variation (TV). (c) Subdomain segmentation and global recovery of mask.

2.1. Forward Imaging Model

Under the assumption of a linear time-invariant system, the scanning imaging process of LDW is assumed to be incoherent imaging. As Figure 3 shows, given an input CAD mask M, the laser power is regulated according to the gray level of the mask, and the focused spot of different power is obtained through electrical modulation. After selective exposure by focusing spots of different power, an approximate aerial image of the photoresist surface can be calculated as

$$I_a = M(x,y) \otimes B(x,y) \tag{1}$$

in which $\otimes$ is the convolution operation and B is the Gaussian spot, which can be calculated by

$$B(x,y) = \frac{2P}{\pi \omega_0^2} e^{-\frac{2(x^2+y^2)}{\omega_0^2}} \tag{2}$$

where P and $\omega_0 = \text{FWHM}/\sqrt{2\ln 2}$ are the total power and the waist of the beam.

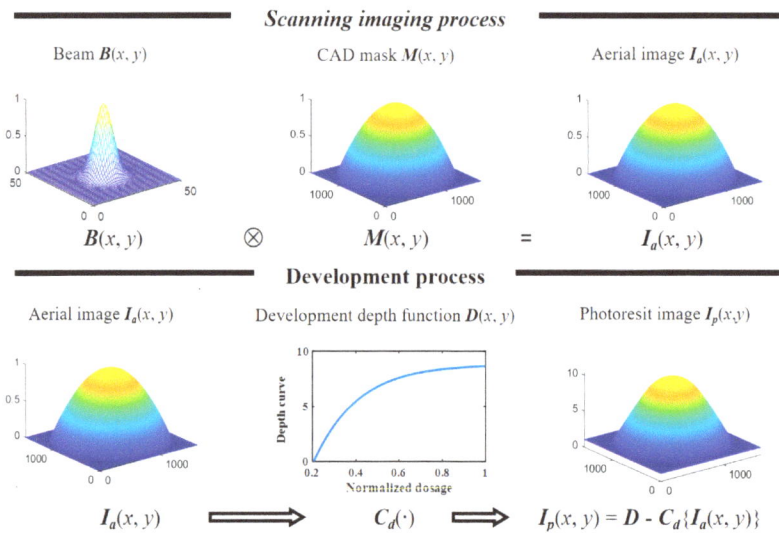

Figure 3. Numerical model of 3D lithography system, where the features on the CAD mask are transferred to the photoresist by a direct exposure and development process.

As a semi-empirical model to describe the exposure process, the Dill exposure model leverages the photochemical reaction through the transport equation and nonlinear kinetic model. Therefore, the Dill model can accurately predict the formation of photoresist patterns and the generation of defects [22–24]. Such predictions can guide the optimization of the exposure parameters and process flow, hence improving the print fidelity. The numerical expression of the Dill exposure model can be described as

$$\frac{\partial I(z,t)}{\partial z} = -I(z,t)[AM_{PAC}(z,t) + B] \tag{3}$$

$$\frac{\partial M_{PAC}(z,t)}{\partial t} = -CI(z,t)M_{PAC}(z,t) \tag{4}$$

in which $I(z,t)$ and $M_{PAC}(z,t)$ are the intensity and photoactive compound concentration (PAC) at the depth z in the resist. A, B, and C are the Dill parameters, which can be calculated by the transmittance of exposed and unexposed photoresists [22]. Although 3D lithography is discontinuous in the exposure time of the resist surface, the quasi-static approximation method based on the surface exposure is also useful [24]. Such prediction allows the optimization of the exposure parameters and process flow, thereby improving the print fidelity. To this end, it is possible to set the boundary condition of $I(0,0) = I_a$ and $M_{PAC}(0,0) = 1$. Assuming that the development time is sufficient, and the properties of the photoresist are not damaged, an accurate development result can be obtained by setting the PAC concentration threshold M_{tr}. Figure 4 depicts the simulation results and experimental results of the Dill model. Figure 4a,c give the normalized PAC computed by different incident intensities I_a, with the resist curves computed by the Dill model and the experiment results given in Figure 4b,d. However, the iteration of partial differential equations based on Equations (3) and (4) is impractical in 3D OPC, because it brings huge computational resource requirements in derivation. Therefore, it is necessary to convert the 3D model into a 2D model.

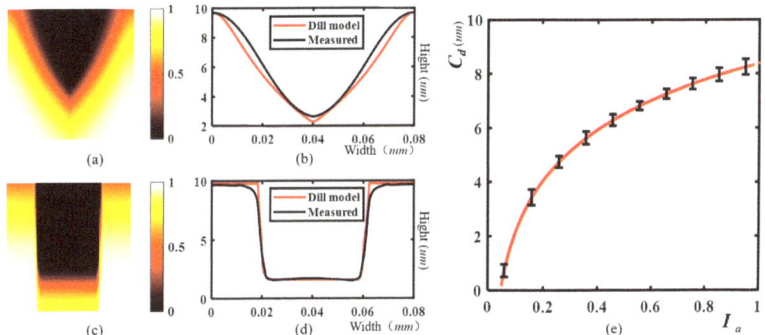

Figure 4. Comparison of Dill model and experiment, where (**a**,**c**) are the normalized PAC concentrations of resists with two different incident intensities I_a, and the corresponding resist curves computed by the Dill model and experiment are given in (**b**,**d**), respectively. (**e**) depicts the nonlinear relationship between normalized aerial image I_a and photoresist depth $C_d(I_a)$.

Fortunately, Equations (3) and (4) were solved analytically by Herrick, and B can often be safely neglected while $A \gg B$ [46]. Thus, the relationship between the intensity and PAC can be approximated as

$$I(z) = I(0) \cdot \frac{1 - M(z)}{1 - M(0)} = I(0) \cdot \frac{M(z)}{M(0)} e^{-Az} \quad (5)$$

Using the threshold method, assume that $I(z) = I_{tr}$ and $M(z) = M_{tr}$ are the development thresholds at depth z. The PAC at the top of photoresist $M(0) = M_{min}$ below the photochemical reaction threshold. Equation (5) is converted into Equation (6) by setting boundary condition $I(0) = I_a$:

$$z = \frac{1}{A} \ln\left[\left(\frac{M_{tr}}{M_{min} I_{tr}}\right) \cdot I_a\right] \quad (6)$$

Considering that the derivation of the photochemical reaction curve uses a series of approximate conditions, we have adopted a more general form in calibration:

$$C_d(I_a) = z = a_1 \cdot \ln(I_a + a_2) + a_3 \quad (7)$$

The quasi-static approximation result can be calibrated as shown in Figure 4e. To this end, the photoresist image after development C_d can be uniformly computed as

$$I_p = \mathcal{T}\{M(x,y)\} = D - C_d(I_a) \quad (8)$$

in which $\mathcal{T}\{\cdot\}$ represents the mapping relationship from the CAD mask and photoresist image, and D is the thickness of the photoresist.

2.2. Inverse Optimization Method with Statistics Subdomain Division

The goal of 3D OPC is to invert the optimal CAD mask $\tilde{M}$ and minimize the dissimilarity between desired target pattern I_0 and photoresist image I_p over all pixels. The score function $\mathcal{S}_{pe}$ employed in convex optimization can be computed by

$$\mathcal{S}_{pe}(M) = \frac{1}{2}\|I_p - I_0\|_2^2 \tag{9}$$

in which $\|\cdot\|_2^2$ means taking the square of the l2 norm.

However, the optimization is impracticable when the pixel number N reaches millions. An effective way to solve this problem is to divide the optimization into subdomains; then, the subdomains are optimized, and the global solution can be obtained by combing these subdomains solutions finally. Considering the uniqueness of the Fresnel design, the symmetrical structure is characterized through a limited area. Therefore, the inverted solutions in a limited subdomain M^{sub} can perceive global solutions $\tilde{M}$. According to Equation (1), the optical proximity effect (OPE) always exists, and the distortion degree of the aerial image in the subdomain I_a^{sub} is increased as subdomain M^{sub} decreases. In Figure 5a, the subdomain mask with width W and height H is denoted as M^{sub}, and the mask intensity in polar coordinates is determined by the radius and independent of the angle. The 3D profile of the subdomain is given in Figure 5b, and the aerial image in the subdomain I_a^{sub} has an obvious distortion at the edges when $H < 0.002$mm. Figure 5c gives the normalized aerial image $\hat{I}_a^{sub}$ computed by

$$\hat{I}_a^{sub}(x,y) = M^{sub}(x,y) \otimes B_{sum}^{-1}(x,y) \tag{10}$$

where $B_{sum}^{-1}(x,y) = B(x,y)/\sum_{(x,y)} B(x,y)$.

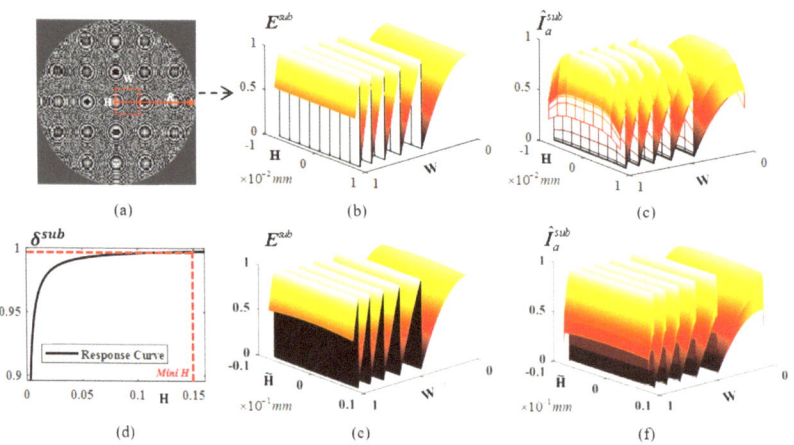

Figure 5. Subdomain division based on perceptual statistics. (**a**) Fresnel lens with radius R = 1 cm. (**b**) is the initial subdomain with the corresponding aerial image in (**c**). (**d**) gives the convolution response between the mask and aerial image. (**e**) shows the global performance of the convolution response. (**f**) is the optimal width and height of the subdomain.

According to the linear characteristic of the convolution operation, $\hat{I}_a^{sub}(x,y) \approx M^{sub}(x,y)$ in the fully convoluted region, and $\hat{I}_a^{sub}(x,y) \ll M^{sub}(x,y)$ in the incompletely convoluted region. Thus, the responses of convolution on subdomain M^{sub} can be computed to describe the overall response relationship. To quantify the extent of distortion, we utilize the linear characteristic of the convolution operation. The response relationship between $\hat{I}_a^{sub}$ and M^{sub} is defined as

$$\delta^{sub} = \frac{1}{kh \cdot kw} \sum_{x=1}^{kh} \sum_{y=1}^{kw} \hat{I}_a^{sub}(x,y) \cdot \left[M^{sub}(x,y) + \varepsilon_\delta\right]^{-1} \tag{11}$$

in which kh and kw represent the pixel number in the x/y direction of subdomain M^{sub}, and Figure 5d shows the relationship between response δ^{sub} and height $\mathbf{H}$. Thus, the optimal height $\widetilde{\mathbf{H}} = 0.16$ mm is determined by $\delta^{sub} = 0.997$ according to the 3δ statistic principle. The subdomain and corresponding aerial image of subdomain $\hat{I}_a^{sub}$ are shown in Figure 5e,f. Similarly, the optimal width $\widetilde{\mathbf{W}} = \mathbf{R} + \widetilde{\mathbf{H}}$ can be obtained from the circular symmetry property. The region to be optimized and the corresponding discrete pixel matrix are compressed to $\widetilde{\mathbf{W}} \cdot \widetilde{\mathbf{H}} \cdot (2 \cdot \mathbf{R})^{-2}$. Hence, when optimizing a Fresnel lens with a diameter of 10 mm, we can effectively reduce the memory requirement to less than 1% by selecting $\widetilde{\mathbf{H}}$ = 0.16 mm and $\mathbf{R}$ = 5 mm as the appropriate parameters. This optimization allows us to strike a favorable balance between computational efficiency and maintaining the desired level of accuracy in the lens design process.

Without loss of generality, the optimization scheme is carried out in the Cartesian coordinate system, and the discontinuity of the optimal $\hat{M}^{sub}$ should be penalized by total variation (TV) in each radial direction. It is worth noting that we use the square of TV to avoid arithmetic errors. Therefore, the optimization problem can be formulated as

$$\begin{aligned} \text{minimize} \quad & \mathcal{S} = \lambda_1 \mathcal{S}_{pe}\left(M^{sub}\right) + \lambda_2 \mathcal{S}_{TV}^2\left(M^{sub}\right) \\ \text{subject to} \quad & M^{sub} = (1 + \cos \omega^{sub})/2 \end{aligned} \tag{12}$$

where $\lambda_1 = \mathcal{S}_{pe}/|\mathcal{S}_{pe}|$ and $\lambda_2 = \mathcal{S}_{TV}^2/|\mathcal{S}_{TV}^2|$ represent the weights of $\mathcal{S}_{pe}$ and $\mathcal{S}_{TV}^2$, and $\mathcal{S}_{TV}^2\left(M^{sub}\right)$ is given by

$$\mathcal{S}_{TV}^2\left(M^{sub}\right) = \|\nabla_x M^{sub}(x,y)\|_2^2 + \|\nabla_y M^{sub}(x,y)\|_2^2 \tag{13}$$

in which $\nabla_x M^{sub}(x,y)$ and $\nabla_y M^{sub}(x,y)$ denote the gradients of $M^{sub}(x,y)$ in the x and y directions:

$$\begin{cases} \nabla_x M^{sub}(x,y) = \frac{M^{sub}(x+1,y) - M^{sub}(x-1,y)}{2} \\ \nabla_y M^{sub}(x,y) = \frac{M^{sub}(x,y+1) - M^{sub}(x,y-1)}{2} \end{cases} \tag{14}$$

Therefore, the optimal subdomain $\hat{M}^{sub}$ with respect to ω can be solved numerically by the steepest gradient descent (SGD) method, with the update rules given by

$$\omega_{t+1}^{sub} = \omega_t^{sub} - \eta \cdot \partial \mathcal{S}_t / \partial \omega_t^{sub} \tag{15}$$

where η is set to a reasonable value and ω will converge under the impetus of $\partial \mathcal{S}/\partial \omega$. The calculation details of $\partial \mathcal{S}/\partial \omega$ can be found in [47]. It is worth noting that Equation (14) is converted into matrix multiplication to realize the solution of the partial differential.

2.3. Accelerated Algorithms

Although Equation (13) gives a stable and reliable optimization scheme, the convergence speed suffers in the case of multi-task optimization (optimization with constraints), which brings inconsistency in time steps η, as we have demonstrated [25,26]. In particular, 3D OPC is an ill-posed non-convex optimization problem. In addition to considering the optimization rate, it is crucial to prevent the optimization from converging to local minima. Thus, more robust Adam optimizer is employed in this paper [48], where the finite difference scheme at time-step $t + 1$ is

$$\omega_{t+1}^{sub} = \omega_{t+1}^{sub} - \eta \cdot \hat{m}_t / \left(\sqrt{\hat{v}_t} + \varepsilon\right) \tag{16}$$

in which $\eta = 0.01$ is the global learning rate. $\hat{m}_t = m_t/(1 - \beta_1^t)$ is the first bias-corrected moment estimate of the first bias moment $m_t = \beta_1 \cdot m_{t-1} + (1 - \beta_1) \cdot g_t$. $v_t = \beta_2 \cdot v_{t-1} + (1 - \beta_2) \cdot g_t^2$ is the second bias-corrected moment estimate of the second bias moment $v_t = \beta_2 \cdot v_{t-1} + (1 - \beta_2) \cdot g_t^2$. The gradient is set as $g_t = \partial \mathcal{S}/\partial \omega$, with the global decay rates $\beta_1 = 0.99$ and $\beta_2 = 0.999$. The additional term $\varepsilon = 10^{-8}$ is included to ensure that the denominator is not zero.

On the other hand, the Fourier transform operation is applied to accelerate the convolution operation:

$$\mathbf{M}^{sub}(x,y) \otimes \mathbf{B}(x,y) = \mathcal{F}^{-1}\left\{\mathcal{F}\left\{\mathbf{M}^{sub}(x,y)\right\} \odot \mathcal{F}\{\mathbf{B}(x,y)\}\right\} \tag{17}$$

in which $\mathcal{F}\{\cdot\}$ and $\mathcal{F}^{-1}\{\cdot\}$ represent the forward and inverse Fourier transform operations, respectively, and $\odot$ denotes the entry-by-entry multiplication operation.

3. Fabrication of the Fresnel Lens

3.1. Design of the Fresnel Lens

Similar to our previous work, a hyperbolic lens is designed in advance [15,16]:

$$\frac{(z-a)^2}{a^2} - \frac{r^2}{b^2} = 1 \tag{18}$$

in which $a = -7.475$ mm and $b = 8.356$ mm are the semi-perimeters of the real and imaginary axes of the hyperbola, respectively. The conical coefficient and radius of curvature being $k = -2.5$ and $R = 9.34$ mm, the focal length and f-number of the lens are $f = 23.35$ mm and $f/\# = 23.35$ mm. Thus, the depth of the Fresnel lens grooves, with a constant height, can be simplified as [35,36]

$$z_{i+1} = R - \sqrt{R^2 - (ip)^2} \tag{19}$$

where i and p are the point coordinate and pitch size of the groove.

3.2. Equipment and Process Parameters

As illustrated in Figure 1a, the lithography device based on the 4096-order electrical modulation is the PicoMaster-100 (Raith/4PICO Litho, Dortmund, Germany). The full width at half maximum (FWHM) of the Gaussian beam profile is 850 nm, the scan speed and step size are v = 100 mm/s and s = 200 nm, and the minimum and maximum exposure doses are 50 mj/cm^2 and 1000 mj/cm^2, respectively. The customized 10-inch glass substrate is covered by the AZ4562 photoresist (PuZhao Display Equipment Co., Ltd., Changsha, China) with a thickness of 10 µm. The development process uses 25% KOH solvent for 120 s at a room temperature of 22 °C.

3.3. Simulation Parameters of 3D OPC

The parameters used for the numerical simulations are as follows: full width at half maxima FWHM = 850 nm; spatial resolution $\Delta x = \Delta y = 200$ nm/pixel; Dill parameters A = 0.45, B = 0.022, C = 0.017, and $M_{tr} = 0.6$; quasi-static approximation coefficient a1 = 3.38, a2 = 47.76, and a3 = -17.6; height of subdomain $\tilde{H} = 0.16$ mm. All computations are performed on an RTX 3090 GPU (NVIDIA, Santa Clara, CA, USA) with 24 GB memory.

3.4. Results and Discussion

The optimization and experimental results are given in Figure 6. Figure 6a–c depict the simulation results and the lithography results before and after 3D OPC. Figure 6(a-i) gives the convergence performance of the 3D OPC method; the score function S is reduced from 8.5 million to 76 thousand. Figure 6(a-ii) shows a cross-section of the CAD mask of the last 10 grooves, where the blue and red lines represent the CAD mask before and after 3D OPC. The CAD mask values based on 3D OPC are increased due to the resist effect, and the discontinuities of the edges are searched to deal with the optical proximity effect (OPE). Figure 6b gives the lithography result before (Figure 6(b-i)) and after (Figure 6(b-ii)) 3D OPC, which was measured by an optical profiler (ZYGO Co., Ltd., Nexview NX2, Middlefield, CT, USA). Figure 6(c-i) illustrates the target structure (black line) and the lithographic structures before (blue line) 3D OPC, where the MSE error (blue horizontal dotted line) between the target one (black line) and the lithographic one (read line) is 0.854 µm. Figure 6(c-ii) illustrates the target structure (black line) and the lithographic structures after (red line) 3D OPC, where the MSE error (red horizontal dotted line) between the target one (black line) and the lithographic one (read line) is 0.171 µm. According to the experiments, the MSE error of the optimized profile is reduced to 79.98% compared with the profile without 3D OPC, which confirms the superiority of the proposed method. It is worth noting that the proposed memory compression method is based on the properties of structural profiles and reverse optimization algorithms, so it is suitable for different lithography processes and lithography models.

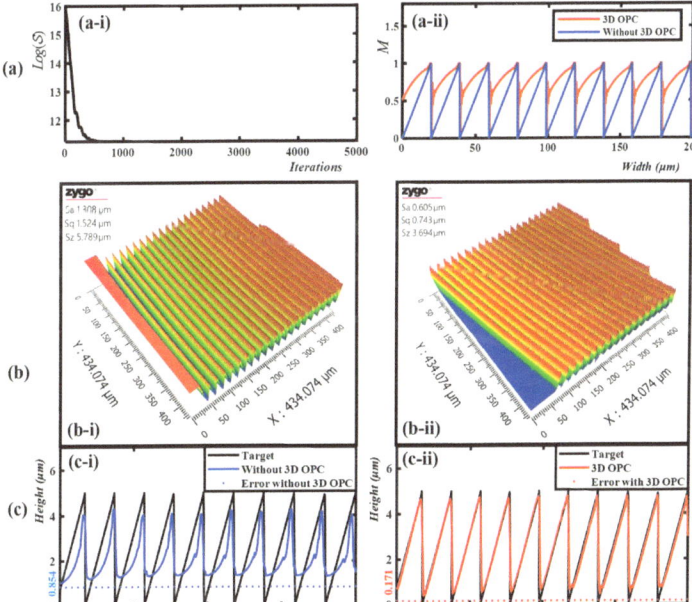

Figure 6. Simulation and experimental results with last 10 grooves. Rows (**a–c**): the simulation results and the lithography results before and after 3D OPC, respectively. (**a-i**) Convergence performance of score function. (**a-ii**) The cross-section profile of CAD mask before (blue line) and after (red line) 3D OPC. (**b-i,b-ii**) shows the lithography result before and after 3D OPC, with the 3D profile measured by an optical profiler. (**c-i**) Schematic illustration of target profile (black line) and the profiles before (blue line) 3D OPC, where the MSE error between the target profile and lithographic profiles (blue horizontal dotted line) is 0.854 µm. (**c-ii**) Schematic illustration of target profile (black line) and the profiles after (red line) 3D OPC, where the MSE error between the target profile and lithographic profiles (read horizontal dotted line) is 0.171 µm.

In the process of spot measurement, it is necessary to move the Fresnel lens continuously in the z direction to determine the focal plane at the best focusing quality, so as to obtain the spot at the focal plane. Figure 7 shows a comparison of the focusing spot of Fresnel lenses (before and after 3D OPC). When the focal length is $f = 32.55$ mm before 3D OPC, there are multiple undesired peaks with considerable intensity in the image. The focal spot image of the Fresnel lens without 3D OPC is given in Figure 7a; the corresponding f-number is $f/\# = 32.55$. Figure 7b depicts the focus spot of the Fresnel lens based on 3D OPC, with a clear peak optimized by 3D OPC, and the improved performance is evidenced by the focus and f-number of $f = 24.25$ mm and $f/\# = 2.425$. The comparison of spot profiles given in Figure 7c, where the simulated profile and the profiles before and after 3D OPC optimization are represented by red solid lines, green dashed lines, and blue dashed lines. The actual focal length of the optimized Fresnel lens is close to the design focal length of $f = 23.35$ mm, and the actual focal length of the optimized Fresnel lens is much longer than the designed focal length because of the concave distortion.

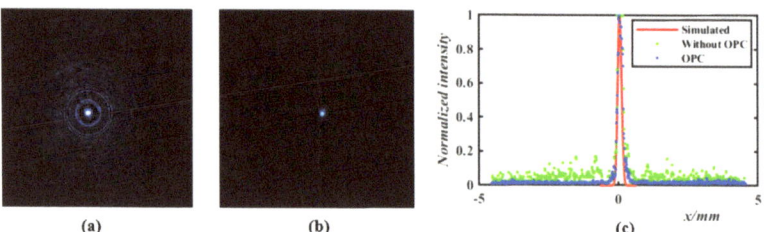

Figure 7. Schematic illustration of focusing spot, with the wavelength 450 nm. (**a**) Focal spot of Fresnel lens without 3D OPC. (**b**) Focusing spot of Fresnel lens based on 3D OPC. (**c**) Intensity profile of focal spot image and simulated intensity profile.

4. Application of the Fresnel Lens

4.1. Fabrication of the Transferred Fresnel Lens

We convert convex Fresnel lenses into concave Fresnel lenses for myopia correction. The transfer process from photoresist to PDMS is divided into three steps. The first step is to mix PDMS (Sylgard 184, Dow Corning, Midland County, MI, USA) with a curing agent at a weight ratio of 1:10 and to remove bubbles through a vacuum environment. The second step is to apply PDMS on the Fresnel lens and use the spin coating method to prepare a mixture film. The spin coating process includes two stages: the film is spun with the speed of 400 rpm for 20 s in the first stage; then, it is spun with the speed of 700 rpm for 40 s in the second stage. In the third step, the PDMS mixture coating on the zoom MLA is fed into an oven at 60 for 7 h. After the baking process, the PDMS mixture is peeled off from the Fresnel lens.

4.2. Vision-Correcting System

In the experiments with the vision-correcting system, a 4× objective lens (Daheng, GCO-2121, Beijing, China) is used to represent the human crystalline lens; a CCD (Daheng, MER-2000-19U3C-L, acquisition frame rate 25 Hz, exposure time 50,000 μs) is used to observe the image, which is simply approximated as an image on the retina. A white light source is selected to imitate natural light imaging and used for testing on the 1951 USAF resolution test card. The color image target is displayed on the OLED screen. The adjustment of the object (mask) and the image distances is realized by two three-dimensional motion platforms (Daheng, GCM-901602 M).

4.3. Results and Discussion

In this paper, we consider the correction of myopia, where the retina receives blurred images in the distance. Therefore, a concave lens needs to be placed in front of the human eye to extend the focal length. With the transfer process provided in Section 3.4, the vision correction lens (transferred Fresnel lens) is used to extend the focal length. Figure 8a shows the optical system for vision correction. The distance between the mask and objective lens is defined as z_1, and the distance between the objective lens and CCD is defined as z_2. In practice, the vision correction approximation experiment can be divided into three steps: normal vision imaging, myopia vision imaging, and vision correction imaging.

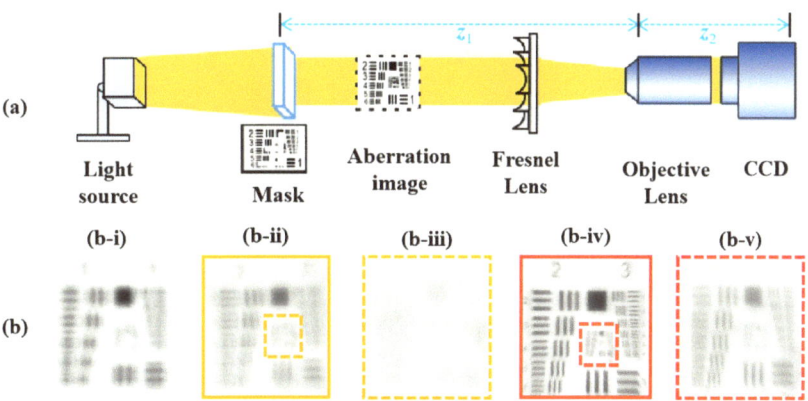

Figure 8. (**a**) Schematic illustration of an experiment for vision correction. (**b**) Comparisons of best imaging performance with 1951 USAF resolution test chart. (**b-i**) is optical image without lens, (**b-ii**) is the optical image corrected by transferred lens (without 3D OPC), (**b-iii**) is the zoomed-in image of group 5 element of (**b-ii**), (**b-iv**) is the optical image corrected by transferred lens (3D OPC), (**b-v**) is the zoomed-in image of group 5 element of (**b-iv**).

The first step is to obtain normal vision imaging without using a vision correction lens: the positions of the mask and objective lens are continuously adjusted to enable the CCD to capture a clear image. During this process, white light illumination is used, and a 1951 USAF resolution test card is used as the mask, resulting in $z_1 = 40$ mm, $z_2 = 90$ mm. The second step is to obtain myopia vision imaging without using a vision correction lens: we stabilize the position of the objective lens and CCD and move the mask away from the objective lens to allow the CCD to receive aberration images. Figure 8(b-i) gives the aberration images received by CCD, where the distance between the mask and objective lens is $z_1 = 80$ mm. The third step is to obtain vision correction imaging by using a vision correction lens: we place a transferred Fresnel lens in front of the objective lens to extend the focal length of the optical path. We continuously adjust the position of the vision correction lens to achieve the best image quality received by the CCD. Based on a visual correction lens without 3D OPC, the corrected image, as shown in Figure 8(b-ii), is blurry, and the zoomed-in image of group 5 is given in Figure 8(b-iii). It can be observed that only the features in element 2 of group 3 can be distinguished. Figure 8(b-iv) shows the imaging correction result of the vision correction lens optimized by 3D OPC, with the zoomed-in image of group 5 elements shown in Figure 8(b-v), and the equivalent corrected resolutions are 14.25 lp/mm (group 3, element 6) and 32 lp/mm (group 5, element 6). Although the blurred grayscale images of group 5 cannot represent the absolute resolution of corrected vision, all of the resolution features are captured.

Through the above operations, the optimal placement of vision correction lenses can be ensured. Thus, colorful image masks can be employed in the imaging system to quantify the imaging performance. The portrait and blocks shown in Figure 9a,b are displayed on the OLED screen. The contrast is calculated as shown in Figure 9c, i.e., the contrast-H and -V of the whole image are averaged horizontally and vertically.

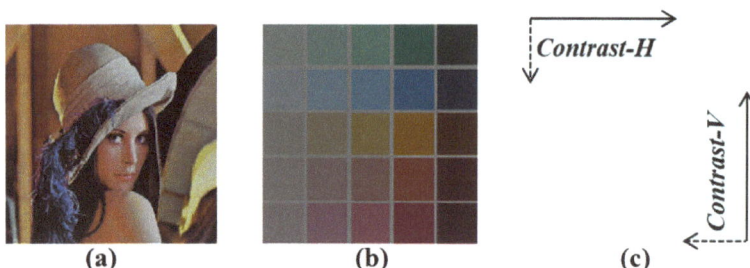

Figure 9. (**a**) The portrait target. (**b**) The color blocks target. (**c**) Schematic diagram of contrast calculation.

Figure 10 depicts the imaging correction results based on vision correction lenses. The columns from left to right show the imaging results at $z1$ = 80, 85, 90, and 95 mm. With the portrait mask employed in rows (a) and (b), the imaging results based on the vision correction lens (before and after 3D OPC) are compared. In row (a) of Figure 10, the imaging results are corrected by the Fresnel lens without 3D OPC, where PSNR = 13.4, 13.2, 13.2, 12.4 dB and contrast = 0.24, 0.24, 0.24, 0.23 at $z1$ = 80, 85, 90 and 95 mm, respectively. Row (b) gives the imaging correction results for the Fresnel lens based on 3D OPC, with the improved PSNR = 16.6, 16.5, 16.2, 15.0 dB and contrast = 0.37, 0.34, 0.32, 0.27. Similarly, the imaging correction results with the block target are depicted in rows (c) and (d). The PSNR = 18.2, 18.3, 18.1, 17.9 dB and contrast = 0.18, 0.18, 0.18, 0.17 based on a Fresnel lens without 3D OPC are given in row (c). The improved performance of corrected imaging is shown in row (d), with the PSNR = 21.8, 21.6, 21.4, 21.2 dB and contrast = 0.26, 0.25, 0.23, 0.22. Although the PSNR and contrast calculated based on different complexity masks are different, the Fresnel lens based on 3D OPC greatly improves the imaging performance of the vision correction system. The average PSNR and the average contrast (computed based on different positions $z1$) before and after 3D OPC are presented in Table 1. Conservatively quantified, we only take smaller values. Compared with the imaging results formed by the vision correction lens without 3D OPC, the average PSNR and average contrast with the 3D OPC-based vision correction lens are improved by 18.92% and 36%, respectively. Without loss of generality, we take the minimum improvement ratio for quantification.

Table 1. Performance of mask imaging based on vision correction lens before and after 3D OPC.

Mask	Vision Correction Lens without 3D OPC		Vision Correction Lens with 3D OPC		Improvement	
	PSNR (dB)	Contrast	PSNR (dB)	Contrast	PSNR	Contrast
Portrait	13.0	0.24	16.1	0.32	23.3%	37.29%
Color Blocks	18.1	0.18	21.5	0.24	18.92%	36%

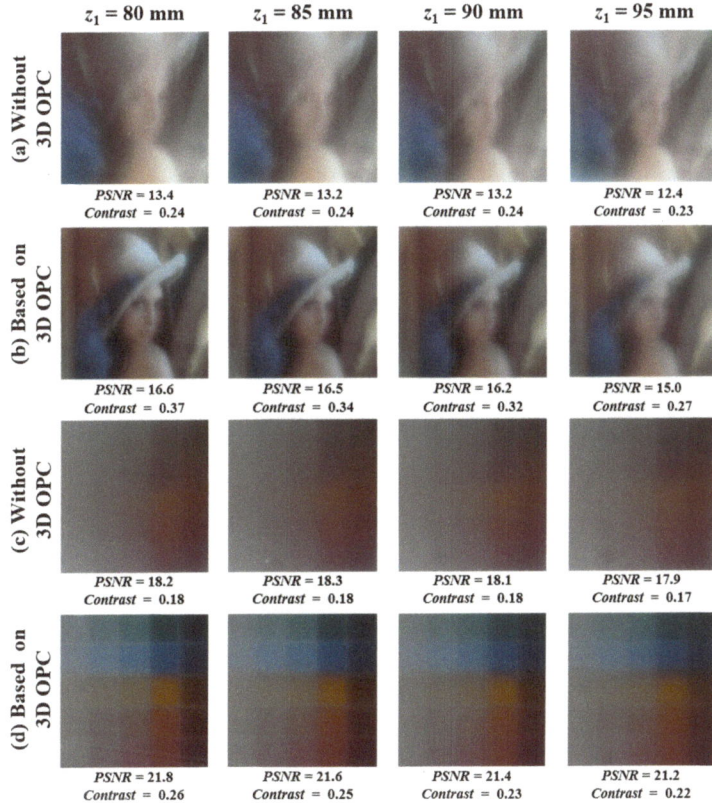

Figure 10. Experimental results of imaging correction for vision correction lenses. Columns from left to right: the imaging results at z1 = 80, 85, 90, and 95 mm, respectively. Rows (**a**,**b**) give the imaging correction results using portrait mask. Rows (**c**,**d**) show the imaging correction results using color block mask.

5. Conclusions

In this paper, an improved 3D OPC method based on statistical principles is proposed, which reduces the memory requirement to less than 1% and realizes 3D OPC in the micron to millimeter scale. With the quasi-static approximated Dill exposure model embedded in 3D OPC, the 3D model is approximated as a 2D model and transformed into a differentiable scheme. Benefitting from the proposed methods, the CAD mask of the Fresnel lens is inverted within 10 min, and the experimental results show that the reduction in the geometric profile error after optimization is 79.98%. With a vision-correcting lens transferred from a 3D OPC-based Fresnel lens by PDMS, the resolution of the vision-correcting system reaches 32 lp/mm. Compared with the transferred vision correction lens without 3D OPC, the PSNR and average image contrast of the vision correction system are improved by 18.92% and 36%, respectively. It should be noted that the image quality of the optimized Fresnel lens is still insufficient. This is due to various factors contributing to the generation of aberrations, as well as inherent limitations in the device's imaging characteristics, particularly evident in Fresnel lenses. This paper compares the image quality of the Fresnel lens before and after optimization in terms of preparation and validates the reliability of the proposed algorithm. Future work will involve employing a 3D OPC algorithm to optimize the achromatic lens, with the goal of enhancing the PSNR and contrast of the imaging. These improvements are envisioned to have potential applications in vision correction.

Author Contributions: Conceptualization, F.P. and C.G.; writing—original draft preparation, F.P.; data measurement C.S.; writing—review H.W. and C.G. All authors have read and agreed to the published version of the manuscript.

Funding: This work was funded by the National Natural Science Foundation of China (No. U20A6004).

Data Availability Statement: The data presented in this study are available from the corresponding author on request. The data are not publicly available due to privacy.

Acknowledgments: The authors acknowledge the support from the Simax Shanghai Company Limited and the Yimeteq Company Limited.

Conflicts of Interest: The authors declare no conflict of interest.

References

1. Trantidou, T.; Friddin, M.S.; Gan, K.B.; Han, L.; Bolognesi, G.; Brooks, N.J.; Ces, O. Mask-free laser lithography for rapid and low-cost microfluidic device fabrication. *Anal. Chem.* **2018**, *90*, 13915–13921. [CrossRef] [PubMed]
2. Melnikov, A.; Köble, S.; Schweiger, S.; Chiang, Y.K.; Marburg, S.; Powell, D.A. Microacoustic metagratings at ultra-high frequencies fabricated by two-photon lithography. *Adv. Sci.* **2022**, *9*, 2198–3844. [CrossRef]
3. Guizar-Sicairos, M.; Johnson, I.; Diaz, A.; Holler, M.; Karvinen, P.; Stadler, H.-C.; Dinapoli, R.; Bunk, O.; Menzel, A. High-throughput ptychography using Eiger: Scanning X-ray nano-imaging of extended regions. *Opt. Express* **2014**, *22*, 14859–14870. [CrossRef]
4. Achenbach, S.; Hengsbach, S.; Schulz, J.; Mohr, J. Optimization of laser writer-based UV lithography with high magnification optics to pattern X-ray lithography mask templates. *Microsyst. Technol.* **2019**, *25*, 2975–2983. [CrossRef]
5. Lin, V.; Liu, X.; Huo, T.; Zheng, G. Design and fabrication of long-focal-length microlens arrays for Shack–Hartmann wavefront sensors. *Micro Nano Lett.* **2011**, *6*, 523–526. [CrossRef]
6. Zhang, L.; Zhou, W.; Naples, N.J.; Yi, A.Y. Fabrication of an infrared Shack–Hartmann sensor by combining high-speed single-point diamond milling and precision compression molding processes. *Appl. Opt.* **2018**, *57*, 3598–3605. [CrossRef]
7. Zhou, X.; Peng, Y.; Peng, R.; Zeng, X.; Zhang, Y.A.; Guo, T. Fabrication of large-Scale microlens arrays based on screen printing for integral imaging 3D display. *ACS Appl. Mater. Interfaces* **2016**, *8*, 24248–24255. [CrossRef]
8. Hong, J.; Kim, Y.; Park, S.; Hong, J.; Min, S.; Lee, S.; Lee, B. 3D/2D convertible projection-type integral imaging using concave half mirror array. *Opt. Express* **2010**, *18*, 20628–20637. [CrossRef]
9. Luan, S.; Xu, P.; Zhang, Y.; Xue, L.; Song, Y.; Gui, C. Flexible Superhydrophobic Microlens Arrays for Humid Outdoor Environment Applications. *ACS Appl. Mater. Interfaces* **2022**, *14*, 53433–53441. [CrossRef]
10. Luan, S.; Cao, H.; Deng, H.; Zheng, G.; Song, Y.; Gui, C. Artificial Hyper Compound Eyes Enable Variable-Focus Imaging on both Curved and Flat Surfaces. *ACS Appl. Mater. Interfaces* **2022**, *14*, 46112–46121. [CrossRef]
11. Su, S.; Liang, J.; Li, X.; Xin, W.; Ye, X.; Xiao, J.; Xu, J.; Chen, L.; Yin, P. Hierarchical Artificial Compound Eyes with Wide Field-of-View and Antireflection Properties Prepared by Nanotip-Focused Electrohydrodynamic Jet Printing. *ACS Appl. Mater. Interfaces* **2021**, *13*, 60625–60635. [CrossRef]
12. Vu, V.; Hasan, S.; Youn, H.; Park, Y.; Lee, H. Imaging performance of an ultra-precision machining-based Fresnel lens in ophthalmic devices. *Opt. Express* **2021**, *29*, 32068–32080. [CrossRef]
13. Vu, V.; Yeon, H.; Youn, H.; Lee, J.; Lee, H. High diopter spectacle using a flexible Fresnel lens with a combination of grooves. *Opt. Express* **2022**, *30*, 38371–38382. [CrossRef]
14. Wu, J.; Zhang, H.; Zhang, W.; Jin, G.; Cao, L.; Barbastathis, G. Single-shot lensless imaging with fresnel zone aperture and incoherent illumination. *Light Sci. Appl.* **2020**, *9*, 53. [CrossRef]
15. Ma, Y.; Wu, J.; Chen, S.; Cao, L. Explicit-restriction convolutional framework for lensless imaging. *Opt. Express* **2022**, *30*, 15266–15278. [CrossRef]
16. Dun, X.; Ikoma, H.; Wetzstein, G.; Wang, Z.; Cheng, X.; Peng, Y. Learned rotationally symmetric diffractive achromat for full-spectrum computational imaging. *Optica* **2020**, *7*, 913–922. [CrossRef]
17. Xiao, X.; Zhao, Y.; Ye, X.; Chen, C.; Lu, X.; Rong, Y.; Deng, J.; Li, G.; Zhu, S.; Li, T. Large-scale achromatic flat lens by light frequency-domain coherence optimization. *Light Sci. Appl.* **2022**, *11*, 323. [CrossRef]
18. Arguello, H.; Pinilla, S.; Peng, Y.; Ikoma, H.; Bacca, J.; Wetzstein, G. Shift-variant color-coded diffractive spectral imaging system. *Optica* **2021**, *8*, 1424–1434. [CrossRef]
19. Sitzmann, V.; Diamond, S.; Peng, Y.; Dun, X.; Boyd, S.; Heidrich, W.; Heide, F.; Wetzstein, G. End-to-end optimization of optics and image processing for achromatic extended depth of field and super-resolution imaging. *ACM Trans. Graph.* **2018**, *37*, 1–13. [CrossRef]
20. Waller, E.H.; Von Freymann, G. Spatio-Temporal Proximity Characteristics in 3D μPrinting via Multi-Photon Absorption. *Polymers* **2016**, *8*, 297. [CrossRef]
21. Lang, N.; Enns, S.; Hering, J.; Freymann, G.V. Towards efficient structure prediction and pre-compensation in multi-photon lithography. *Opt. Express* **2022**, *30*, 28805–28816. [CrossRef]

22. Yesilkoy, F.; Choi, K.; Dagenais, M.; Peckerar, M. Implementation of e-beam proximity effect correction using linear programming techniques for the fabrication of asymmetric bow-tie antennas. *Solid State Electron.* **2010**, *54*, 1211–1215. [CrossRef]
23. Bolten, J.; Wahlbrink, T.; Schmidt, M.; Gottlob, H.D.; Kurz, H. Implementation of electron beam grey scale lithography and proximity effect correction for silicon nanowire device fabrication. *Microelectron. Eng.* **2011**, *88*, 1910–1912. [CrossRef]
24. Dill, F.H.; Hornberger, W.P.; Hauge, P.S.; Shaw, J.M. Characterization of positive photoresist. *IEEE Trans. Electron. Dev.* **1975**, *22*, 445–452. [CrossRef]
25. Luan, S.; Peng, F.; Zheng, G.; Gui, C.; Song, Y.; Liu, S. High-speed, large-area and high-precision fabrication of aspheric micro-lens array based on 12-bit direct laser writing lithography. *Light Adv. Manuf.* **2022**, *3*, 47. [CrossRef]
26. Yang, Z.; Peng, F.; Luan, S.; Wan, H.; Song, Y.; Gui, C. 3D OPC method for controlling the morphology of micro structures in laser direct writing. *Opt. Express* **2023**, *31*, 3212–3226. [CrossRef]
27. Fleming, A.J.; Wills, A.G.; Routley, B.S. Exposure optimization in scanning laser lithography. *IEEE Potentials* **2016**, *35*, 33–39. [CrossRef]
28. Ghalehbeygi, O.T.; O'Connor, J.; Routley, B.S.; Fleming, A.J. Iterative Deconvolution for Exposure Planning in Scanning Laser Lithography. In Proceedings of the 2018 Annual American Control Conference (ACC), Milwaukee, WI, USA, 27–29 June 2018; pp. 6684–6689.
29. Ghalehbeygi, O.T.; Wills, A.G.; Routley, B.S.; Fleming, A.J. Gradient-based optimization for efficient exposure planning in maskless lithography. *J. Micro Nanolithogr. MEMS MOEMS* **2017**, *16*, 033507. [CrossRef]
30. Fleming, A.J.; Ghalehbeygi, O.T.; Routley, B.S.; Wills, A.G. Scanning laser lithography with constrainedquadratic exposure optimization. *IEEE Trans. Control Syst. Technol.* **2019**, *27*, 2221–2228. [CrossRef]
31. Sun, X.; Yin, S.; Jiang, H.; Zhang, W.; Gao, M.; Du, J.; Du, C. U-Net convolutional neural network-based modification method for precise fabrication of three-dimensional microstructures using laser direct writing lithography. *Opt. Express* **2021**, *29*, 6236. [CrossRef]
32. Lv, W.; Liu, S.; Xia, Q.; Wu, X.; Shen, Y.; Lam, E.Y. Level-set-based inverse lithography for mask synthesis using the conjugate gradient and an optimal time step. *J. Vac. Sci. Technol. B* **2013**, *31*, 041605. [CrossRef]
33. Li, J.; Lam, E.Y. Robust source and mask optimization compensating for mask topography effects in computational lithography. *Opt. Express* **2014**, *22*, 9471–9485. [CrossRef]
34. Li, J.; Liu, S.; Lam, E.Y. Efficient source and mask optimization with augmented lagrangian methods in optical lithography. *Opt. Express* **2013**, *21*, 8076–8090. [CrossRef]
35. Ma, X.; Shi, D.; Wang, Z.; Li, Y.; Arce, G.R. Lithographic source optimization based on adaptive projection compressive sensing. *Opt. Express* **2017**, *25*, 7131–7149. [CrossRef]
36. Ma, X.; Wang, Z.; Li, Y.; Arce, G.R.; Dong, L.; Garcia-Frias, J. Fast optical proximity correction method based on nonlinear compressive sensing. *Opt. Express* **2018**, *26*, 14479–14498. [CrossRef]
37. Shen, Y.; Peng, F.; Zhang, Z. Efficient optical proximity correction based on semi-implicit additive operator splitting. *Opt. Express* **2019**, *27*, 1520–1528. [CrossRef]
38. Shen, Y.; Peng, F.; Zhang, Z. Semi-implicit level set formulation for lithographic source and mask optimization. *Opt. Express* **2019**, *27*, 29659–29668. [CrossRef]
39. Ma, X.; Zhao, Q.; Zhang, H.; Wang, Z.; Arce, G.R. Model-driven convolution neural network for inverse lithography. *Opt. Express* **2018**, *26*, 32565–32584. [CrossRef]
40. Ma, X.; Zheng, X.; Arce, G.R. Fast inverse lithography based on dual-channel model-driven deep learning. *Opt. Express* **2020**, *28*, 20404–20421. [CrossRef]
41. Peng, F.; Yang, Z.; Song, Y. 3D grayscale lithography based on exposure optimization. In Proceedings of the International Workshop on Advanced Patterning Solutions (IWAPS), Foshan, China, 12–13 December 2021; pp. 1–3.
42. Jidling, C.; Fleming, A.J.; Wills, A.G.; Schön, T.B. Memory efficient constrained optimization of scanning-beam lithography. *Opt. Express* **2022**, *30*, 20564–20579. [CrossRef]
43. Yuan, K.; Yu, B.; Pan, D.Z. E-beam lithography stencil planning and optimization with overlapped characters. *IEEE Trans. Comput. Aided Des. Integr. Circuits Syst.* **2012**, *31*, 167–179. [CrossRef]
44. Li, F.; Xie, Y.; Sun, Q.; Cao, Z.; Lu, Z.; Wang, Z. Analyzing of line profile for laser direct writing lithograph. *Acta Photonica Sin.* **2004**, *33*, 136–139.
45. Du, C.; Dong, X.; Qiu, C.; Deng, Q.; Zhou, C. Profile control technology for high-performance microlens array. *Opt. Eng.* **2004**, *43*, 2595–2602. [CrossRef]
46. Mack, C.A. *Fundamental Principles of Optical Lithography: The Science of Microfabrication*; John Wiley & Sons, Ltd.: New York, NY, USA, 2007.
47. Jia, N.; Edmund, Y.L. Pixelated source mask optimization for process robustness in optical lithography. *Opt. Express* **2011**, *19*, 19384–19398. [CrossRef]
48. Shen, Y.; Peng, F.; Huang, X.; Zhang, Z. Adaptive gradient-based source and mask co-optimization with process awareness. *Chin. Opt. Lett.* **2019**, *17*, 121102. [CrossRef]

Disclaimer/Publisher's Note: The statements, opinions and data contained in all publications are solely those of the individual author(s) and contributor(s) and not of MDPI and/or the editor(s). MDPI and/or the editor(s) disclaim responsibility for any injury to people or property resulting from any ideas, methods, instructions or products referred to in the content.

Review

Scanning Strategies in Laser Surface Texturing: A Review

Denys Moskal, Jiří Martan *[ID] and Milan Honner [ID]

New Technologies Research Centre (NTC), University of West Bohemia, Univerzitní 8, 30100 Plzeň, Czech Republic; moskal@ntc.zcu.cz (D.M.); honner@ntc.zcu.cz (M.H.)
* Correspondence: jmartan@ntc.zcu.cz

Abstract: Laser surface texturing (LST) is one of the most promising technologies for controllable surface structuring and the acquisition of specific physical surface properties needed in functional surfaces. The quality and processing rate of the laser surface texturing strongly depend on the correct choice of a scanning strategy. In this paper, a comparative review of the classical and recently developed scanning strategies of laser surface texturing is presented. The main attention is paid to maximal processing rate, precision and existing physical limitations. Possible ways of further development of the laser scanning strategies are proposed.

Keywords: laser machining; laser ablation; processing rate; scanning techniques; heat accumulation; plasma shielding

Citation: Moskal, D.; Martan, J.; Honner, M. Scanning Strategies in Laser Surface Texturing: A Review. *Micromachines* 2023, *14*, 1241. https://doi.org/10.3390/mi14061241

Academic Editors: Yao Liu and Jinjie Zhou

Received: 28 April 2023
Revised: 2 June 2023
Accepted: 8 June 2023
Published: 12 June 2023

Copyright: © 2023 by the authors. Licensee MDPI, Basel, Switzerland. This article is an open access article distributed under the terms and conditions of the Creative Commons Attribution (CC BY) license (https://creativecommons.org/licenses/by/4.0/).

1. Introduction

Laser pulse surface ablation leads to the formation of a crater with a depth of several nanometres up to micrometres. The application of consecutive laser pulses with sequential laser beam movement facilitates the creation of regular laser-formed surface structures, from simple periodical dimples or segments up to polygonal or hierarchical structures [1–4]. Such a process of periodical objects laser formation is named laser surface texturing (LST), and it has a wide range of applications: in tribology, material engineering, wettability, brazing, medicine, optics and other areas [5–8]. The development of laser surface scanning techniques is needed for increasing the processing rate of LST. Involving the fastest scanning techniques, for example, the application of polygon scanners, acousto-optic beam deflection or sample rotation, has several limitations, such as dead processing time, low deflection angle, sample geometry specification and others. At higher laser surface scanning speeds, it becomes more and more difficult to maintain the high precision of LST, due to the synchronization loop between the laser and the scanning system [9–11]. The application of high-frequency lasers with ultrashort laser pulses meets the physical limitations of LST, such as heat accumulation between laser pulses and ablated plasma shielding effects [12–16]. The speed limitation for the application of high-frequency lasers becomes important for texturing of 3D bent surfaces [17,18]. There are alternative techniques, such as using low-frequency lasers with high pulse energy for processing entire wide areas at once with a low-speed laser beam or sample movement [19–24]. For such a technique, plasma shielding and heat accumulation do not have such an important role, but on the other hand, the flexibility and scalability of the laser-processed area is limited by a spot size in comparison with scanning technologies.

The correct choice of an appropriate LST method needs to have an overview of the actual scanning techniques and their limitations. There are a lot of reviews for laser surface scanning strategies for selective melting in 3D printing technology [25–27]. However, there are no systematic reviews of the existing laser beam scanning strategies for LST with a comparative analysis of their scanning parameters, such as processing rate or maximal scanning speed. In this paper, such a review of the classical and recently developed laser surface scanning strategies with a comparative analysis of several scanning parameters of

LST is presented. The scanning strategy is defined as a spatial and temporal arrangement of laser paths on the scanning surface, which can be applied in LST.

Different laser surface scanning strategies will be more suitable for different materials, especially considering the laser source limitations [13]. The resulting efficiency, roughness or minimal periodicity of LST structure are significantly influenced by the laser source used. There are a lot of other laser surface processing parameters which significantly affect initiated physical processes of material ablation, such as laser pulse fluence, duration, wavelength, polarization, air pressure or blowing speed or even pulse parity [28–36]. The formation of periodically distributed micro-objects on the laser surface is additionally affected by the limitations of the laser beam scanning system and desired structure of the laser-formed texture. The full description of all passible physical mechanisms, which affect the resulting surface structure and effectivity of laser surface ablation, goes far beyond one article.

The goal of this paper is to describe existing strategies of laser surface texturing beam scanning techniques, which principally differ only in laser beam path ordering and filling arrangement. For that reason, in this review, only technical parameters of laser beam scanning strategies in LST were compared: processing rate, surface structures intervals, precision and scanning speed. The main physical limitations, which are mainly affected by the chosen scanning strategy of LST, are also discussed.

2. Physical Limitations of Laser Surface Texturing

There are a lot of physical effects which are activated during the interaction of a laser pulse with the material, depending on the intensity and pulse duration of the laser pulses: from slight surface heating up to intense Coulomb explosion [37]. However, the main physical effects, which are affected by the chosen scanning strategy, are heat accumulation and plasma shielding effects. Although some scanning parameters exist, where optimal heat accumulation and even plasma shielding effects have a positive role in achieving higher efficiency or better quality of processed surface [38,39], generally, they play a negative role in the biggest cases [40–42]. In this section, the mentioned physical effects are discussed as principally influenced by the choice of the scanning strategy.

2.1. Plasma Shielding Effect

The ablation plasma plume spreads over the laser-processed surface, and the next laser pulses will be partially or fully blocked by the plume. The time of the plasma plume surface shielding depends on laser pulse parameters, such as pulse duration, pulse energy or focused spot size. Understanding the main principles of the ablation plasma plume evolution helps in the correct choice of an optimal scanning strategy for LST.

The laser pulse absorption initiates the surface temperature reaching the region of overcritical fluid formation [37,43–48]. It was shown that ultrafast solid-to-liquid phase transitions already appear in the first few hundred femtoseconds [46]. Following this, material expansion is detected in 10–20 picoseconds after laser pulse absorption [46,47]. Exposed materials are able to achieve great speeds, more than 8–10 km/s [49–52]. The ablation plume is ejected at the highest speed in the phase explosion regime. The speed of the laser-ablated plume at the very start of the explosion can be described by a kinetic equation of adiabatic expansion [53]:

$$v_f \sim (E/M)^{1/2} \qquad (1)$$

where M is the mass of ablated plume, and E is the total energy. In ambient gas, the free expansion of the ablated products will be changed by breaking the movement of the plasma plume with a generation of a frontal shock wave [49,53,54]. The following part of the ablated plume has slower medium-size clusters (up to 10,000 atoms) that have ejection velocities of less than 4 km/s and droplets that are slower than 3 km/s [49]. The pressure in the front shock wave is typically 100–200 atm and decays to close to ambient pressure in 100–200 ns [55]. In the latter time, the shock wave slows down to the speed of sound and travels forward as a sound wave [56]. The formation of this shock wave expanding starts in a short time period (0.2–0.5 ns), when the mass of the shock wave becomes comparable with the plume mass [57,58]. The following movement of the plasma plume in ambient gas can be described by the drag model [35,56,59]:

$$R = R_0(1 - exp(-\beta t)) \quad (2)$$

where R is the position of the plasma plume front during expansion, R_0 is the stopping distance of the plume, and β is the deceleration coefficient. The stopping position of the ablated plume at normal pressure of ambient gas above the scanned surface is about 1–2 mm and higher (Figure 1) [35,60]. The post-ablation products contain clusters and droplets, which continue to have high enough speeds $\sim$ 100 m/s and inertial mass for achieving bigger distances above the laser-processed surface [61,62]. Such ablation products can cover several centimetres above the processed area for a long time, up to tens of microseconds [34,63]. Plasma and particle shielding effects significantly limit the laser repetition rate and effectiveness of laser surface processing [64].

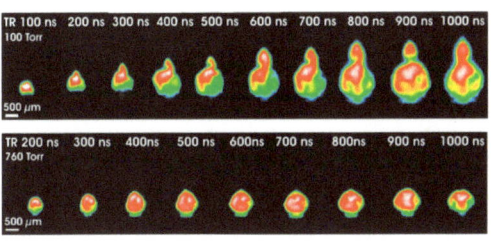
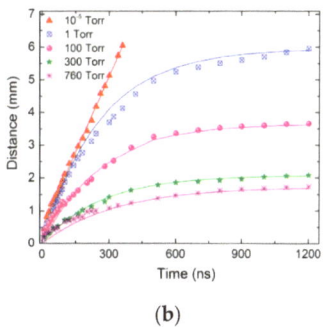

(a) (b)

Figure 1. Explosion of ablation plasma plume: (a) plasma plume evolution [60,65]; (b) R-t plots obtained from ICCD images are given for various pressure levels [35].

The plasma plume transparency dynamic was studied in detail by J. König et al. [44]. The transparency of the plasma plume above the metallic target was measured by a probe laser beam, which was parallel to the sample surface. It was found that in the first moment of the plasma plume expansion, its transparency decreases down to 10%. The transparency of the plasma plume stays reduced by a certain value till 2–3 µs; during this time, the ablated material expands by hundreds of micrometres [35,56]. Such a highly optically dense plasma plume works as a shield above the processed surface for every subsequent laser pulse, if it comes before the plasma plume dissipates. Overcoming plasma shielding effects can be achieved via the application of a long enough time delay between laser pulses (MHz or lower frequencies), and then the plasma plume optical density becomes low enough before every next laser pulse. Such laser pulse frequencies are applicable for optical systems with laser beam scanning speeds up to 10–20 m/s, which are needed for optimal laser spots overlapping 30–50% [10,13].

An opposite way of overcoming the shielding effects can be achieved via the application of several laser pulses within a short time interval, shorter than that of a plasma plume, which will be expanded above the surface [15,66]. The plasma plume expanding cannot be avoided by the application laser pulse with a short time interval, down to 1–2 ps, but the energy conversion in such fast processes is able to change the ablation mechanism and suppress the shielding effects [15,66]. Such a shielding-suppressing regime is very dependent on processed materials, laser pulse intervals and even pairing of the laser pulses [28]. The frequency of laser pulses for such a mechanism of plasma plume suppression should be in the hundreds of GHz or in THz and can be realized in burst regimes [15,67–70]. Such a short interval between laser pulses can be achieved with special techniques of pulse dividing, such as crystals array or optical branches [28,71,72]. A more practical approach is usually to use long enough time delays when the plasma is gone upon the arrival of the next pulse.

Another alternative is application of a scanning strategy with special distribution of laser spots, where the distance between laser spots will be bigger than the plasma plume size (more than 200–300 µm). It can be multibeam surface processing with low frequency of laser pulses or high-speed surface scanning equidistant distribution of laser spots [9,73,74]. Several existing scanning techniques of distant laser spot distribution will be discussed in the Section 3 of this review.

2.2. Heat Accumulation in Pulsed Laser Surface Processing

There are a lot of papers in which the important role of heat accumulation during laser surface processing was discussed [40,75–77]. It was shown that in some optimal conditions, heat accumulation becomes a positive factor for increasing the ablation rate or improving the quality of the machined surface [31,42,78–84]. However, undesired high heat accumulation leads to thermal surface degradation with material boiling, intense oxidation and uncontrollable splash formation [40,75,76,85–87]. The evaluation of the heat accumulation level can be performed by thermal field modelling or by experimental [32,76,86,88–90].

After ablation, a thin remelted layer with a number of point defects and dislocations, produced by ultrafast cooling and shock wave, remains in the laser spot area [91,92]. The output laser-irradiated surface will be contaminated by precipitation of the ablated products and associated oxidation of the upper surface layers [93–96]. In most cases, the surface ablation process has a semi-thermal character ($k_B T_i \geq \varepsilon_b$). For example, at Gaussian distribution of energy in the laser spot, the laser pulse residual heat stays in the subsurface layers and in the near-ablated zone. The residual heat in subsurface layers can be brought by secondary effects: heat conductivity from ablated layers, ballistic and diffusion effects, convective and radiating energy exchange between the ambient and solid target, and shock wave spreading [37,49,93,97,98]. These residual effects have a great influence on the quality and efficiency of LST in a high-repetition multi-pulse regime [31,76,99]. Of course, even a single laser pulse can produce many thermal effects: material boiling, oxidation and splash formation. These thermal effects have short-term character, and in the biggest cases, high-temperature fields dissipate within 2–5 µs [41,76]. However, in the case of multi-pulse laser surface processing, the residual heat is rising from pulse to pulse in the laser-affected zone. Such heat accumulation is able to prolong and intensify the thermal processes to an undesired level, although this occurs in the case whereby one single pulse is unable to initiate significant thermal processes.

The value of heat accumulation can be predicted theoretically as a sum of the residual heat from all laser pulses in a laser pulse sequence [40,41]:

$$T(r,t) = \sum_{i=0}^{N_f} \Delta T_i \left(r, t + i \cdot \frac{1}{f_{pulse}} \right) \qquad (3)$$

where $T(r,t)$ is the temperature in the point r at the time moment t, N_f is the full number of the laser pulses in the pulse sequence, ΔT_i is the residual temperature after i-th laser pulse and f_{pulse} is the laser pulse generation frequency. In this equation, the residual temperature ΔT_i after absorption of a discrete laser pulse can be approximately defined from a 3D model with an instance heat point source [41,100,101]. A more detailed study of the heat accumulation effects under moving Gaussian laser spots was conducted by Bauer et al. [40].

For evaluation of heat accumulation under a laser-scanned surface in a fixed subsurface point, a semi-planar thermal model can be applied [76,102]:

$$\Delta T(r,t) = \frac{F \cdot e^{-\frac{2r_x^2}{w_0^2}}}{\rho \cdot c \cdot \sqrt{4 \cdot \pi \cdot \alpha \cdot t}} \qquad (4)$$

where x is the coordinate of a fixed surface point, r_x is the distance from the centre of the laser spot to the fixed surface point x, α is the thermal diffusivity, ρ is the material density and c is the specific heat. It has been found that the maximal heat accumulation in a thin subsurface layer is achieved within a specific time interval when the laser beam central point has already passed the fixed subsurface point [76,102]. In the exemplary work of F. Bauer et al. [40], the critical temperature for heat accumulation was defined through experiments involving offset temperature shifting in the laser-scanned surface. The defined offset temperature shift was compared with the corresponding temperature shift in thermal simulations. It was shown that the critical temperature for heat accumulation in steel surface processing has a value near 607 °C. Oxidation and surface degradation were detected at higher temperatures of the scanned surface preheating, even when other laser scanning parameters were not changed [40,76].

There are several other approximation methods for the evaluation of the heat accumulation level, for example, post-control of laser-processed surface by microscopy and profilometry [11,40,103–107] or energy-dispersive X-ray spectroscopy (EDX) of oxidation level [11,40,41,84,108]. It gives us the possibility of defining some limitations of the applied scanning strategies, but such methods are not able to determine temperature changes, which appear during the laser surface processing.

For the direct control of temperature changes under laser beam scanning, a contact method of temperature changes can be applied [68,85]. In this case, the measurements are not affected by optical effects, such as emissivity changes, plasma shielding or undesired influence of laser beam reflection on optical measurements. The disadvantages of such methods are the volumetric character of the achieved data and long response time of the measurements.

Non-contact distant detection of temperature changes can be performed via the application of thermal IR cameras [86]. The application of an IR camera gives a mean value of temperature changes (maximal frame rate 1–2 kHz only [109]), and it does not recognise thermal changes after every individual laser pulse (Figure 2). A similar technique was applied for the detection of heat accumulation during the direct laser interference patterning (DLIP) process, but the camera was installed above the surface [110]. Such a solution is useful for the detection of surface distribution and the long-term dynamic of laser-induced heat accumulation.

For non-contact measurements with sub-nanosecond resolution and a frame rate reaching tens of GHz, the IR detectors can be applied [102,111,112] (Figure 3a). Such IR measurements are fast enough for detection of pulse-to-pulse heat accumulation, and they can be applied for comparing several types of LST: straight line grooving [76], surface microobjects formation by different scanning strategies [113–115], LIPSS (laser-induced periodic surface structures) creation by laser multibeam processing [116] and other [117,118]. Such fast in-process IR measurements of heat accumulation were used for detecting thermal regimes during laser surface processing (Figure 3a) [76,112]. The heat accumulation was evaluated as a subtraction of the background level signal from the thermal IR signal (Figure 3b). In a similar way, an optimal regime of laser surface processing, LIPSS formation, phase changes or other laser-initiated effects can be evaluated [111,116,119–121]. The application of the multibeam technique decreases heat accumulation by dividing of thermal load into several separated fluxes, when instead of one gigantic laser pulse, the surface is irradiated by an array of laser spots with low energy [90,122].

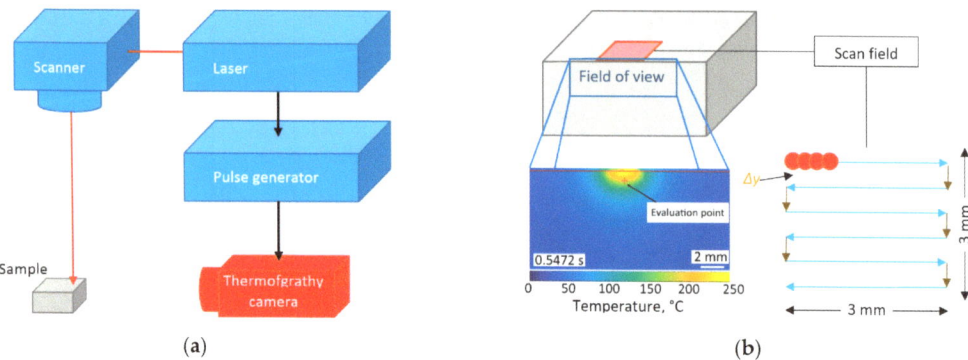

Figure 2. Heat accumulation detection with IR camera: (**a**) IR detection with thermal camera of heat accumulation in laser surface scanning process; (**b**) temperature distribution on side surface under scanning field [86].

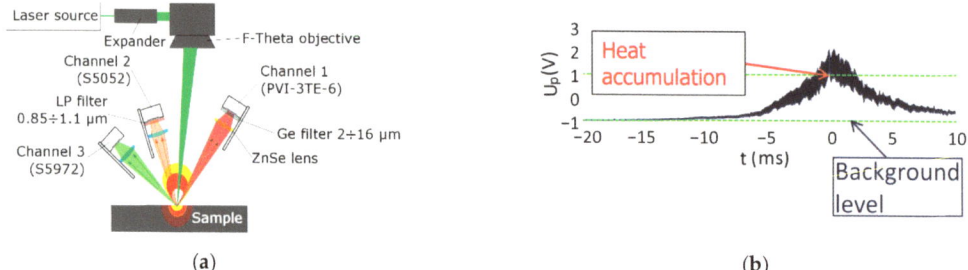

Figure 3. Fast detection of laser-induced surface heating: (**a**) IR measurement set-up with three IR photodiodes [102]; (**b**) evaluation of heat accumulation by subtraction algorithm [76].

In the case of LIPSS formation, an additional mechanism of heat conversion appears when the absorbed energy (after the laser pulse) is divided between different thermal processes, and it can affect the resulting heat accumulation. Such concurrent phases' affection of heat accumulation during LIPSS formation was detected as a thermal double-maxima signature in fast IR-thermal measurements [116].

An interesting and original effect of the influence of temperature regime on LST processing and LIPSS formation was presented in the newest work by W. Gao et al. [123]. In this work, it was shown that decreasing the laser-processed surface temperature to below the freezing point has the potential to dramatically change the LST process. It is mentioned that the local frost layer around the laser-irradiated spot melts into water, helping to boost ablation efficiency, suppress the recast layer and reduce the heat-affected zone, while the remaining frost layer can prevent ablation debris from adhering to the target surface. The frost layer eliminates the debris deposition and recast layer, and LIPSS formation has a mechanism similar to high-spatial-frequency (HSF) LIPSS formation in water [124,125].

The newest results in experiments on the detection of heat accumulation have shown that the highest effectivity is achieved when the temperature of the surface reaches ~800 °C [76]. At the same time, it was shown that the most effective temperature regime does not lead to the best quality of surface processing, but it corresponds to a lower value of heat accumulation around 600 °C [40,76,83].

The correct choice of scanning strategy and optimization of laser beam parameters brings the possibility to overcome the aforementioned physical limitations, especially heat accumulation. The next development of the IR in-process fast measurements will be in on-fly control of the laser beam parameters during laser surface scanning. The on-fly IR control of the laser surface processing in combination with high-speed scanning technique is a way to achieve the high standards of Industry 4.0.

3. Scanning Techniques of LST

The need for increasing the throughput of LST technologies stimulates the development of new strategies for high-speed laser surface processing. There are several well-known scanning technologies for high-speed laser beam deflection: galvo scanners, polygon scanners, piezo scanners, static and resonant scanners, micro-lens scanners, electro-optic deflectors (EOD) and acousto-optic deflectors (AOD) [21,103,126]. The inertial scanning systems have a maximal deflection angle and a number of resolvable spots on the scanned surface [126]. There are two traditional techniques for high-speed laser surface machining with a large processing area: galvanometer beam scanning and polygon optical scanning systems (Figure 4) [127,128]. The maximal scanning speed of the available galvanometer scanners lies in the range of 10–40 m/s, whereas the polygon scanner is able to achieve a scanning speed of more than 1000 m/s [127,129,130]. The higher speed of the polygon scanners is of great benefit in the fast provision of LST in large areas. However, polygon scanners do not provide the smooth wall profiles of vector scans for cutting a circumference or trepanning large holes greater than 50–150 µm. The laser beam deflection in polygon scanners should be corrected by an additional galvanometer scanner. For LST, when the processed area is smaller than 15% of the working field, the polygon scanners are not cost-efficient, and alternative techniques will be more suitable [127]. The correct choice of scanning strategy helps to improve the laser-processed surface quality and precision of the laser pulse delivery.

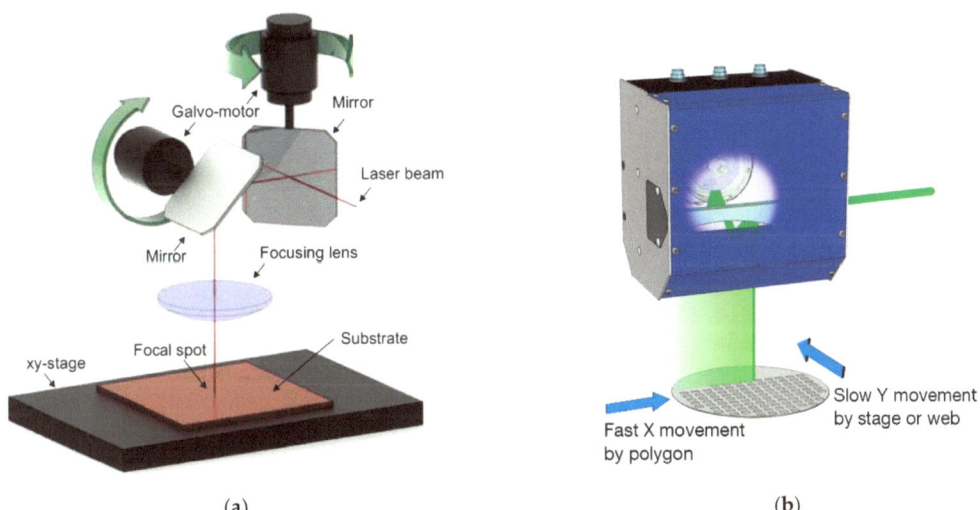

Figure 4. Inertial laser scanning systems: (**a**) galvanometer scanner; (**b**) polygon scanning technique (adapted from [126,127]).

The output processing rate of LST depends on a favourable choice of the combination of the scanning strategy with different scanning techniques (Table 1, rows 1 and 2). The maximal processing rate of 9000 cm^2/min for DLIP was found by a team from Fraunhofer Institute IWS [131–134]. The period and distribution of laser surface textured objects with DLIP are directly dependent on the wavelength of the laser beam [135]. This limitation makes it difficult to apply the DLIP for texturing surfaces in cases of irregular structure or complex nonsymmetrical textures, for example, for super hydrophilic surfaces, high optical absorbance and hydrodynamic effects. The process of the formation of such a structure is based on self-organized effects, cone-like structure formation or programmable direct laser machining with spot size resolution [136–138]. For the laser-controlled formation of fine complex submicron structures, a combination of the DLIP technique with dynamic systems has been provided. The processing rate of such a technique is about 0.7–10 cm^2/min [132,139]. The benefit of such a combination of DLIP with regular micro-texturing techniques gives the possibility to create unique hierarchical structures [132,140].

Submicron surface structures might be created by laser scanning with the self-organized formation of LIPSS [116,141]. In this case, the laser scanning parameters, such as laser spot overlapping and scanning speed, have a key role for highly regular LIPSS (Table 1, rows 3 and 16). The achieved processing rate for LIPSS formation directly depends on the applied scanning speed. For example, I. Gnilitskyi et al. [142] have reported a processing rate of LIPSS forming on stainless steel equal to 6.3 cm^2/min with a scanning speed of 3 m/s. The last study predicts several times higher processing rates with polygon scanners [77]. LIPSS forming is competitive with industrial standards of nano-manufacturing (~1 cm^2 in 10 s) [143]. Like the DILP technique, LIPSS might be applied for the formation of hierarchical surface structures in combination with micro-texturing. The benefit of such a solution is that the period of the upper LIPSS can be smaller than half of the laser wavelength [143].

Another highly productive LST strategy is via forming an array of micro-objects by dividing the laser beam with diffraction or shadow masks [144–147] (Table 1, rows 12 and 17). This scanning strategy could be used in cases when the pulse energy is high enough to be divided into multi-beams [90]. The distance between laser spots is given by mask parameters or diffraction light distribution. In the case of the application of a solid-state static mask, the spot distance is constant, and part of the laser beam energy will be lost. The application of spatial light modulators (SLMs) gives us the possibility to change the laser spot distribution of the scanning process, but the average power will be limited to under 300 W [148]. However, a processing rate of up to 1800–5400 cm^2/min in a multibeam scanning solution can be achieved [148,149].

LST by straight hatch lines is the most suitable strategy for the polygon scanner technology (Table 1, rows 13–16, 20, 22). In this case, the laser beam scanning speed achieves high values, up to 800–2000 m/s [42]. The polygon scanner has a throughput several times higher in comparison to galvanometer scanners [150]. However, providing LST with a polygon scanner needs to involve a correlation between mirror position and laser pulse generation for the precise formation of micro-objects. This imposes a restriction on the scanning speed for LST. The maximal processing rates up to 7680 cm^2/min were achieved for a linear texture, and this was performed at 320 m/s [77]. However, for polygon scanners, the problem of processing arrays of micro-objects with specific geometry remains unresolved [151]. It is difficult to control the laser drilling of micro-objects with high-speed scanning, because there are substantial data in relation to large arrays with small objects or micro-objects, up to 800 MB per second [130]. Additionally, there is not enough time for precise control of laser spot distribution inside every micro-object in the array. Moreover, ultra-high-speed laser beam processing with polygon scanning involves artefacts such as jitter, banding, bow and other problems characteristic of these systems. These artefacts involve two components: periodical and random. There are several hardware techniques for reducing polygon scanner artefacts, but known classic methods of laser beam processing of the array of objects in ultra-fast scanning systems do not have a fully finished solution to the mentioned problems, and scanning techniques must still be improved [151,152].

The galvanometer scanner can create curved lines purposefully with the fast swinging of two deflection mirrors (Figure 4a). This technique was applied in direct laser formation of an array of micro-objects with different structures (Table 1, rows 6–11). The galvanometer scanner is able to achieve a processing rate of up to 428 cm^2/min for forming an array of micro-objects with a one-beam simple scanning technique at one pulse per object [153]. The high precision formation of surface structures with hatch filling of more complex micro-objects reduces the processing rate to 25 cm^2/min [38]. Noticeably higher processing rates with galvanometer scanners might be achieved by splitting the laser beam into several spots. In this case, a processing rate of up to 5400 cm^2/min is reached [148]. A multibeam solution has potential for industrial applications, especially where there is a need to create a wide array of periodical surface microstructures [116,148].

Table 1. The surface processing rate of different scanning strategies.

	Scanning Strategy	Scanning Technique	Structure Period (μm)	Processing Rate (cm^2/min)	Scanning Speed (m/s)	Reference
1	DLIP-head (ps-laser)	Sample movement	0.343–1.064	100–9000	1	[131–134]
2	DLIP-head (ps-laser)	Galvanometer scanner	0.532–1.064	0.7–10	16·10^{-3}–6.8	[132,139]
3	LIPSS (fs-laser)	Galvanometer scanner	0.9	6.3	3	[142]
4	Path cutting (fs-laser)	Sample movement	0.8	0.01	0.3	[154]
5	Unidirectional scan (ps-laser)	Sample rotation and acousto-optic beam deflection	250	~46.8	1.5 (rotation) 40 (AOD scanning)	[155]
6	Hatch filling (ns-laser)	Galvanometer scanner	12.5–200	1.8–428	0.25–4	[153]
7	Point-by-point ablation (fs-laser)	Galvanometer scanner	30–40	0.4–0.75	~0.05–0.2	[156,157]
8	Hatch filling (ps-laser)	Galvanometer scanner	2000	0.15–0.20	0.5	[158]
9	Path writing (0.1 μs laser)	Galvanometer scanner	50	1.2	0.4–2	[159]
10	Hatch filling (fs-laser)	Galvanometer scanner	4	8–25	4.5–17.1	[38]
11	Interlaced (ps-laser)	Galvanometer scanner	1.2–6	0.017–2	0.024–0.6	[160]
12	Hatch filling (ps-laser)	Multibeam galvanometer scanner	500	5400	20	[148]
13	Hatch filling	Polygon scanner	14.5–40	148–7680	60–800	[77,127,129]
14	Hatch filling (ps-laser)	Polygon scanner	10	840	10–200	[42]
15	Hatch filling (fs-laser)	Polygon scanner	1–12	0.03, approx. 60	25	[161]
16	Hatch filling (fs, ps-laser)	Polygon scanner	40	43	15	[162]
17	Laser pulse pattern	Sample movement with mask	20	1800	–	[149,163]
18	Shifted path (ps-laser)	Galvanometer scanner	200	17.4	8	[102,164]
19	Shifted burst (ps-laser)	Galvanometer scanner	60–570	160	8	[102,113]
20	Unidirectional hatch	Polygon scanner and self-organizing	≲0.5	1510	560	[165]
21	Hatch filling with multibeam	Galvanometer scanner with DOE	~0.4	1910	9	[116]
22	Hatch filling with ns-laser	Polygon scanner	50	1386	200	[166]

4. Scanning Strategies

4.1. Classic Methods of Laser Beam Scanning

In the classic methods, the precision of laser surface machining is reached by continuous control of the laser beam movement. Mirrors' inertia in galvanometer scan heads requires additional time for acceleration and deceleration. Incorrect delays in laser switching on and off lead to floating of the overlapping at the edges of the scanning paths (Figure 5a). Corrections on the laser path edges are provided by sky-writing or by additional synchronization between laser mirror position and laser pulse generation. A strong correlation between laser pulse delivery and scanning mirrors' position improves the precision of LST close to 1–2 µm [103,129] (Figure 5b,c). On the other hand, every additional correction of the scanning parameters may lead to an escalation of the processing time up to 50% [129,167].

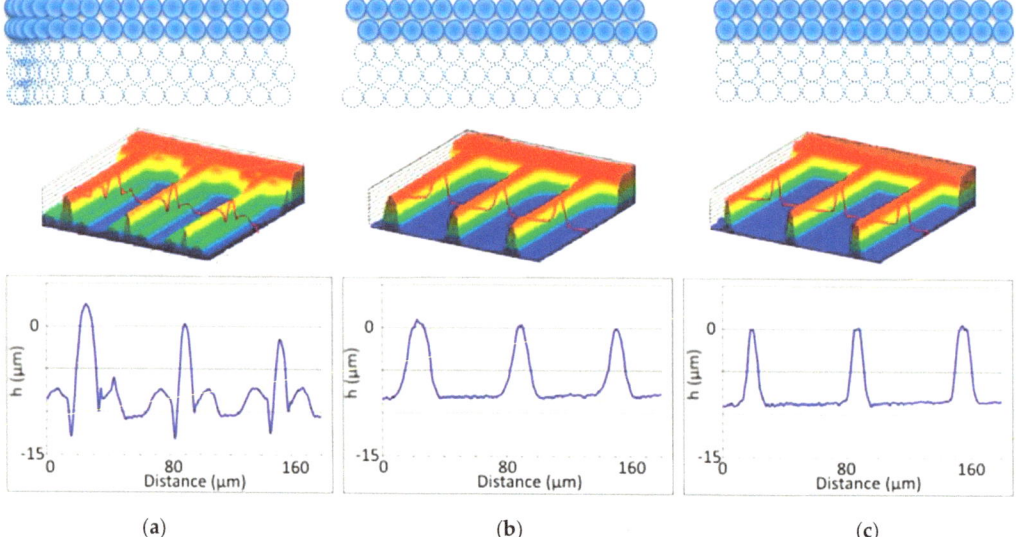

Figure 5. Precision and quality of bar texturing (from top to bottom: laser spots disposition, 3D and section profilometry): (**a**) without sky-writing; (**b**) with sky-writing; (**c**) synchronized system (10 ps, diameter 5.7 µm, 120 mW, 300 kHz, 1 µm pitch, 60 layers) [10,11].

The quality of laser surface machining also depends on the applied laser beam paths' arrangement [161,168,169]. There are several main scanning strategies for filling the laser-textured objects with laser spots: straight hatching, path filling, interlaced filling, criss-cross texturing, unidirectional or bidirectional scanning, angular hatching, wobble scanning and their combinations [103,108,132,160,170–177] (Figure 6).

The correct choice of scanning method significantly affects the efficiency and quality of laser material processing. Dold [176] has provided a detailed study of the influence of the different scanning strategies on the ablation rate, roughness and processing time of the laser surface machining (Figure 7).

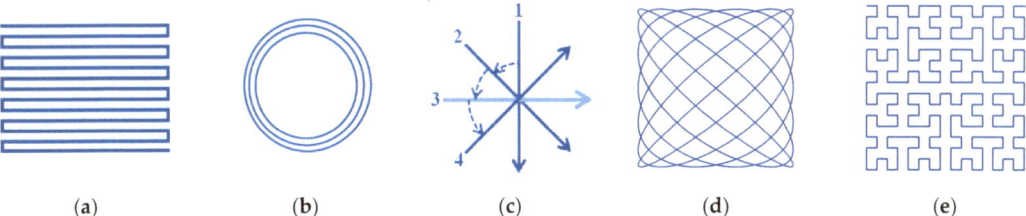

Figure 6. Scanning strategies for laser surface machining with different laser beam movement arrangements (adapted from references): (**a**) straight line hatching [177–179], (**b**) path filling [172], (**c**) angular hatching [175], (**d**) filling by Lissajous carves [176], (**e**) filling by Hilbert curves hatching [178].

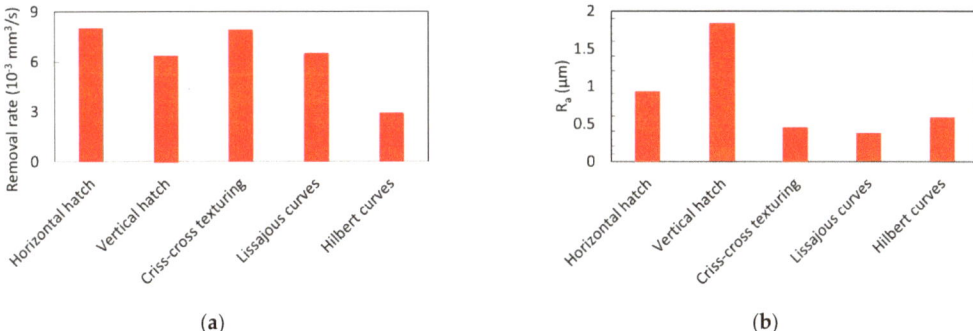

Figure 7. Evaluation of different hatch geometries: (**a**) matter removal rate; (**b**) average roughness analysis (4.5 W, 800 kHz, τ_p = 10 ps, diameter 34 µm, 0.63 J/cm^2, hatch overlap 1.7 µm, 25 scans. Adapted from [176]).

It was shown that the highest efficiency of laser surface processing with scanning by straight hatching lines has an efficiency more than two times higher in comparison to fractal filling. The quality of a laser-processed surface depends not just on the laser beam filling strategy, but on the direction of the scanned lines, i.e., whether it is in a vertical or horizontal direction. This distinguishing feature of horizontal and vertical directions can be explained by the difference in the dynamics of X-scanning and Y-scanning galvanometer mirrors.

All of the presented scanning strategies have their advantages and disadvantages. A discussion of a full long list of their variants and combinations would be inefficient. In the next part of this paper, the scanning strategies, which are widely used in the experimental realizations of LST, are discussed. Additionally, some special scanning strategies are discussed, which were developed for overcoming undesired heat accumulation or for achieving special conditions for LIPSS formation.

4.2. Classic Strategies of Micro-Object Formation in LST

In classic LST, generally, there are two most known scanning strategies of micro-object formation: path and hatch filling (Table 1). The first classic-path filling strategy of a micro-object array formation is similar to helical scanning with several concentric circles (Figure 8a). The dimple texturing is performed consecutively, as every next dimple in the array is formed only after all the filling paths inside the previous dimple have been completely finished. The short length of the laser beam paths inside every micro-object can increase the interline heat accumulation in addition to the inline heat accumulation.

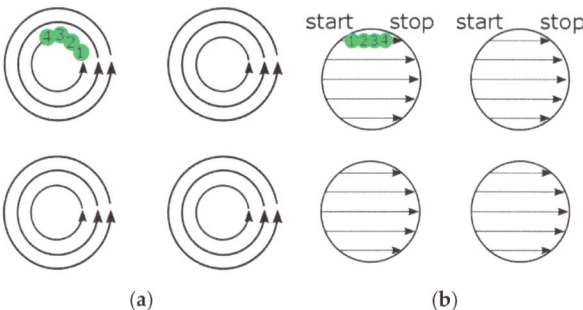

Figure 8. Laser pulse distribution for classic scanning strategies for laser writing of micro-objects: (**a**) path filling; (**b**) hatching. Numbers 1–4 show successive laser pulses.

The second classic strategy is hatch scanning of micro-objects in an LST array. The micro-objects on the textured surface are formed by straight scanning of the laser beam through all the micro-objects in one row (Figure 8b). This means that one hatching line belongs to the several dimples in the scanned LST array in one horizontal direction. The next hatching line is started only after fully finishing the writing of the previous hatching line in a row. Inline laser pulse generation should be stopped after writing one hatching segment and started again on the next dimple hatching segment.

In both these classic strategies, the processing rate v_{PR} is decreased by "on-fly" synchronization between mirrors' position and laser pulse generation. In the case of the classic path filling method, the inertia of the galvanometer mirrors becomes a principal limitation of scanning speed [180]. The application of the classic LST methods with high repetition rate lasers is additionally limited by physical effects, such as heat accumulation, plasma shielding effects and non-effective laser spots distance. Moreover, there are some technical limitations of the application of the classic LST methods at high scanning speed: low precision and a large amount of data needing to be processed in a short time [103,130,180].

4.3. Interlaced Method of Laser Beam Scanning

The distance between hatching lines and their ordering becomes important in the case of laser machining of temperature sensitive materials, for example, in selective laser melting (SLM) technology, composite materials treatment and the formation of biocompatible structures [25,41,181–183]. Heat accumulation between inline laser spots is not the only aspect responsible for thermally damaged results [184,185]. There are several types of heat accumulation leading to material damage: pulse-to-pulse-, rerun-, and geometry-density heat accumulation [14,186,187]. Overcoming the heat accumulation at high repetition rates and high speeds of laser surface processing can be achieved with the interlaced method. In this interlaced method, the scanning lines do not have sequential ordering (Figure 9) [108,122,160,188,189]. The application of the interlaced scanning method is able to improve the ablation rate from 4 up to 13 times in comparison to the classic sequential method (Figure 9a) [160].

The interlaced method of laser beam surface scanning of stainless steel was studied by Neuenschwander et al. [108]. With the interlaced method, the time interval between overlapping scanning lines is not equal to the period of the laser lines scanning, but it is given by the full scanning time of one area. The classic sequential surface scanning method initiates heat accumulation, and the processed surface is damaged by cavity formation (Figure 10b). Unlike the classic method, a surface machined via the interlaced method with the same spot distance between two neighbouring spots shows a good surface quality (Figure 10c).

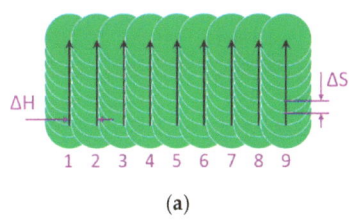 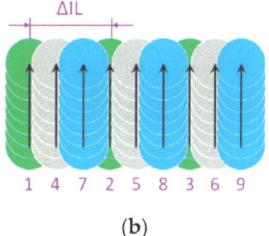

(a)   (b)

Figure 9. Straight line laser scanning strategies: (**a**) classic unidirectional sequential; (**b**) unidirectional interlaced. Meaning of symbols: ΔH—hatch distance; ΔS—distance between pulses in the laser beam scan direction; ΔIL—interlace distance; Numbers 1–9 show successive laser paths [160].

Figure 10. Application of the interlaced scanning method with a polygon scanner: (**a**) increasing the ablation rate above four times as compared to the classic sequential method (adapted from [160]); (**b**) surface machined with a pitch 4.9 μm with the sequential method; (**c**) surface machined with a pitch 9.8 μm with four interlaced patterns (reproduced from [108], with the permission of AIP Publishing).

It can be concluded that the interlaced method has great potential for laser surface processing with high repetition rates of several tens of MHz and high scanning speeds. For overcoming heat accumulation, the lateral distribution of the laser spots should be comparable to the laser spot diameter. It requires great scanning speeds, i.e., several hundred meters per second. At such scanning speeds, laser switching for controllable pulse picking will definitely not be possible anymore [108,173].

4.4. Shifted Laser Surface Texturing

The physical and technical limitations of the classic and interlaced strategies [160,173] can be overcome by using an asynchronous surface scanning method. Shifted laser surface texturing (sLST) is an asynchronous scanning strategy which was developed for faster laser writing of a wide array of repeating micro-objects [114,164]. The algorithm of sLST can be explained in an example with an LST array of triangles. Scanning is performed on straight lines, and asynchronous laser pulsing has a continuous character during the processing of the whole scanning line. In this approach, the laser pulses are rapidly rasterized on the whole processed surface by applying only one laser spot per one micro-object in the array (Figure 11a). The scanning raster is slightly shifted on the surface at every next application of the raster (Figure 11b). The sequence of shifts along a triangular shape produces an array of triangular micro-objects (Figure 11c).

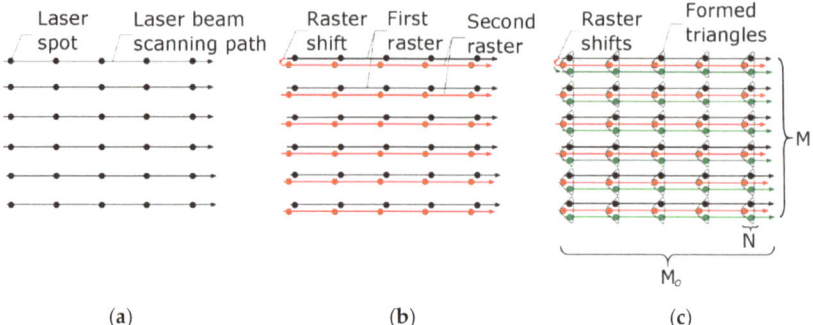

Figure 11. The shifted LST method of triangle array formation: (**a**) linear raster with one spot per one micro-object location; (**b**) one small shift in the linear raster to the next position; (**c**) formation of triangular objects in an array by two sequenced raster shifts.

In the case of larger objects forming, a longer series of laser pulses can be applied instead of pulse-by-pulse laser surface processing. In this case, one laser scan raster contains an area of straight line segments (Figure 12). The classical strategies of LST need to have continuous synchronization between mirrors and laser switching for every segment in the scanned line (Figure 12a). Unlike this, in the asynchronous burst sLST method, the position of the straight segments is determined indirectly by the scanning speed of the laser beam and by the period between the bursts (Figure 12b). The form of the laser-written micro-objects is defined by the sequence of shifts in the linear rasters, which is similar to the one-spot-per-one-object strategy sLST (Figure 11) [114,164]. K. Ratautas et al. [190] have shown that the shifted method of laser spot distribution decreases heat accumulation several times and that such a method is suitable for high-temperature-sensitive materials.

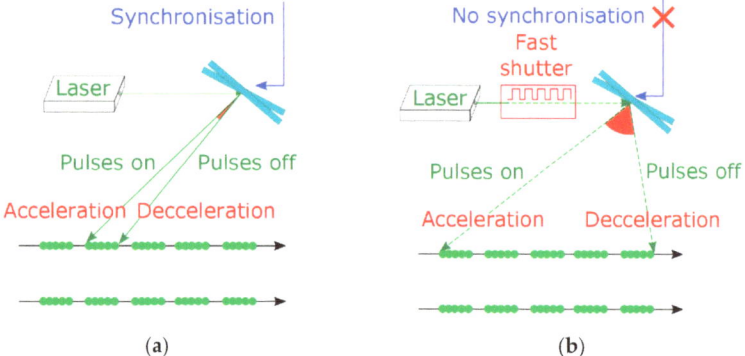

Figure 12. Equidistant straight segments formed by: (**a**) classic hatch strategy, in which it is necessary to control the mirror position and laser switching for every object in the array; (**b**) shifted LST in burst regime, in which there is no need to know the position of every object. Synchronization between laser and mirrors is provided only at the start and finish positions on the scanning field [113].

It was shown by M. Gafner, S. M. Remund et al. [122] that the synchronized scanning method is able to achieve a higher precision at the cost of scanning speed. Higher scanning speeds with high precision can be achieved via the simultaneous application of the synchronized mirrors' movement on raster line ends with laser spot distribution according to the shifted method (Figure 12b) [102,113,122,190]. However, in this case, feedback communication between the laser source and scanning system for every laser spot position puts a limit on the maximal scanning speed. The novelty of the sLST method lies in the combination of two principal properties of the laser scanning system: the inertia of the deflection mirrors and the stable frequency of the laser pulse generator. In this method, the laser works as an asynchronous source, and correction of the first laser spot position is needed only at the starting position of the scanning mirror.

4.5. Scanning Strategies Combined with LIPSS Formation

Laser surface processing with application wave-optical effects such as multibeam interference and diffraction beam modulation in LST gives us the possibility of achieving a resolution at a level of detail close to the optical limit (Figure 13a) [73,132,133,140,191,192]. DLIP has provided one of the best results, showing the possibility to form a surface pattern with 180 nm detailing. It is lower than the wavelength of the applied laser (266 nm), and it is already very close to the theoretical Abbe optical limitation of $\lambda/2$ [193]. Through the application of immersion techniques, detailing of features with a size down to $\lambda/4n$ is possible, where n is the refractive index of the surrounding medium [194].

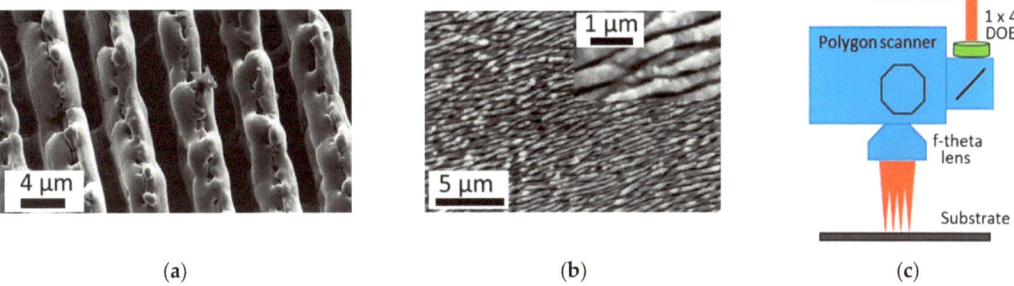

Figure 13. Fast laser surface processing: (**a**) close to the optical limit (bidirectional hatch) [168], (**b**) self-organized nanostructure [195], (**c**) multi-beam high-speed scanning technology [165].

The application of laser-initiated self-organized processes allows us to overcome the wavelength principal limit by achieving surface nano-structuring and a remarkable increase in the processing rate (Figure 13b) [116,165,195–197]. The combination of wave-optical effects with one of the known laser surface scanning techniques has great potential for the application of LST on a wide surface area (Figure 13c) [165,198,199]. For example, the papers of Lasagni et al. [131–134,200] introduced a combination of hatch-scanning techniques and interference patterning, where the processing rate achieved 0.9 m^2/min. A higher processing rate and smaller period of processed relief can be achieved in experiments on surface texturing by forming self-organized LIPSS structures [195,201]. Schiele et al. reported the achievement of a processing rate up to 1.5 m^2/min with self-organized LIPSS nanostructures on metallic surfaces [42,165].

The application of a multispot technique in combination with scanning technology is a helpful method for decreasing the thermal load on material with an expanding processing area [122,165]. There are two main methods of multibeam forming: Diffractive Optic Element (DOE) and Spatial Light Modulation (SLM) [202]. In the case of the application of DOE, the periodicity of multibeam spot distribution is fixed, and the flexibility of an applied laser scanning strategy becomes limited by diffraction spot distribution [73,122]. On the other hand, SLM units are able to change laser spot distribution during scanning processes with a frame rate of 60–180 Hz [202,203]. However, the application of SLM units has limitations if the maximal laser beam power is above 200 W [148]. Improving the existing scanning strategies with LIPSS and their combination with multibeam technology looks like the most promising way for reaching industrial scales of LST [116,122,165,204].

5. Conclusions

In this paper, classical and recently developed scanning strategies utilizing laser surface texturing (LST) were compared. In the beginning, plasma and particle shielding effects and heat accumulation were described as basic physical limitations of short and ultrashort LST. Different methods of heat accumulation evaluation for the optimization of scanning strategies were briefly discussed. Several methods of laser beam movement arrangements contained in the scanning strategies were discussed, such as straight hatching, path filing, criss-cross texturing, and Lissajous or Hilbert curves filling. The main attention was paid to the classical path and straight hatch filling strategies coupled with galvanometer and polygon scanning techniques as the most known strategies. It was observed that classical scanning strategies of LST are limited in terms of processing rate by the mentioned physical limitations. Alternative, recently developed scanning strategies with high-processing-rate LST were discussed, namely, the interlaced method and shifted LST. It was shown that a combination of several techniques, such as multibeam processing, asynchronous shifted LST strategy and LIPSS formation, can offer a way to achieve a higher processing rate in LST. The next step in achieving the high standards of Industry 4.0 can be the application of the on-fly non-contact IR control of the temperature regime in laser surface processing coupled with a high-speed scanning technique.

Author Contributions: Conceptualization, J.M. and D.M.; methodology, D.M.; investigation, D.M.; validation, J.M. and M.H.; formal analysis, D.M.; writing—original draft preparation, D.M.; writing—review and editing, J.M. and D.M.; visualization, D.M.; supervision, J.M. and M.H.; project administration, J.M. and M.H.; funding acquisition, J.M. and M.H. All authors have read and agreed to the published version of the manuscript.

Funding: The work has been supported by the Technology Agency of the Czech Republic (M-era.Net project ADVENTURE, No. TH75020001) and the Ministry of Education, Youth and Sports of the Czech Republic (OP RDE program, LABIR-PAV project, No. CZ.02.1.01/0.0/0.0/18_069/0010018).

Data Availability Statement: No new data were created or analyzed in this study. Data sharing is not applicable to this article.

Conflicts of Interest: Authors D. Moskal and J. Martan are inventors of the patent: M. Kucera, D. Moskal, J. Martan, Method of laser beam writing with shifted laser surface texturing, patent, US10160229, WO2016189344, CZ308932, DE112015006574, 2015.

References

1. He, F.; Liao, Y.; Lin, J.; Song, J.; Qiao, L.; Cheng, Y.; Sugioka, K. Femtosecond Laser Fabrication of Monolithically Integrated Microfluidic Sensors in Glass. *Sensors* **2014**, *14*, 19402–19440. [CrossRef] [PubMed]
2. Yang, T.; Lin, H.; Jia, B. Two-dimensional material functional devices enabled by direct laser fabrication. *Front. Optoelectron.* **2018**, *11*, 2–22. [CrossRef]
3. Etsion, I.; Halperin, G.; Becker, E. The effect of various surface treatments on piston pin scuffing resistance. *Wear* **2006**, *261*, 785–791. [CrossRef]
4. Schille, J.; Loeschner, U.; Ebert, R.; Scully, P.; Goddard, N.; Exner, H. Laser micro processing using a high repetition rate femto second laser. In Proceedings of the International Congress on Applications of Lasers and Electro-Optics, Wuhan, China, 23–25 March 2010; pp. 1491–1499. [CrossRef]
5. Zhang, L.; Lin, N.; Zou, J.; Lin, X.; Liu, Z.; Yuan, S.; Yu, Y.; Wang, Z.; Zeng, Q.; Chen, W.; et al. Super-hydrophobicity and corrosion resistance of laser surface textured AISI 304 stainless steel decorated with Hexadecyltrimethoxysilane (HDTMS). *Opt. Laser Technol.* **2020**, *127*, 106146. [CrossRef]
6. Chen, X.; Lei, Z.; Chen, Y.; Jiang, M.; Jiang, N.; Bi, J.; Lin, S. Enhanced wetting behavior using femtosecond laser-textured surface in laser welding-brazing of Ti/Al butt joint. *Opt. Laser Technol.* **2021**, *142*, 107212. [CrossRef]
7. He, C.; Yang, S.; Zheng, M. Analysis of synergistic friction reduction effect on micro-textured cemented carbide surface by laser processing. *Opt. Laser Technol.* **2022**, *155*, 108343. [CrossRef]
8. Luo, X.; Tian, Z.; Chen, C.; Jiang, G.; Hu, X.; Wang, L.; Peng, R.; Zhang, H.; Zhong, M. Laser-textured High-throughput Hydrophobic/Superhydrophobic SERS platform for fish drugs residue detection. *Opt. Laser Technol.* **2022**, *152*, 108075. [CrossRef]
9. Martan, J.; Moskal, D.; Smeták, L.; Honner, M. Performance and Accuracy of the Shifted Laser Surface Texturing Method. *Micromachines* **2020**, *11*, 520. [CrossRef]

10. Jaeggi, B.; Neuenschwander, B.; Hunziker, U.; Zuercher, J.; Meier, T.; Zimmermann, M.; Hennig, G. High precision and high throughput surface structuring by synchronizing mechanical axes with an ultra short pulsed laser system in MOPA arrangement. In Proceedings of the International Congress on Applications of Lasers and Electro-Optics, Orlando, FL, USA, 14–18 October 2012; pp. 1046–1053. [CrossRef]
11. Jaeggi, B.; Neuenschwander, B.; Meier, T.; Zimmermann, M.; Hennig, G. High precision surface structuring with ultra-short laser pulses and synchronized mechanical axes. *Phys. Procedia* **2013**, *41*, 319–326. [CrossRef]
12. Račiukaitis, G.; Brikas, M.; Gečys, P.; Voisiat, B.; Gedvilas, M. Use of high repetition rate and high power lasers in microfabrication: How to keep the efficiency high? *J. Laser Micro/Nanoeng.* **2009**, *4*, 186–191. [CrossRef]
13. Neuenschwander, B.; Jaeggi, B.; Schmid, M.; Hennig, G. Surface structuring with ultra-short laser pulses: Basics, limitations and needs for high throughput. *Phys. Procedia* **2014**, *56*, 1047–1058. [CrossRef]
14. Faas, S.; Bielke, U.; Weber, R.; Graf, T. Prediction of the surface structures resulting from heat accumulation during processing with picosecond laser pulses at the average power of 420 W. *Appl. Phys. A* **2018**, *124*, 612. [CrossRef]
15. Förster, D.J.; Faas, S.; Gröninger, S.; Bauer, F.; Michalowski, A.; Weber, R.; Graf, T. Shielding effects and re-deposition of material during processing of metals with bursts of ultra-short laser pulses. *Appl. Surf. Sci.* **2018**, *440*, 926–931. [CrossRef]
16. Itina, T.; Povarnitsyn, M.; Khishchenko, K. *Laser Ablation: Effects and Applications*; Black, S.E., Ed.; Springer Science & Business Media: Berlin/Heidelberg, Germany, 2011; pp. 99–125.
17. Xu, D.; Wu, W.; Malhotra, R.; Chen, J.; Lu, B.; Cao, J. Mechanism investigation for the influence of tool rotation and laser surface texturing (LST) on formability in single point incremental forming. *Int. J. Mach. Tools Manuf.* **2013**, *73*, 37–46. [CrossRef]
18. Batal, A.; Michalek, A.; Penchev, P.; Kupisiewicz, A.; Dimov, S. Laser processing of freeform surfaces: A new approach based on an efficient workpiece partitioning strategy. *Int. J. Mach. Tools Manuf.* **2020**, *156*, 103593. [CrossRef]
19. Sugioka, K. Progress in ultrafast laser processing and future prospects. *Nanophotonics* **2017**, *6*, 393–413. [CrossRef]
20. Khan, A.; Wang, Z.; Sheikh, M.A.; Whitehead, D.J.; Li, L. Laser micro/nano patterning of hydrophobic surface by contact particle lens array. *Appl. Surf. Sci.* **2011**, *258*, 774–779. [CrossRef]
21. Li, L.; Guo, W.; Wang, Z.B.; Liu, Z.; Whitehead, D.; Luk'yanchuk, B. Large-area laser nano-texturing with user-defined patterns. *J. Micromech. Microeng.* **2009**, *19*, 054002. [CrossRef]
22. Lai, N.D.; Lin, J.H.; Huang, Y.Y.; Hsu, C.C. Fabrication of two- and three-dimensional quasi-periodic structures with 12-fold symmetry by interference technique. *Opt. Express* **2006**, *14*, 10746. [CrossRef]
23. Georgiev, D.G.; Baird, R.J.; Avrutsky, I.; Auner, G.; Newaz, G. Controllable excimer-laser fabrication of conical nano-tips on silicon thin films. *Appl. Phys. Lett.* **2004**, *84*, 4881–4883. [CrossRef]
24. Simon, P.; Ihlemann, J. Ablation of submicron structures on metals and semiconductors by femtosecond UV-laser pulses. *Appl. Surf. Sci.* **1997**, *109–110*, 25–29. [CrossRef]
25. Marattukalam, J.J.; Karlsson, D.; Pacheco, V.; Beran, P.; Wiklund, U.; Jansson, U.; Hjörvarsson, B.; Sahlberg, M. The effect of laser scanning strategies on texture, mechanical properties, and site-specific grain orientation in selective laser melted 316L SS. *Mater. Des.* **2020**, *193*, 108852. [CrossRef]
26. Mugwagwa, L.; Dimitrov, D.; Matope, S.; Yadroitsev, I. Evaluation of the impact of scanning strategies on residual stresses in selective laser melting. *Int. J. Adv. Manuf. Technol.* **2019**, *102*, 2441–2450. [CrossRef]
27. Shipley, H.; McDonnell, D.; Culleton, M.; Coull, R.; Lupoi, R.; O'Donnell, G.; Trimble, D. Optimisation of process parameters to address fundamental challenges during selective laser melting of Ti-6Al-4V: A review. *Int. J. Mach. Tools Manuf.* **2018**, *128*, 1–20. [CrossRef]
28. Bornschlegel, B.; Finger, J. In-Situ Analysis of Ultrashort Pulsed Laser Ablation with Pulse Bursts. *J. Laser Micro/Nanoeng.* **2019**, *14*, 88–94. [CrossRef]
29. Ivanov, D.S.; Rethfeld, B. The effect of pulse duration on the interplay of electron heat conduction and electron-phonon interaction: Photo-mechanical versus photo-thermal damage of metal targets. *Appl. Surf. Sci.* **2009**, *255*, 9724–9728. [CrossRef]
30. Mannion, P.; Magee, J.; Coyne, E.; O'Connor, G.; Glynn, T. The effect of damage accumulation behaviour on ablation thresholds and damage morphology in ultrafast laser micro-machining of common metals in air. *Appl. Surf. Sci.* **2004**, *233*, 275–287. [CrossRef]
31. Schille, J.; Schneider, L.; Loeschner, U. Process optimization in high-average-power ultrashort pulse laser microfabrication: How laser process parameters influence efficiency, throughput and quality. *Appl. Phys. A* **2015**, *120*, 847–855. [CrossRef]
32. Moskal, D.; Martan, J.; Honner, M.; Beltrami, C.; Kleefoot, M.-J.; Lang, V. Inverse Dependence of Heat Accumulation on Pulse Duration in Laser Surface Processing with Ultra-Short Pulses. *SSRN Electron. J.* **2022**, *213*, 124328. [CrossRef]
33. Wang, J.; Ma, Y.; Liu, Y.; Yuan, W.; Song, H.; Huang, C.; Yin, X. Experimental investigation on laser ablation of C/SiC composites subjected to supersonic airflow. *Opt. Laser Technol.* **2019**, *113*, 399–406. [CrossRef]
34. Wang, H.; Liu, J.; Xu, Y.; Wang, X.; Ren, N.; Ren, X.; Hu, Q. Experimental characterization and real-time monitoring for laser percussion drilling in titanium alloy using transverse electric field assistance and/or lateral air blowing. *J. Manuf. Process.* **2021**, *62*, 845–858. [CrossRef]
35. Farid, N.; Harilal, S.S.; Ding, H.; Hassanein, A. Dynamics of ultrafast laser plasma expansion in the presence of an ambient. *Appl. Phys. Lett.* **2013**, *103*, 191112. [CrossRef]

36. Mitko, V.S.; Römer, G.R.B.E.; Huis in 't Veld, A.J.; Skolski, J.Z.P.; Obona, J.V.; Ocelík, V.; De Hosson, J.T.M. Properties of High-Frequency Sub-Wavelength Ripples on Stainless Steel 304L under Ultra Short Pulse Laser Irradiation. *Phys. Procedia* **2011**, *12*, 99–104. [CrossRef]
37. Gamaly, E.G.; Rode, A. V Physics of ultra-short laser interaction with matter: From phonon excitation to ultimate transformations. *Prog. Quantum Electron.* **2013**, *37*, 215–323. [CrossRef]
38. Schille, J. Highspeed Laser Micro Processing using Ultrashort Laser Pulses. *J. Laser Micro/Nanoeng.* **2014**, *9*, 161–168. [CrossRef]
39. Neuenschwander, B.; Jaeggi, B.; Zimmermann, M.; Hennig, G. Influence of particle shielding and heat accumulation effects onto the removal rate for laser micromachining with ultra-short pulses at high repetition rates. In Proceedings of the International Congress on Applications of Lasers and Electro-Optics, Orlando, FL, USA, 14–18 October 2018; pp. 218–226. [CrossRef]
40. Bauer, F.; Michalowski, A.; Kiedrowski, T.; Nolte, S. Heat accumulation in ultra-short pulsed scanning laser ablation of metals. *Opt. Express* **2015**, *23*, 1035–1043. [CrossRef] [PubMed]
41. Weber, R.; Graf, T.; Berger, P.; Onuseit, V.; Wiedenmann, M.; Freitag, C.; Feuer, A. Heat accumulation during pulsed laser materials processing. *Opt. Express* **2014**, *22*, 11312. [CrossRef]
42. Schille, J.; Schneider, L.; Streek, A.; Kloetzer, S.; Loeschner, U. High-throughput machining using high average power ultrashort pulse lasers and ultrafast polygon scanner. In Proceedings of the Laser-Based Micro- and Nanoprocessing X, San Francisco, CA, USA, 13–18 February 2016; Klotzbach, U., Washio, K., Arnold, C.B., Eds.; Volume 9736, p. 97360R. [CrossRef]
43. Rethfeld, B.; Sokolowski-Tinten, K.; Von Der Linde, D.; Anisimov, S.I. Timescales in the response of materials to femtosecond laser excitation. *Appl. Phys. A Mater. Sci. Process.* **2004**, *79*, 767–769. [CrossRef]
44. König, J.; Nolte, S.; Tünnermann, A. Plasma evolution during metal ablation with ultrashort laser pulses. *Opt. Express* **2005**, *13*, 10597. [CrossRef]
45. Mildner, J.; Sarpe, C.; Götte, N.; Wollenhaupt, M.; Baumert, T. Applied Surface Science Emission signal enhancement of laser ablation of metals (aluminum and titanium) by time delayed femtosecond double pulses from femtoseconds to nanoseconds. *Appl. Surf. Sci.* **2014**, *302*, 291–298. [CrossRef]
46. Carrasco-García, I.; Vadillo, J.M.; Javier Laserna, J. Visualization of surface transformations during laser ablation of solids by femtosecond pump–probe time-resolved microscopy. *Spectrochim. Acta Part B At. Spectrosc.* **2015**, *113*, 30–36. [CrossRef]
47. Carrasco-García, I.; Vadillo, J.M.; Javier Laserna, J.; Laserna, J.J. Monitoring the dynamics of the surface deformation prior to the onset of plasma emission during femtosecond laser ablation of noble metals by time-resolved reflectivity microscopy. *Spectrochim. Acta Part B At. Spectrosc.* **2017**, *131*, 1–7. [CrossRef]
48. Rethfeld, B.; Sokolowski-Tinten, K.; Temnov, V.V.; Kudryashov, S.I.; Bialkowski, J.; Cavalleri, A.; von der Linde, D. Ablation dynamics of solids heated by femtosecond laser pulses. In *Nonresonant Laser-Matter Interaction (NLMI-10)*; Libenson, M.N., Ed.; SPIE: San Francisco, CA, USA, 2001; Volume 4423, pp. 186–196. [CrossRef]
49. Wu, C.; Zhigilei, L.V. Microscopic mechanisms of laser spallation and ablation of metal targets from large-scale molecular dynamics simulations. *Appl. Phys. A Mater. Sci. Process.* **2014**, *114*, 11–32. [CrossRef]
50. Amoruso, S.; Bruzzese, R.; Pagano, C.; Wang, X. Features of plasma plume evolution and material removal efficiency during femtosecond laser ablation of nickel in high vacuum. *Appl. Phys. A Mater. Sci. Process.* **2007**, *89*, 1017–1024. [CrossRef]
51. Lorazo, P.; Lewis, L.J.; Meunier, M. Thermodynamic pathways to melting, ablation, and solidification in absorbing solids under pulsed laser irradiation. *Phys. Rev. B* **2006**, *73*, 134108. [CrossRef]
52. Sankar, P.; Shashikala, H.D.; Philip, R. Effect of laser beam size on the dynamics of ultrashort laser-produced aluminum plasma in vacuum. *Phys. Plasmas* **2019**, *26*, 013302. [CrossRef]
53. Arnold, N.; Gruber, J.; Heitz, J. Spherical expansion of the vapor plume into ambient gas: An analytical model. *Appl. Phys. A Mater. Sci. Process.* **1999**, *69*, S87–S93. [CrossRef]
54. Trusso, S.; Barletta, E.; Barreca, F.; Fazio, E.; Neri, F. Time resolved imaging studies of the plasma produced by laser ablation of silicon in O2/Ar atmosphere. *Laser Part. Beams* **2005**, *23*, 149–153. [CrossRef]
55. Choi, T.Y.; Grigoropoulos, C.P. Plasma and ablation dynamics in ultrafast laser processing of crystalline silicon. *J. Appl. Phys.* **2002**, *92*, 4918–4925. [CrossRef]
56. Tański, M.; Barbucha, R.; Kocik, M.; Garasz, K.; Mizeraczyk, J. Investigation of the Laser Generated Ablation Plasma Plume Dynamics and Plasma Plume Sound Wave Dynamics. In *Laser Technology 2012: Applications of Lasers*; Woliński, W.L., Jankiewicz, Z., Romaniuk, R.S., Eds.; SPIE: San Francisco, CA, USA, 2013; Volume 8703, p. 87030O. [CrossRef]
57. Gacek, S.; Wang, X. Plume splitting in pico-second laser-material interaction under the influence of shock wave. *Phys. Lett. Sect. A Gen. At. Solid State Phys.* **2009**, *373*, 3342–3349. [CrossRef]
58. Gacek, S.; Wang, X. Dynamics evolution of shock waves in laser-material interaction. *Appl. Phys. A Mater. Sci. Process.* **2009**, *94*, 675–690. [CrossRef]
59. Misra, A.; Thareja, R.K. Investigation of laser ablated plumes using fast photography. *IEEE Trans. Plasma Sci.* **1999**, *27*, 1553–1558. [CrossRef]
60. Diwakar, P.K.; Harilal, S.S.; Phillips, M.C.; Hassanein, A. Characterization of ultrafast laser-ablation plasma plumes at various Ar ambient pressures. *J. Appl. Phys.* **2015**, *118*, 043305. [CrossRef]
61. Keller, W.J.; Shen, N.; Rubenchik, A.M.; Ly, S.; Negres, R.; Raman, R.N.; Yoo, J.-H.; Guss, G.; Stolken, J.S.; Matthews, M.J.; et al. Physics of picosecond pulse laser ablation. *J. Appl. Phys.* **2019**, *125*, 085103. [CrossRef]

62. Olsson, R.; Powell, J.; Palmquist, A.; Brånemark, R.; Frostevarg, J.; Kaplan, A.F.H. Production of osseointegrating (bone bonding) surfaces on titanium screws by laser melt disruption. *J. Laser Appl.* **2018**, *30*, 042009. [CrossRef]
63. Ullmann, F.; Loeschner, U.; Hartwig, L.; Szczepanski, D.; Schille, J.; Gronau, S.; Knebel, T.; Drechsel, J.; Ebert, R.; Exner, H. Highspeed Laser Ablation Cutting of Metal. In *High-Power Laser Materials Processing: Lasers, Beam Delivery, Diagnostics, and Applications II*; Dorsch, F., Ed.; SPIE: San Francisco, CA, USA, 2013; Volume 8603, p. 860311. [CrossRef]
64. Semerok, A.; Sallé, B.; Wagner, J.-F.; Petite, G. Femtosecond, picosecond, and nanosecond laser microablation: Laser plasma and crater investigation. *Laser Part. Beams* **2002**, *20*, 67–72. [CrossRef]
65. Diwakar, P.K.; Harilal, S.S.; Hassanein, A.; Phillips, M.C. Expansion dynamics of ultrafast laser produced plasmas in the presence of ambient argon. *J. Appl. Phys.* **2014**, *116*, 133301. [CrossRef]
66. Povarnitsyn, M.E.; Itina, T.E.; Khishchenko, K.V.; Levashov, P.R. Suppression of ablation in femtosecond double-pulse experiments. *Phys. Rev. Lett.* **2009**, *103*, 195002. [CrossRef] [PubMed]
67. Letan, A.; Audouard, E.; Mishchik, K.; Hönninger, C.; Mottay, E. Use of Bursts for Femtosecond Ablation Efficiency Increase. In *Lasers in Manufacturing*; John Wiley & Sons: Hoboken, NJ, USA, 2019; pp. 1–6.
68. Kerse, C.; Kalaycloĝ Lu, H.; Elahi, P.; Çetin, B.; Kesim, D.K.; Akçaalan, Ö.; Yavaş, S.; Aşlk, M.D.; Öktem, B.; Hoogland, H.; et al. Ablation-cooled material removal with ultrafast bursts of pulses. *Nature* **2016**, *537*, 84–88. [CrossRef]
69. Gaudiuso, C.; Giannuzzi, G.; Volpe, A.; Lugarà, P.M.; Choquet, I.; Ancona, A. Incubation during laser ablation with bursts of femtosecond pulses with picosecond delays. *Opt. Express* **2018**, *26*, 8958–8968. [CrossRef]
70. Moskal, D.; Martan, J.; Smazalová, E.; Houdková, Š. Influence of initial surface state on laser surface texturing result. In Proceedings of the METAL 2016-25th Anniversary International Conference on Metallurgy and Materials, Brno, Czech Republic, 25–27 May 2016.
71. Fang, Z.; Zhou, T.; Perrie, W.; Bilton, M.; Schille, J.; Löschner, U.; Edwardson, S.; Dearden, G. Pulse Burst Generation and Diffraction with Spatial Light Modulators for Dynamic Ultrafast Laser Materials Processing. *Materials* **2022**, *15*, 9059. [CrossRef] [PubMed]
72. Mur, J.; Petkovšek, R. Near-THz bursts of pulses—Governing surface ablation mechanisms for laser material processing. *Appl. Surf. Sci.* **2019**, *478*, 355–360. [CrossRef]
73. Bruening, S.; Du, K.; Jarczynski, M.; Jenke, G.; Gillner, A. Ultra-fast laser micro processing by multiple laser spots. *Procedia CIRP* **2018**, *74*, 573–580. [CrossRef]
74. Kuang, Z.; Perrie, W.; Leach, J.; Sharp, M.; Edwardson, S.P.; Padgett, M.; Dearden, G.; Watkins, K.G. High throughput diffractive multi-beam femtosecond laser processing using a spatial light modulator. *Appl. Surf. Sci.* **2008**, *255*, 2284–2289. [CrossRef]
75. Weber, R.; Graf, T.; Freitag, C.; Feuer, A.; Kononenko, T.; Konov, V.I. Processing constraints resulting from heat accumulation during pulsed and repetitive laser materials processing. *Opt. Express* **2017**, *25*, 3966. [CrossRef]
76. Martan, J.; Prokešová, L.; Moskal, D.; Ferreira de Faria, B.C.; Honner, M.; Lang, V. Heat accumulation temperature measurement in ultrashort pulse laser micromachining. *Int. J. Heat Mass Transf.* **2021**, *168*, 120866. [CrossRef]
77. Loeschner, U.; Schille, J.; Streek, A.; Knebel, T.; Hartwig, L.; Hillmann, R.; Endisch, C. High-rate laser microprocessing using a polygon scanner system. *J. Laser Appl.* **2015**, *27*, S29303. [CrossRef]
78. Fraggelakis, F.; Mincuzzi, G.; Lopez, J.; Manek-Hönninger, I.; Kling, R. Texturing metal surface with MHz ultra-short laser pulses. *Opt. Express* **2017**, *25*, 18131. [CrossRef]
79. Jaeggi, B.; Neuenschwander, B.; Schmid, M.; Muralt, M.; Zuercher, J.; Hunziker, U. Influence of the pulse duration in the ps-regime on the ablation efficiency of metals. *Phys. Procedia* **2011**, *12*, 164–171. [CrossRef]
80. Kramer, T. Increasing the Specific Removal Rate for Ultra Short Pulsed Laser-Micromachining by Using Pulse Bursts. *J. Laser Micro/Nanoeng.* **2017**, *12*, 107–114. [CrossRef]
81. Schille, J.; Schneider, L.; Hartwig, L.; Loeschner, U. High-rate laser processing of metals using high-average power ultrashort pulse lasers. In Proceedings of the 38th International MATADOR Conference, Yunlin, Taiwan, 28–30 March 2015; pp. 135–152.
82. Jaeggi, B.; Remund, S.; Streubel, R.; Goekce, B. Laser Micromachining of Metals with Ultra-Short Pulses: Factors Limiting the Laser Micromachining of Metals with Ultra-Short Pulses: Factors Limiting the Scale-Up Process. *J. Laser Micro/Nanoeng.* **2017**, *12*. [CrossRef]
83. Moskal, D.; Martan, J.; Kučera, M.; Houdková, Š.; Kromer, R. Picosecond laser surface cleaning of AM1 superalloy. *Phys. Procedia* **2016**, *83*, 249–257. [CrossRef]
84. Di Niso, F.; Gaudiuso, C.; Sibillano, T.; Mezzapesa, F.P.; Ancona, A.; Lugarà, P.M. Role of heat accumulation on the incubation effect in multi-shot laser ablation of stainless steel at high repetition rates. *Opt. Express* **2014**, *22*, 12200–12210. [CrossRef]
85. Förster, D.J.; Weber, R.; Graf, T. Heat accumulation effects on efficiency during laser drilling of metals. In Proceedings of the Lasers in Manufacturing Conference, München, Germany, 26–29 June 2017.
86. Bornschlegel, B.; Köller, J.; Finger, J. In-situ analysis of heat accumulation during ultrashort pulsed laser ablation. *J. Laser Micro/Nanoeng.* **2020**, *15*, 56–62. [CrossRef]
87. Ancona, A.; Röser, F.; Rademaker, K.; Limpert, J.; Nolte, S.; Tunnermann, A. High speed laser drilling of metals using a high repetition rate, high average power ultrafast fiber CPA system. *Opt. Express* **2008**, *16*, 8958–8968. [CrossRef]
88. Taylor, L.L.; Scott, R.E.; Qiao, J. Integrating two-temperature and classical heat accumulation models to predict femtosecond laser processing of silicon. *Opt. Mater. Express* **2018**, *8*, 648. [CrossRef]

89. Bulgakova, N.M.; Zhukov, V.P.; Mirza, I.; Meshcheryakov, Y.P.; Tomáštík, J.; Michálek, V.; Haderka, O.; Fekete, L.; Rubenchik, A.M.; Fedoruk, M.P.; et al. Ultrashort-pulse laser processing of transparent materials: Insight from numerical and semi-analytical models. In Proceedings of the Laser Applications in Microelectronic and Optoelectronic Manufacturing (LAMOM) XXI, San Francisco, CA, USA, 13–18 February 2016; Volume 9735, p. 97350N. [CrossRef]
90. Zhanwen, A.; Zou, G.; Wu, Y.; Wu, Y.; Feng, B.; Xiao, Y.; Huo, J.; Jia, Q.; Du, C.; Liu, L. Temporal and spatial heat input regulation strategy for high-throughput micro-drilling based on multi-beam ultrafast laser. *Opt. Laser Technol.* **2022**, *155*, 108424. [CrossRef]
91. Shugaev, M.V.; Gnilitskyi, I.; Bulgakova, N.M.; Zhigilei, L.V. Mechanism of single-pulse ablative generation of laser-induced periodic surface structures. *Phys. Rev. B* **2017**, *96*, 205429. [CrossRef]
92. Wu, C.; Christensen, M.S.; Savolainen, J.M.; Balling, P.; Zhigilei, L.V. Generation of subsurface voids and a nanocrystalline surface layer in femtosecond laser irradiation of a single-crystal Ag target. *Phys. Rev. B Condens. Matter Mater. Phys.* **2015**, *91*, 035413. [CrossRef]
93. Bulgakova, N.; Panchenko, A.; Zhukov, V.; Kudryashov, S.; Pereira, A.; Marine, W.; Mocek, T.; Bulgakov, A. Impacts of Ambient and Ablation Plasmas on Short- and Ultrashort-Pulse Laser Processing of Surfaces. *Micromachines* **2014**, *5*, 1344–1372. [CrossRef]
94. Lee, J.M.; Watkins, K.G. In-process monitoring techniques for laser cleaning. *Opt. Lasers Eng.* **2000**, *34*, 429–442. [CrossRef]
95. Verdier, M.; Costil, S.; Coddet, C.; Oltra, R.; Perret, O. On the topographic and energetic surface modifications induced by laser treatment of metallic substrates before plasma spraying. *Appl. Surf. Sci.* **2003**, *205*, 3–21. [CrossRef]
96. Křenek, T.; Pola, J.; Docheva, D.; Stich, T.; Fajgar, R.; Kovářík, T.; Pola, M.; Martan, J.; Moskal, D.; Jandová, V.; et al. Porous micro/nano structured oxidic titanium surface decorated with silicon monoxide. *Surf. Interfaces* **2021**, *26*, 101304. [CrossRef]
97. Afanasiev, Y.V.; Chichkov, B.N.; Demchenko, N.N.; Isakov, V.A.; Zavestovskaya, I.N. Ablation of metals by ultrashort laser pulses: Theoretical modeling and computer simulations. *J. Russ. Laser Res.* **1999**, *20*, 89–115. [CrossRef]
98. Dausinger, F.; Hugel, H.; Konov, V.I. Micromachining with ultrashort laser pulses: From basic understanding to technical applications. In Proceedings of the ALT'02 International Conference on Advanced Laser Technologies, Adelboden, Switzerland, 15–20 September 2002; Weber, H.P., Konov, V.I., Graf, T., Eds.; Volume 5147, pp. 106–115. [CrossRef]
99. Sedao, X.; Lenci, M.; Rudenko, A.; Faure, N.; Pascale-Hamri, A.; Colombier, J.P.; Mauclair, C. Influence of pulse repetition rate on morphology and material removal rate of ultrafast laser ablated metallic surfaces. *Opt. Lasers Eng.* **2019**, *116*, 68–74. [CrossRef]
100. Rykalin, N. *Rascety Teplovych Processov Pri Svarke*; Jerohin, A., Ed.; MASGIZ: Moscow, Russia, 1951.
101. Carslaw, H.S.; Jaeger, J. *Conduction of Heat in Solids*, 2nd ed.; Oxford at the Clarendon Press: London, UK, 1959.
102. Moskal, D. Thermo-Physical Processes and Ultrashost Pulse Laser Scanning Methods in Surface Texturing. Ph.D. Thesis, University of West Bohemia, Plzeň, Czech Republic, 2019.
103. Zimmermann, M.; Jaeggi, B.; Neuenschwander, B. Improvements in ultra-high precision surface structuring using synchronized galvo or polygon scanner with a laser system in MOPA arrangement. In Proceedings of the Laser Applications in Microelectronic and Optoelectronic Manufacturing (Lamom) XX, San Francisco, CA, USA, 7–12 February 2015; Roth, S., Nakata, Y., Neuenschwander, B., Xu, X., Eds.; Volume 9350, p. 935016. [CrossRef]
104. Bizi-Bandoki, P.; Valette, S.; Audouard, E.; Benayoun, S. Time dependency of the hydrophilicity and hydrophobicity of metallic alloys subjected to femtosecond laser irradiations. *Appl. Surf. Sci.* **2013**, *273*, 399–407. [CrossRef]
105. Dunn, A.; Carstensen, J.V.; Wlodarczyk, K.L.; Hansen, E.B.; Gabzdyl, J.; Harrison, P.M.; Shephard, J.D.; Hand, D.P. Nanosecond laser texturing for high friction applications. *Opt. Lasers Eng.* **2014**, *62*, 9–16. [CrossRef]
106. Scorticati, D.; Römer, G.-W.; de Lange, D.F.; Huis in 't Veld, B. Ultra-short-pulsed laser-machined nanogratings of laser-induced periodic surface structures on thin molybdenum layers. *J. Nanophotonics* **2012**, *6*, 063528. [CrossRef]
107. Schnell, G.; Duenow, U.; Seitz, H. Effect of Laser Pulse Overlap and Scanning Line Overlap on Femtosecond Laser-Structured Ti6Al4V Surfaces. *Materials* **2020**, *13*, 969. [CrossRef]
108. Neuenschwander, B.; Jaeggi, B.; Zimmermann, M.; Markovic, V.; Resan, B.; Weingarten, K.; de Loor, R.; Penning, L. Laser surface structuring with 100 W of average power and sub-ps pulses. *J. Laser Appl.* **2016**, *28*, 022506. [CrossRef]
109. White, J. *High Speed Infrared Cameras Enable Demanding Thermal Imaging Applications*; Scientific Imaging White Paper High; Electrophysics Resource Center: Fairfield, CT, USA, 2010.
110. Schröder, N.; Vergara, G.; Voisat, B.; Lasagni, A.F. Monitoring the Heat Accumulation during Fabrication of Surface Micropatterns on Metallic Surfaces Using Direct Laser Interference Patterning. *J. Laser Micro/Nanoeng.* **2020**, *15*, 150–157. [CrossRef]
111. Kučera, M.; Martan, J.; Franc, A. Time-resolved temperature measurement during laser marking of stainless steel. *Int. J. Heat Mass Transf.* **2018**, *125*, 1061–1068. [CrossRef]
112. Martan, J.; Moskal, D.; Prokešová, L.; Honner, M. Detection of Heat Accumulation in Laser Surface Texturing by Fast Infrared Detectors. In *The Laser in Manufacturing (LiM2019)*; JLPS-Japan Laser Processing Society: Osaka, Japan, 2019; pp. 1–7.
113. Moskal, D.; Martan, J.; Kučera, M. Shifted Laser Surface Texturing (sLST) in Burst Regime. *J. Laser Micro/Nanoeng.* **2019**, *14*, 179–185. [CrossRef]
114. Moskal, D.; Martan, J.; Kučera, M. Scanning strategy of high speed shifted laser surface texturing. In Proceedings of the Lasers in Manufacturing Conference, LIM 2017, Munich, Germany, 26–29 June 2017; pp. 1–5.
115. Houdková, Š.; Šperka, P.; Repka, M.; Martan, J.; Moskal, D. Shifted laser surface texturing for bearings applications. *J. Phys. Conf. Ser.* **2017**, *843*, 012076. [CrossRef]

116. Hauschwitz, P.; Martan, J.; Bičišťová, R.; Beltrami, C.; Moskal, D.; Brodsky, A.; Kaplan, N.; Mužík, J.; Štepánková, D.; Brajer, J.; et al. LIPSS-based functional surfaces produced by multi-beam nanostructuring with 2601 beams and real-time thermal processes measurement. *Sci. Rep.* **2021**, *11*, 22944. [CrossRef]
117. Martan, J.; Semmar, N.; Cibulka, O. Precise nanosecond time resolved infrared radiometry measurements of laser induced silicon phase change and melting front propagation. *J. Appl. Phys.* **2008**, *103*, 3–5. [CrossRef]
118. Martan, J.; Cibulka, O.; Semmar, N. Nanosecond pulse laser melting investigation by IR radiometry and reflection-based methods. *Appl. Surf. Sci.* **2006**, *253*, 1170–1177. [CrossRef]
119. Purtonen, T.; Kalliosaari, A.; Salminen, A. Monitoring and adaptive control of laser processes. *Phys. Procedia* **2014**, *56*, 1218–1231. [CrossRef]
120. Xu, X.; Grigoropoulos, C.P.; Russo, R.E. Measurement of solid-liquid interface temperature during pulsed excimer laser melting of polycrystalline silicon films. *Appl. Phys. Lett.* **1994**, *65*, 1745–1747. [CrossRef]
121. Martan, J.; Semmar, N.; Boulmer-Leborgne, C. IR radiometry optical system view factor and its application to emissivity investigations of solid and liquid phases. *Int. J. Thermophys.* **2007**, *28*, 1342–1352. [CrossRef]
122. Gafner, M.; Remund, S.M.; Chaja, M.W.; Neuenschwander, B. High-rate laser processing with ultrashort laser pulses by combination of diffractive elements with synchronized galvo scanning. *Adv. Opt. Technol.* **2021**, *10*, 333–352. [CrossRef]
123. Gao, W.; Zheng, K.; Liao, Y.; Du, H.; Liu, C.; Ye, C.; Liu, K.; Xie, S.; Chen, C.; Chen, J.; et al. High-Quality Femtosecond Laser Surface Micro/Nano-Structuring Assisted by A Thin Frost Layer. *Adv. Mater. Interfaces* **2023**, *10*, 2201924. [CrossRef]
124. Piccolo, L.; Wang, Z.; Lucchetta, G.; Shen, M.; Masato, D. Ultrafast Laser Texturing of Stainless Steel in Water and Air Environment. *Lasers Manuf. Mater. Process.* **2022**, *9*, 434–453. [CrossRef]
125. Miyaji, G.; Miyazaki, K.; Zhang, K.; Yoshifuji, T.; Fujita, J. Mechanism of femtosecond-laser-induced periodic nanostructure formation on crystalline silicon surface immersed in water. *Opt. Express* **2012**, *20*, 14848. [CrossRef]
126. Römer, G.R.B.E.; Bechtold, P. Electro-optic and Acousto-optic Laser Beam Scanners. *Phys. Procedia* **2014**, *56*, 29–39. [CrossRef]
127. De Loor, R.; Penning, L.; Slagle, R. A need for speed in laser processing and micromachining. *Laser Tech. J.* **2014**, *3*, 32–34. [CrossRef]
128. Žemaitis, A.; Gaidys, M.; Gečys, P.; Račiukaitis, G.; Gedvilas, M. Rapid high-quality 3D micro-machining by optimised efficient ultrashort laser ablation. *Opt. Lasers Eng.* **2019**, *114*, 83–89. [CrossRef]
129. De Loor, R. Polygon Scanner System for Ultra Short Pulsed Laser Micro-Machining Applications. *Phys. Procedia* **2013**, *41*, 544–551. [CrossRef]
130. Streek, A.; Lee, M. Ultrafast Material Processing with High-Brightness Fiber Lasers. *Laser Tech. J.* **2017**, *14*, 22–25. [CrossRef]
131. Lasagni, A.; Benke, D.; Kunze, T.; Bieda, M.; Eckhardt, S.; Roch, T.; Langheinrich, D.; Berger, J. Bringing the direct laser interference patterning method to industry: A one tool-complete solution for surface functionalization. *J. Laser Micro/Nanoeng.* **2015**, *10*, 340–344. [CrossRef]
132. Aguilar-Morales, A.I.; Alamri, S.; Lasagni, A.F. Micro-fabrication of high aspect ratio periodic structures on stainless steel by picosecond direct laser interference patterning. *J. Mater. Process. Technol.* **2018**, *252*, 313–321. [CrossRef]
133. Lang, V.; Roch, T.; Lasagni, A.F. High-Speed Surface Structuring of Polycarbonate Using Direct Laser Interference Patterning: Toward 1 m^2 min^{-1} Fabrication Speed Barrier. *Adv. Eng. Mater.* **2016**, *18*, 1342–1348. [CrossRef]
134. Lang, V.; Rank, A.; Lasagni, A.F. Large Area One-Step Fabrication of Three-Level Multiple-Scaled Micro and Nanostructured Nickel Sleeves for Roll-to-Roll Hot Embossing. *Adv. Eng. Mater.* **2017**, *19*, 1700126. [CrossRef]
135. Burrowg, G.M.; Gaylord, T.K. Multi-beam interference advances and applications: Nano-electronics, photonic crystals, metamaterials, subwavelength structures, optical trapping, and biomedical structures. *Micromachines* **2011**, *2*, 221–257. [CrossRef]
136. Razi, S.; Madanipour, K.; Mollabashi, M. Laser surface texturing of 316L stainless steel in air and water: A method for increasing hydrophilicity via direct creation of microstructures. *Opt. Laser Technol.* **2016**, *80*, 237–246. [CrossRef]
137. Mutlak, F.A.H.; Ahmed, A.F.; Nayef, U.M.; Al-zaidi, Q.; Abdulridha, S.K. Improvement of absorption light of laser texturing on silicon surface for optoelectronic application. *Optik* **2021**, *237*, 166755. [CrossRef]
138. Ancona, A.; Joshi, G.; Volpe, A.; Scaraggi, M.; Lugarà, P.; Carbone, G. Non-Uniform Laser Surface Texturing of an Un-Tapered Square Pad for Tribological Applications. *Lubricants* **2017**, *5*, 41. [CrossRef]
139. Furlan, V.; Demir, A.G.; Pariani, G.; Bianco, A.; Previtali, B. A new approach to Direct Laser Interference Patterning with scanner optics for high productivity. In Proceedings of the European Society for Precision Engineering and Nanotechnology, Conference Proceedings-18th International Conference and Exhibition, EUSPEN 2018, Venice, Italy, 4–8 June 2018; pp. 49–50.
140. Cardoso, J.T.; Aguilar-Morales, A.I.; Alamri, S.; Huerta-Murillo, D.; Cordovilla, F.; Lasagni, A.F.; Ocaña, J.L. Superhydrophobicity on hierarchical periodic surface structures fabricated via direct laser writing and direct laser interference patterning on an aluminium alloy. *Opt. Lasers Eng.* **2018**, *111*, 193–200. [CrossRef]
141. Sandra, H.; Kirner, S.V.; Rosenfeld, A. Laser-Induced Periodic Surface Structures—A Scientific Evergreen. *IEEE J. Sel. Top. Quantum Electron.* **2017**, *23*, 9000615. [CrossRef]
142. Gnilitskyi, I.; Derrien, T.J.-Y.; Levy, Y.; Bulgakova, N.M.; Mocek, T.; Orazi, L. High-speed manufacturing of highly regular femtosecond laser-induced periodic surface structures: Physical origin of regularity. *Sci. Rep.* **2017**, *7*, 8485. [CrossRef]
143. Martínez-Calderon, M.; Rodríguez, A.; Dias-Ponte, A.; Morant-Miñana, M.C.; Gómez-Aranzadi, M.; Olaizola, S.M. Femtosecond laser fabrication of highly hydrophobic stainless steel surface with hierarchical structures fabricated by combining ordered microstructures and LIPSS. *Appl. Surf. Sci.* **2016**, *374*, 81–89. [CrossRef]

144. Mottay, E.; Delaigue, M.; Schöps, B.; Dalla-Barba, G.; Audouard, E.; Hönninger, C.; Bernard, O.; Mishchik, K. Efficient micro processing with high power femtosecond lasers by beam engineering and modelling. *Procedia CIRP* **2018**, *74*, 310–314. [CrossRef]
145. Myles, D.T.E.; Ziyenge, M.; Shephard, J.D.; Milne, D.C. Scanned mask imaging solid state laser tool for cost effective flip chip—Chip scale package manufacture. *J. Laser Micro/Nanoeng.* **2015**, *10*, 106–109. [CrossRef]
146. Chang, Y.H.; Lin, Y.C.; Liu, Y.S.; Liu, C.Y. Light-extraction enhancement by cavity array-textured n-polar GaN surfaces ablated using a KrF laser. *IEEE Photonics Technol. Lett.* **2012**, *24*, 2013–2015. [CrossRef]
147. Horn, A.; Kalmbach, C.C.; Moreno, J.G.; Schütz, V.; Stute, U.; Overmeyer, L. Laser-Surface-Treatment for Photovoltaic Applications. *Phys. Procedia* **2012**, *39*, 709–716. [CrossRef]
148. Zhu, G.; Whitehead, D.; Perrie, W.; Allegre, O.J.; Olle, V.; Li, Q.; Tang, Y.; Dawson, K.; Jin, Y.; Edwardson, S.P.; et al. Thermal and optical performance characteristics of a spatial light modulator with high average power picosecond laser exposure applied to materials processing applications. *Procedia CIRP* **2018**, *74*, 594–597. [CrossRef]
149. Abbott, C.; Allott, R.M.; Bann, B.; Boehlen, K.L.; Gower, M.C.; Rumsby, P.T.; Stassen Boehlen, I.; Sykes, N. New techniques for laser micromachining MEMS devices. In Proceedings of the High-Power Laser Ablation IV, Taos, NM, USA, 21–26 April 2002; Phipps, C.R., Ed.; Volume 4760, p. 281. [CrossRef]
150. Lopez, J.; Mishchik, K.; Mincuzzi, G.; Audouard, E.; Mottay, E. Efficient Metal Processing Using High Average Power Ultrafast Laser. *J. Laser Micro/Nanoeng.* **2017**, *12*, 1–8. [CrossRef]
151. Marshall, G.F. *Handbook of Optical and Laser Scanning*; Marshall, G.F., Stutz, G.E., Eds.; CRC Press: Boca Raton, FL, USA, 2018; ISBN 9781315218243. [CrossRef]
152. Jaeggi, B.; Neuenschwander, B.; Zimmermann, M.; Loor, R.D.; Penning, L. High throughput ps-laser micro machining with a synchronized polygon line scanner. In Proceedings of the 8th International Conference on Photonic Technologies LANE 2014, Fürth, Germany, 8–11 September 2014; pp. 1–8.
153. Dunn, A.; Wlodarczyk, K.L.; Carstensen, J.V.; Hansen, E.B.; Gabzdyl, J.; Harrison, P.M.; Shephard, J.D.; Hand, D.P. Laser surface texturing for high friction contacts. *Appl. Surf. Sci.* **2015**, *357*, 2313–2319. [CrossRef]
154. Dmitriev, P.A.; Makarov, S.V.; Milichko, V.A.; Mukhin, I.S.; Samusev, A.K.; Krasnok, A.E.; Belov, P.A. Direct Femtosecond Laser Writing of Optical Nanoresonators. *J. Phys. Conf. Ser.* **2016**, *690*, 012021. [CrossRef]
155. Bruening, S.; Hennig, G.; Eifel, S.; Gillner, A. Ultrafast Scan Techniques for 3D-μm Structuring of Metal Surfaces with high repetitive ps-laser pulses. *Phys. Procedia* **2011**, *12*, 105–115. [CrossRef]
156. Lin, Y.; Han, J.; Cai, M.; Liu, W.; Luo, X.; Zhang, H.; Zhong, M. Durable and robust transparent superhydrophobic glass surfaces fabricated by a femtosecond laser with exceptional water repellency and thermostability. *J. Mater. Chem. A* **2018**, *6*, 9049–9056. [CrossRef]
157. Abbott, M.; Cotter, J. Optical and electrical properties of laser texturing for high-efficiency solar cells. *J. Optoelectron. Adv. Mater.* **2006**, *14*, 225–235. [CrossRef]
158. Sampedro, J.; Ferre, R.; Fernández, E.; Pérez, I.; Cárcel, B.; Molina, T.; Ramos, J.A. Surface Functionalization of AISI 316 Steel by Laser Texturing of Shaped Microcavities with Picosecond Pulses. *Phys. Procedia* **2012**, *39*, 636–641. [CrossRef]
159. Liu, Z.; Xu, W.; Hou, Z.; Wu, Z. A Rapid Prototyping Technique for Microfluidics with High Robustness and Flexibility. *Micromachines* **2016**, *7*, 201. [CrossRef]
160. Wlodarczyk, K.L.; Lopes, A.A.; Blair, P.; Maroto-Valer, M.M.; Hand, D.P.; Wlodarczyk, K.L.; Lopes, A.A.; Blair, P.; Maroto-Valer, M.M.; Hand, D.P. Interlaced Laser Beam Scanning: A Method Enabling an Increase in the Throughput of Ultrafast Laser Machining of Borosilicate Glass. *J. Manuf. Mater. Process.* **2019**, *3*, 14. [CrossRef]
161. Mincuzzi, G.; Gemini, L.; Faucon, M.; Kling, R. Extending ultra-short pulse laser texturing over large area. *Appl. Surf. Sci.* **2016**, *386*, 65–71. [CrossRef]
162. Schille, J.; Schneider, L.; Ullmann, F.; Mauersberger, S.; Löschner, U. Bio-inspirierte Funktionalisierung von Technischen Oberflächen Durch Hochrate- Lasermikrostrukturierung. In *ThGOT Thementage Grenz- und Oberflächentechnik und 5. Kolloquium Dünne Schichten in der Optik*; INNOVENT e.V.: Jena, Germany, 2017; Available online: https://www.innovent-jena.de/en/about-us/contact (accessed on 27 April 2023).
163. Riveiro, A.; Maçon, A.L.B.; del Val, J.; Comesaña, R.; Pou, J. Laser Surface Texturing of Polymers for Biomedical Applications. *Front. Phys.* **2018**, *6*, 16. [CrossRef]
164. Martan, J.; Moskal, D.; Kučera, M. Laser surface texturing with shifted method—Functional surfaces at high speed. *J. Laser Appl.* **2019**, *31*, 022507. [CrossRef]
165. Schille, J.; Schneider, L.; Mauersberger, S.; Szokup, S.; Höhn, S.; Pötschke, J.; Reiß, F.; Leidich, E.; Löschner, U. High-Rate laser surface texturing for advanced tribological functionality. *Lubricants* **2020**, *8*, 33. [CrossRef]
166. Roessler, F.; Streek, A. Accelerating laser processes with a smart two-dimensional polygon mirror scanner for ultra-fast beam deflection. *Adv. Opt. Technol.* **2021**, *10*, 297–304. [CrossRef]
167. Delgado, M.A.O.; Lasagni, A.F. Minimizing stitching errors for large area laser surface processing. *J. Laser Micro/Nanoeng.* **2016**, *11*, 185–191. [CrossRef]
168. Aguilar-Morales, A.I.; Alamri, S.; Kunze, T.; Lasagni, A.F. Influence of processing parameters on surface texture homogeneity using Direct Laser Interference Patterning. *Opt. Laser Technol.* **2018**, *107*, 216–227. [CrossRef]
169. Machado, L.M.; Samad, R.E.; Freitas, A.Z.; Vieira, N.D.; De Rossi, W. Microchannels direct machining using the femtosecond smooth ablation method. *Phys. Procedia* **2011**, *12*, 67–75. [CrossRef]

170. Romoli, L. Flattening of surface roughness in ultrashort pulsed laser micro-milling. *Precis. Eng.* **2018**, *51*, 331–337. [CrossRef]
171. Abdo, B.M.A.; Anwar, S.; El-Tamimi, A.M.; Alahmari, A.M.; Abouel Nasr, E. Laser micro-milling of bio-lox forte ceramic: An experimental analysis. *Precis. Eng.* **2018**, *53*, 179–193. [CrossRef]
172. Wang, X.C.; Zheng, H.Y.; Chu, P.L.; Tan, J.L.; Teh, K.M.; Liu, T.; Ang, B.C.Y.; Tay, G.H. Femtosecond laser drilling of alumina ceramic substrates. *Appl. Phys. A* **2010**, *101*, 271–278. [CrossRef]
173. Neuenschwander, B.; Jaeggi, B.; Zimmermann, M.; Markovic, V.; Resan, B.; Weingarten, K.; de Loor, R.; Penning, L. Laser Surface Structuring with 100W of Average Power and Sub-ps Pulses. In Proceedings of the International Congress on Applications of Lasers and Electro-Optics (ICALEO), Atlanta, GA, USA, 18–22 October 2015; pp. 14–22.
174. Dobrzański, L.A.; Drygała, A.; Gołombek, K.; Panek, P.; Bielańska, E.; Zieba, P. Laser surface treatment of multicrystalline silicon for enhancing optical properties. *J. Mater. Process. Technol.* **2008**, *201*, 291–296. [CrossRef]
175. Daniel, C.; Manderla, J.; Hallmann, S.; Emmelmann, C. Influence of an Angular Hatching Exposure Strategy on the Surface Roughness During Picosecond Laser Ablation of Hard Materials. *Phys. Procedia* **2016**, *83*, 135–146. [CrossRef]
176. Dold, C.A. Picosecond Laser Processing of Diamond Cutting Edges. Ph.D. Thesis, Institut für Werkzeugmaschinen und Fertigung, Zurich, Switzerland, 2013. [CrossRef]
177. Hao, X.; Xu, W.; Chen, M.; Wang, C.; Han, J.; Li, L.; He, N. Laser hybridizing with micro-milling for fabrication of high aspect ratio micro-groove on oxygen-free copper. *Precis. Eng.* **2021**, *70*, 15–25. [CrossRef]
178. Ma, L.; Bin, H. Temperature and stress analysis and simulation in fractal scanning-based laser sintering. *Int. J. Adv. Manuf. Technol.* **2007**, *34*, 898–903. [CrossRef]
179. Chen, L.; Zhang, Y.; Kodama, S.; Xu, S.; Shimada, K.; Mizutani, M.; Kuriyagawa, T. Picosecond laser-induced nanopillar coverage of entire mirror-polished surfaces of Ti6Al4V alloy. *Precis. Eng.* **2021**, *72*, 556–567. [CrossRef]
180. Moskal, D.; Kučera, M.; Smazalová, E.; Houdková, S.; Kromer, R. Application of shifted laser surface texturing. In Proceedings of the METAL 2015-24th International Conference on Metallurgy and Materials, Brno, Czech Republic, 3–5 June 2015; pp. 1016–1021.
181. Kruth, J.; Badrossamay, M.; Yasa, E.; Deckers, J.; Thijs, L.; Van Humbeeck, J. Part and material properties in selective laser melting of metals. In Proceedings of the 16th International Symposium on Electromachining, Shanghai, China, 19–23 April 2010; pp. 1–12.
182. Salama, A.; Yan, Y.; Li, L.; Mativenga, P.; Whitehead, D.; Sabli, A. Understanding the self-limiting effect in picosecond laser single and multiple parallel pass drilling/machining of CFRP composite and mild steel. *Mater. Des.* **2016**, *107*, 461–469. [CrossRef]
183. Kumar, N.; Prakash, S.; Kumar, S. Studies of laser textured Ti-6Al-4V wettability for implants. *IOP Conf. Ser. Mater. Sci. Eng.* **2016**, *149*, 012056. [CrossRef]
184. Liu, B.; Jiang, G.; Wang, W.; Mei, X.; Wang, K.; Cui, J.; Wang, J. Porous microstructures induced by picosecond laser scanning irradiation on stainless steel surface. *Opt. Lasers Eng.* **2016**, *78*, 55–63. [CrossRef]
185. Ahmmed, K.M.T.; Ling, E.J.Y.; Servio, P.; Kietzig, A.M. Introducing a new optimization tool for femtosecond laser-induced surface texturing on titanium, stainless steel, aluminum and copper. *Opt. Lasers Eng.* **2015**, *66*, 258–268. [CrossRef]
186. Schonlau, T.; Hebel, R.; Pause, S.; Mayer, G. Machining strategies for versatile ultra-short pulse laser applications. In Proceedings of the 8th International Conference on Photonic Technologies LANE 2014, Furth, Germany, 8–11 September 2014.
187. Faas, S.; Bielke, U.; Weber, R.; Graf, T. Scaling the productivity of laser structuring processes using picosecond laser pulses at average powers of up to 420 W to produce superhydrophobic surfaces on stainless steel AISI 316L. *Sci. Rep.* **2019**, *9*, 1933. [CrossRef]
188. Mohanty, S.; Hattel, J. Cellular scanning strategy for selective laser melting: Capturing thermal trends with a low-fidelity, pseudo-analytical model. *Math. Probl. Eng.* **2014**, *2014*, 715058. [CrossRef]
189. Valente, E.H.; Gundlach, C.; Christiansen, T.L.; Somers, M.A.J. Effect of scanning strategy during selective laser melting on surface topography, porosity, and microstructure of additively manufactured Ti-6Al-4V. *Appl. Sci.* **2019**, *9*, 5554. [CrossRef]
190. Ratautas, K.; Vosylius, V.; Jagminienė, A.; Stankevičienė, I.; Norkus, E.; Račiukaitis, G. Laser-induced selective electroless plating on pc/abs polymer: Minimisation of thermal effects for supreme processing speed. *Polymers* **2020**, *12*, 2427. [CrossRef]
191. Lasagni, A.F.; Roch, T.; Langheinrich, D.; Bieda, M.; Perez, H.; Wetzig, A.; Beyer, E. Large area direct fabrication of periodic arrays using interference patterning. In Proceedings of the Laser-based Micro- and Nanopackaging and Assembly VI, San Francisco, CA, USA, 21–26 January 2012; Volume 8244, pp. 1–10. [CrossRef]
192. Bruening, S.; Jarczynski, M.; Mitra, T.; Du, K.; Fornaroli, C.; Gillner, A. Ultra-Fast Multi-Spot-Parallel Processing of Functional Micro- and Nanoscale Structures on Embossing Dies with Ultrafast Lasers. In Proceedings of the German Scientific Laser Society (WLT), in Lasers in Manufacturing (LiM), Munich, Germany, 26–29 June 2017.
193. Roch, T.; Beyer, E.; Lasagni, A. Surface modification of thin tetrahedral amorphous carbon films by means of UV direct laser interference patterning. *Diam. Relat. Mater.* **2010**, *19*, 1472–1477. [CrossRef]
194. Brueck, S.R.J. Optical and Interferometric Lithography-Nanotechnology Enablers. *Proc. IEEE* **2005**, *93*, 1704–1721. [CrossRef]
195. Wu, B.; Zhou, M.; Li, J.; Ye, X.; Li, G.; Cai, L. Superhydrophobic surfaces fabricated by microstructuring of stainless steel using a femtosecond laser. *Appl. Surf. Sci.* **2009**, *256*, 61–66. [CrossRef]
196. Hermens, U.; Kirner, S.V.; Emonts, C.; Comanns, P.; Skoulas, E.; Mimidis, A.; Mescheder, H.; Winands, K.; Krüger, J.; Stratakis, E.; et al. Mimicking lizard-like surface structures upon ultrashort laser pulse irradiation of inorganic materials. *Appl. Surf. Sci.* **2017**, *418*, 499–507. [CrossRef]

197. Jorge-Mora, A.; Imaz, N.; Garcia-Lecina, E.; O'Connor, G.M.; Gómez-Vaamonde, R.; Alonso-Pérez, A.; Franco-Trepat, E.; García-Santiago, C.; Pino-Minguez, J.; Nieto, D. In vitro response of bone marrow mesenchymal stem cells (hBMSCs) on laser-induced periodic surface structures for hard tissue replacement: Comparison between tantalum and titanium. *Opt. Lasers Eng.* **2018**, *111*, 34–41. [CrossRef]
198. Huerta-Murillo, D.; Aguilar-Morales, A.I.; Alamri, S.; Cardoso, J.T.; Jagdheesh, R.; Lasagni, A.F.; Ocaña, J.L. Fabrication of multi-scale periodic surface structures on Ti-6Al-4V by direct laser writing and direct laser interference patterning for modified wettability applications. *Opt. Lasers Eng.* **2017**, *98*, 134–142. [CrossRef]
199. Giannuzzi, G.; Gaudiuso, C.; Franco, C.D.; Scamarcio, G.; Lugarà, P.M.; Ancona, A. Large area laser-induced periodic surface structures on steel by bursts of femtosecond pulses with picosecond delays. *Opt. Lasers Eng.* **2019**, *114*, 15–21. [CrossRef]
200. Soldera, M.; Alamri, S.; Sürmann, P.A.; Kunze, T.; Lasagni, A.F. Microfabrication and surface functionalization of soda lime glass through direct laser interference patterning. *Nanomaterials* **2021**, *11*, 129. [CrossRef]
201. Florian, C.; Kirner, S.V.; Krüger, J.; Bonse, J. Surface functionalization by laser-induced periodic surface structures. *J. Laser Appl.* **2020**, *32*, 022063. [CrossRef]
202. Gillner, A.; Finger, J.; Gretzki, P.; Niessen, M.; Bartels, T.; Reininghaus, M. High power laser processing with ultrafast and multi-parallel beams. *J. Laser Micro/Nanoeng.* **2019**, *14*, 129–137. [CrossRef]
203. Lamprecht, B.; Satzinger, V.; Schmidt, V.; Peharz, G.; Wenzl, F.P. Spatial light modulator based laser microfabrication of volume optics inside solar modules. *Opt. Express* **2018**, *26*, A227. [CrossRef]
204. Indrišiūnas, S.; Gedvilas, M. Control of the wetting properties of stainless steel by ultrashort laser texturing using multi-parallel beam processing. *Opt. Laser Technol.* **2022**, *153*, 108187. [CrossRef]

Disclaimer/Publisher's Note: The statements, opinions and data contained in all publications are solely those of the individual author(s) and contributor(s) and not of MDPI and/or the editor(s). MDPI and/or the editor(s) disclaim responsibility for any injury to people or property resulting from any ideas, methods, instructions or products referred to in the content.

MDPI AG
Grosspeteranlage 5
4052 Basel
Switzerland
Tel.: +41 61 683 77 34

Micromachines Editorial Office
E-mail: micromachines@mdpi.com
www.mdpi.com/journal/micromachines

Disclaimer/Publisher's Note: The title and front matter of this reprint are at the discretion of the Guest Editors. The publisher is not responsible for their content or any associated concerns. The statements, opinions and data contained in all individual articles are solely those of the individual Editors and contributors and not of MDPI. MDPI disclaims responsibility for any injury to people or property resulting from any ideas, methods, instructions or products referred to in the content.

www.ingramcontent.com/pod-product-compliance
Lightning Source LLC
LaVergne TN
LVHW072323090526
838202LV00019B/2340